Gallium Nitride Power Devices

Gallium Nitride Power Devices

edited by

Hongyu Yu

Tianli Duan

Pan Stanford Publishing

Published by

Pan Stanford Publishing Pte. Ltd.
Penthouse Level, Suntec Tower 3
8 Temasek Boulevard
Singapore 038988

Email: editorial@panstanford.com
Web: www.panstanford.com

British Library Cataloguing-in-Publication Data
A catalogue record for this book is available from the British Library.

Gallium Nitride Power Devices

ISBN 978-981-4774-09-3 (Hardcover)
ISBN 978-1-315-19662-6 (eBook)

Printed in Canada

Contents

Preface

As a third-generation semiconductor, GaN has found broad technological applications in high-power devices by virtue of its high breakdown field and high electron mobility. Therefore, the development of GaN technology is regarded as important national strategic research for many countries. The industry has made phenomenal growth in GaN electronics. Moreover, many GaN-related papers have been published to report the progress in the fundamental concepts and performances of GaN devices.

To access both fundamental knowledge and advanced novel development, a book that could give information about comprehensive material physics and device structure as well as device operational principles was needed. To meet such a need, *Gallium Nitride Power Devices* was prepared. This book comprises nine chapters which discuss the growth technology of GaN wafers and the characteristics of polarization effects of GaN material and device process on device performance, reliability, and packaging. It is a textbook for undergraduate students, and it is also a reference book for graduate students as well as engineers and scientists in GaN research.

In the course of writing this book, I was fortunate to obtain the help of Wang Zhigang and Wang Bing from Southwest Jiaotong University, Cheng Kai from Enkris Semiconductor, Liu Zhihong from the Singapore-MIT Alliance for Research and Technology, and Jiang Lingli, Xia Pengkun, and Dongbin from the South University of Science and Technology of China.

Tianli Duan

Chapter 1

The Growth Technology of High-Voltage GaN on Silicon

Peng Xiang, Liyang Zhang, and Kai Cheng
Enkris Semiconductor, Inc., Suzhou, China
Peng.Xiang@enkris.com

1.1 Introduction

Due to the limited availability and relatively high cost of sufficiently large GaN substrates for homoepitaxial growth, GaN films are normally grown on foreign substrates, such as sapphire, SiC, and Si. Table 1.1 shows properties of III nitrides and these foreign substrates. SiC has the smallest lattice mismatch and thermal expansion coefficient (TEC) mismatch to GaN as compared to others, but the high price and limited diameter (normally ≤6 inches) of a SiC substrate make it unaffordable for power applications. GaN-on-sapphire substrate technology is very mature and is the mainstream in the light-emitting diode (LED) market, but it is unsuitable for power applications due to the poor thermal conductivity of sapphire.

Gallium Nitride Power Devices
Edited by Hongyu Yu and Tianli Duan

ISBN 978-981-4774-09-3 (Hardcover), 978-1-315-19662-6 (eBook)
www.panstanford.com

Si is the most widely used semiconductor nowadays, and using a Si substrate has many advantages over SiC and sapphire, such as a large diameter, low cost, and a ready-made process. The diameter of a Si substrate can be scaled to 12 inches, and the metal-organic chemical vapor deposition (MOCVD) reactor is available to grow 8-inch GaN epilayers on 8-inch Si substrates. In contrast, the SiC and sapphire substrate size is typically smaller than 6 inches. The cost of GaN power devices can be further reduced by using the ready-made 6- and 8-inch complementary metal-oxide-semiconductor (CMOS) process line. A Si substrate also shows high crystalline quality and minimized defect density, thanks to decades of development of the Si semiconductor industry. It has been well accepted that GaN on large-size silicon substrates is one cost-effective way to achieve high-volume production of GaN power devices. Figure 1.1 shows the epitaxial relationship of the GaN(0001) plane (c-plane) on the Si(111) plane. Usually, a Si(111) substrate is utilized for the epitaxial growth of c-plane GaN because of their same trigonal symmetry.

Table 1.1 Properties of III nitrides and substrates [1, 2]

	Material	Lattice parameters (Å)	Lattice mismatch[a] to GaN	TEC[b] (10^{-6} K^{-1})	Thermal conductivity (W/cmK)	Diameter (inches)
III nitrides	AlN	a = 3.112 c = 4.982	2.48%	4.2	2.0	/
	InN	a = 3.548 c = 5.7034	−10.1%	5.7	0.8	/
	GaN	a = 3.1891 c = 5.1855	0	5.59	1.3	/
Substrates	Al_2O_3	a = 4.765 c = 12.982	13.9% Rotated 30°	7.5	0.3	2–6
	SiC	a = 3.081 c = 15.117	3.51%	4.2	4.9	2–6
	Si	a = 5.431	−17%	3.59	1.3	2–12

[a]Lattice mismatch $=(a_{GaN}-a_{eff.sub})/a_{sub}$
[b]TEC: Thermal expansion coefficient

However, the growth of GaN on Si is challenging owing to issues such as melt-back etching and a large thermal and lattice mismatch [4]. A common problem in the growth of GaN on Si is melt-back etching. The origin of melt-back etching is the alloying reaction between Ga and Si at high temperature (HT). Ga has high solubility in Si at high temperature. When Ga comes in contact with Si at high temperature, the alloy of Ga and Si forms and produces hollows and swellings on the substrate, which seriously destroys the epilayers (Fig. 1.2). Once the melt-back-etching process starts, it will expand during the following growth and cannot be stopped. Therefore it is extremely important to prevent melt-back for growing GaN on Si. Ga-free layers, such as AlN, SiC [5], AlAs [6], and Al_2O_3 [7], were employed to prevent Ga from contacting Si. Ascribed with the advantages of high thermal stability and convenience, AlN is normally used as a starting buffer layer for GaN on Si.

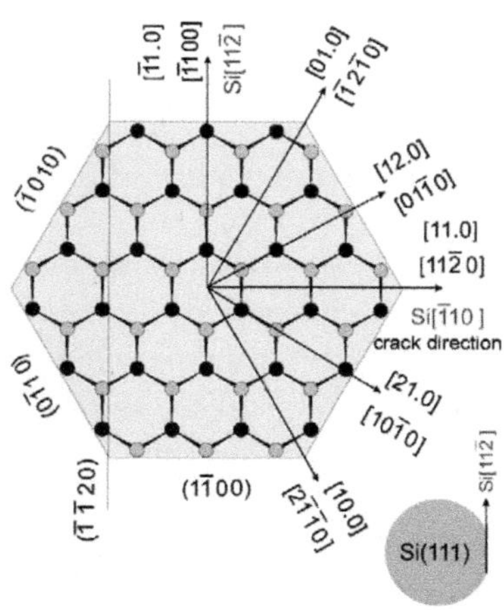

Figure 1.1 Epitaxial relationship GaN(0001) on Si(111). Reprinted from Ref. [3], Copyright (2002), with permission from Elsevier.

One of the biggest challenges is the large stress caused by the thermal mismatch between GaN epilayers and Si substrates. The TEC of GaN is 5.59×10^{-6} K^{-1}, which is much larger than that of Si, 3.59×10^{-6} K^{-1}. A GaN epilayer is usually grown at a high temperature of 1000°C using MOCVD. When GaN is cooling down from growth temperature to room temperature, a large tensile stress of about 1 Gpa is generated in the GaN film because the TEC of the Si substrate is significantly smaller than that of GaN. Consequently, there is a much greater risk that the GaN epilayers will crack if the thickness of GaN exceeds several hundred nanometers, as shown in Fig. 1.3. A GaN layer with a thickness of several hundred nanometers is insufficient for producing high-quality films and devices. GaN electronics on

Si can easily break down from the Si substrate because of the low critical electric field of Si. To achieve a high breakdown voltage, a film thickness of several micrometers is necessary for a GaN-on-Si power device.

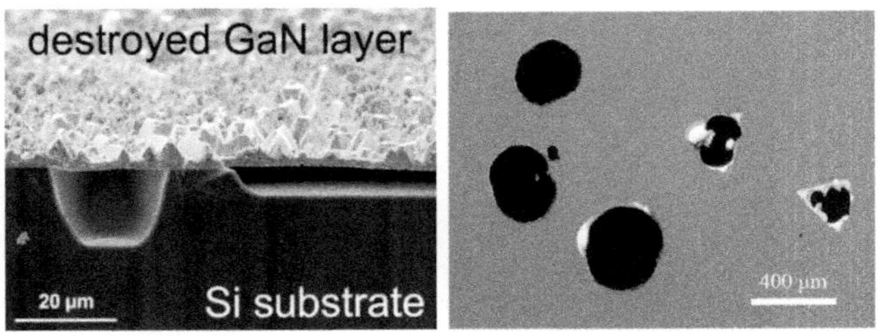

(a) (b)

Figure 1.2 Destroyed GaN layer due to melt-back etching: (a) SEM image of a vertical view. Reprinted from Ref. [8], Copyright (2003), with permission from Elsevier. (b) Optical microscope image of a plan view.

Stress also introduces large wafer bow, which deteriorates the uniformity of the epilayer and causes failure during wafer handling and passing of the stepper in the lithography process, making the material unsuitable for device manufacturing. The wafer curvature κ can be obtained from the Stoney equation:

$$\kappa = \frac{6M_f\varepsilon_m h_f}{M_s h_s^2} \tag{1.1}$$

where ε_m, M_f, M_s, h_f, and h_s denote the biaxial mismatch strain, the biaxial moduli of the film, the biaxial moduli of the substrate, the thickness of the film, and the thickness of the substrate, respectively. Assuming the diameter of the substrate is D, the wafer bow can be obtained by

$$B = \frac{3M_f D^2}{4M_s h_s^2}\varepsilon_m h_f \tag{1.2}$$

The wafer bow is proportional to the square of the substrate diameter, indicating a larger value when a large-diameter Si substrate is utilized.

To grow crack-free GaN on Si with low wafer bow, stress engineering has to be executed. Compressive stress is introduced during growth by inserting Al-rich layers such as AlN [9–13], AlGaN [14–24], and AlN/GaN superlattices (SLs) [25–29] to counterbalance the tensile stress. Nevertheless, this compressive stress also should be carefully controlled. This compressive stress causes curvature of the wafer during growth, which will deteriorate the wafer temperature uniformity and cause plastic deformation (Fig. 1.4) if a too large curvature appears [30]. Thus stress engineering is of great importance and is challenging for the growth of a GaN-on-Si wafer.

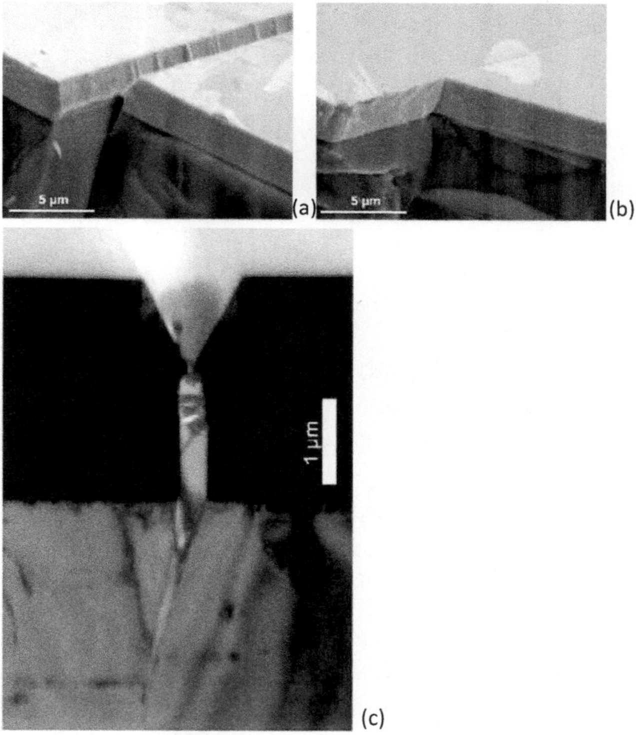

Figure 1.3 Two types of cracks for GaN on Si: (a) an open groove with facets occurs during growth and (b) a closed crack during cooling down; (c) cross section of an open crack. Reprinted from Ref. [3], Copyright (2002), with permission from Elsevier.

The large lattice mismatch between GaN and Si will introduce a high density of dislocations, which are usually larger than $10^8/cm^2$. High dislocation densities deteriorate the crystalline quality of GaN and the performance of power devices.

Apart from the large tensile stress and high dislocation density, GaN on Si has also suffered from high buffer leakage, which should be minimized for power applications. Compared to Si and GaAs,

GaN grown on a heterosubstrate is a very imperfect crystal system with a high density of defects. Usually, the density of the background donor in unintentionally doped GaN is 10^{16}–$10^{17}/cm^3$ due to the existence of oxygen and silicon impurities and nitrogen vacancies, which introduces 10^{16}–$10^{17}/cm^3$ background electrons. Thus leakage current from the buffer layer will occur. Additionally, the Si substrate and the AlN/Si interface are conducting. A GaN buffer with a high background donor will also introduce leakage current from the Si substrate. Compositional doping [32–39] and removal of the Si substrate [40] are effective in reducing the buffer leakage and increasing the breakdown voltage of GaN on Si.

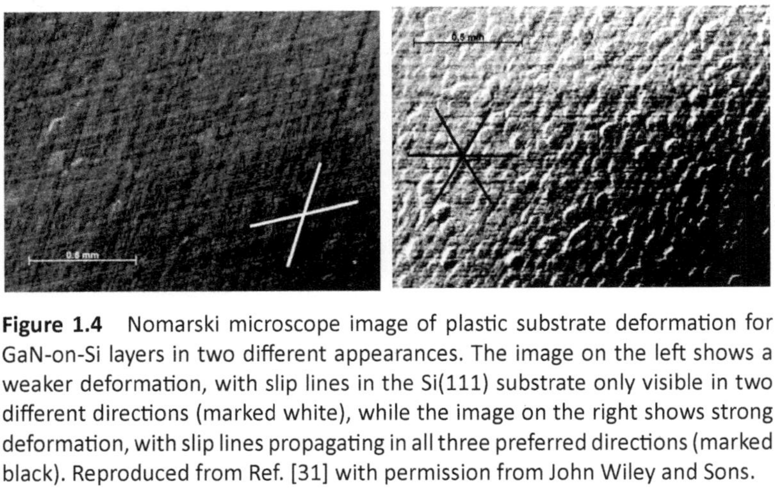

Figure 1.4 Nomarski microscope image of plastic substrate deformation for GaN-on-Si layers in two different appearances. The image on the left shows a weaker deformation, with slip lines in the Si(111) substrate only visible in two different directions (marked white), while the image on the right shows strong deformation, with slip lines propagating in all three preferred directions (marked black). Reproduced from Ref. [31] with permission from John Wiley and Sons.

1.2 The Nucleation Layer Growth

Due to the melt-back etching phenomenon mentioned earlier, GaN cannot be directly grown on a Si substrate at high temperature. AlN [20, 41–49], SiC [5], AlAs [6], and Al_2O_3 [7] were applied as the seeding layer for GaN grown on Si. AlN is the most universal nucleation layer that supports high-quality GaN on a Si substrate. With high thermal stability and good wettability on Si, the AlN buffer prevents the melt-back reaction and favors the subsequent growth of the GaN layer. In addition, compressive stress can be generated in GaN films grown on AlN because of the smaller lattice parameters of AlN, which can counterbalance the tensile stress when cooling down.

Before the deposition of an AlN nucleation layer, an oxide-free layer is preferred since the Si surface can be easily oxidized in air. Wet etching with buffered HF prior to epitaxy or in situ thermal cleaning under hydrogen is usually utilized. A cleaned Si surface can easily react with the ammonia or active nitrogen, leading to a thin amorphous Si_xN_y layer. This amorphous layer will passivate the surface and degrade the epilayers above. A way to avoid this amorphous layer formation is to predeposit a few Al monolayers on the Si surface before introducing the N source [12, 50, 51]. However, ammonia preflow is also adopted by some groups to produce a thin amorphous Si_xN_y layer, which is claimed to act as a nanomask and enlarge the AlN seeds, resulting in a lower dislocation density [52].

The AlN surface is very sensitive to the Al predeposition amount, which determines the thickness of the Al monolayer. With an appropriate Al predeposition time, a clear step structure surface was shown (Fig. 1.5b). However, long-time Al predeposition can result in a discontinued and rough surface (Fig. 1.5c) as too much Al accumulates on the substrate surface [53].

Apart from the Si surface, growth conditions of AlN nucleation also strongly affect the growth of the GaN epilayer, including the growth temperature [41, 54, 55], the V/III ratio [56, 57], and pressure and AlN buffer thickness [43, 45, 58]. There is a strong pre-reaction in the gas phase between the trimethylaluminum (TMAl) and ammonia, which starts with adduct formation and produces nanoparticles [59]. This parasitic reaction will degrade the surface morphology and crystalline quality of AlN and AlGaN and reduce the growth rate of AlN and the efficiency of the precursors. This reaction can be reduced by a low V/III ratio and pressure and careful chamber design.

The surface mobility of Al atoms is much smaller than that of Ga due to the high bond energy of AlN. High temperature can increase the Al atom mobility, resulting in a superior crystalline quality than that grown at low temperature (LT). In 1996, Meng et al. studied the structural characteristics of AlN on a Si substrate grown by ultrahigh-vacuum reactive sputter deposition [42]. And Raman scattering, ion beam channeling, and transmission electron microscopy (TEM) indicated epitaxial growth of wurtzitic AlN thin films on Si(111) at temperatures above 600°C. Zamir and coworkers studied the influence of the AlN growth temperature on the morphology and preferred orientation of GaN films [41] and reported enhanced crystalline quality with increasing buffer growth temperature (Fig. 1.6). LT/HT bilayer AlN [20] and HT/LT/HT multilayer AlN [47] nucleation layers were also utilized to achieve high-quality GaN on a Si substrate.

Figure 1.5 The surface of AlN buffer layers on a Si(1 1 1) substrate with preseeding Al for (a) 0s, (b) 5s, and (c) 10s. Reprinted from Ref. [53], Copyright (2001), with permission from Elsevier.

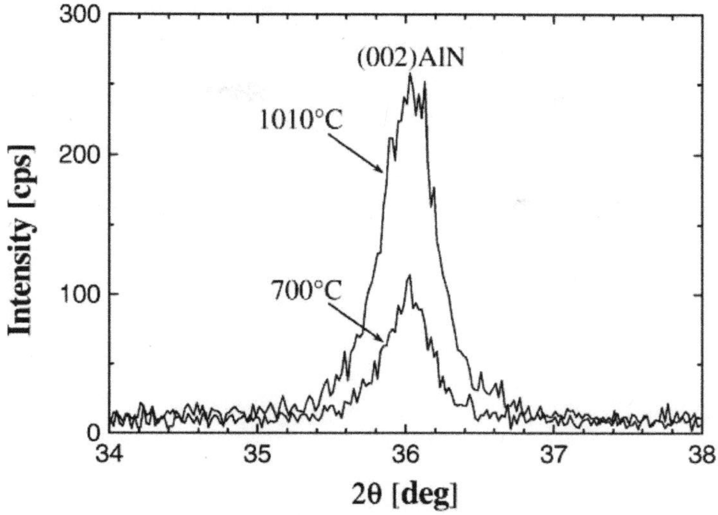

Figure 1.6 AlN 002 $2\theta/\omega$ scan of AlN on Si grown at 700°C and 1010°C. Reprinted from Ref. [41], Copyright (2000), with permission from Elsevier.

The influences of the ammonia flow rate on the property of AlN nucleation layers were studied by Krishnan and coworkers. A smooth AlN surface was obtained at a lower ammonia flow rate, and with a high growth temperature and a low V/III ratio, an X-ray diffraction full-width at half-maximum (XRD FWHM) between 796 and 998 arcsec of AlN was achieved [57].

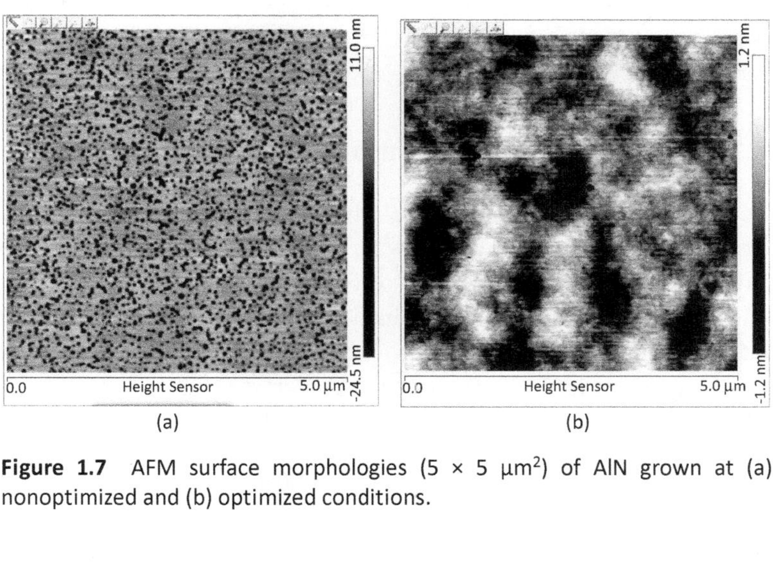

Figure 1.7 AFM surface morphologies (5 × 5 μm^2) of AlN grown at (a) nonoptimized and (b) optimized conditions.

By optimizing the growth process, an AlN nucleation layer with a smooth surface (Fig. 1.7) and high crystalline quality can be obtained on a large-diameter Si substrate [57, 60], which is the promise of a high-quality GaN layer.

1.3 Stress Engineering

The critical thickness of crack-free GaN when grown directly on a HT-AlN buffer layer is only about several hundred nanometers due to the large tensile thermal stress introduced upon cooling from growth temperature to room temperature, which is caused by the large thermal mismatch between GaN and Si. This is far too thin to achieve high-performance, high-voltage electronic devices, where at least a crack-free GaN film several micrometers thick is required.

To achieve thick, crack-free GaN on Si with low wafer bow, stress engineering has to be introduced. The main idea of stress engineering is to introduce sufficient compressive stress by using buffer layers or interlayers, which counterbalance the tensile stress when cooling down from growth temperature to room temperature. Because of the lattice mismatch between GaN and AlN, GaN grown on Al-rich layers such as AlN, AlGaN, and AlN/GaN SLs is under compressive stress during growth. As reported, the LT-Al(Ga)N interlayer [9–13], the AlGaN buffer layer [14–24], and Al(Ga)N/GaN SLs [25–29] are popular stress engineering methods. Besides, a patterned silicon substrate [61], engineered Si substrates [50, 62, 63], the GaN/AlN sequence [64], and the Ga-rich GaN buffer [65] were also used to release the tensile stress.

1.3.1 The LT-Al(Ga)N Interlayer

The LT-Al(Ga)N interlayer is an effective method to grow crack-free GaN on Si. This method was first used for GaN grown on sapphire. Amano and coworkers [66] found that the crack density and dislocation of GaN on sapphire were significantly reduced by introducing a thin AlN layer grown at low temperature between two GaN layers. In 2000, Dadgar and coworkers applied this LT-AlN interlayer to grow GaN on a Si substrate and achieved >1 μm thick crack-free GaN layers [9]. In 2006, by inserting four LT-AlN layers in the GaN, a 5.6 μm crack-free GaN film was grown on a 6-inch Si substrate [10]. Figure 1.8 shows the SEM image of 14 μm thick crack-free GaN layers on a Si substrate with six LT-AlN interlayers [31].

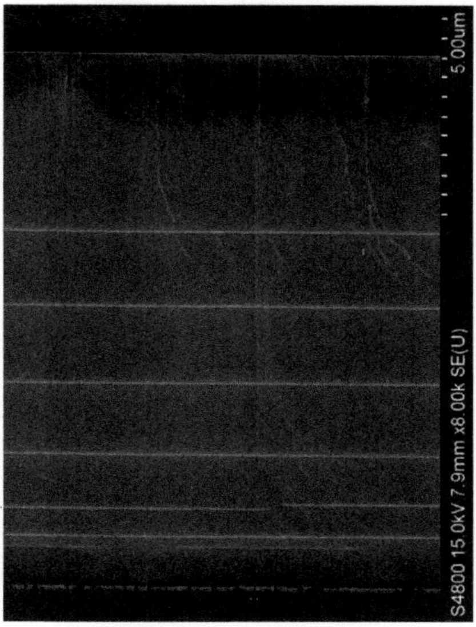

Figure 1.8 SEM image of 14 μm thick crack-free GaN layers on a Si substrate with six LT-AlN interlayers. Reproduced from Ref. [31] with permission from John Wiley and Sons.

The mechanism and influences of growth conditions of an LT-AlN interlayer were investigated substantially [10–13]. Similar to the AlGaN buffer layer, GaN grown on an LT-AlN interlayer was compressively strained due to the lattice mismatch, which counterbalances the tensile thermal stress when cooling down, indicated by the in situ curvature measurements [10], as shown in Fig. 1.9.

The growth temperature and thickness of the AlN interlayer have a strong influence on the stress in GaN films. As reported by Krost et al. [67], when the growth temperature of the AlN interlayer is 600°C to 1000°C, a tensile stress of less than 0.2 GPa was achieved in a ~1.3 μm thick GaN layer with one AlN interlayer, as shown in Fig. 1.10. When the AlN interlayer growth temperature increased to 1100°C and 1200°C, the tensile stress increased rapidly to 0.35 GPa and 0.65 GPa, respectively. They suggested that the LT-AlN interlayers have a decoupling effect on the upper GaN layers since the AlN grown at low temperature was highly defective.

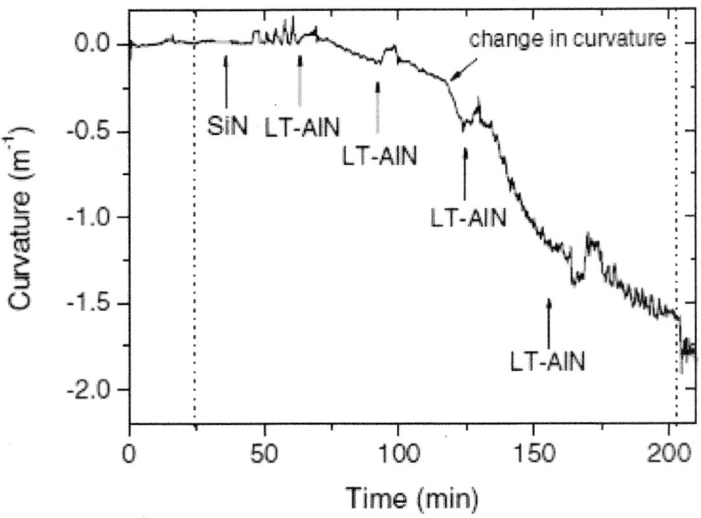

Figure 1.9 In situ curvature date of GaN on a Si substrate with LT-AlN interlayers. Reproduced from Ref. [13] with permission from John Wiley and Sons.

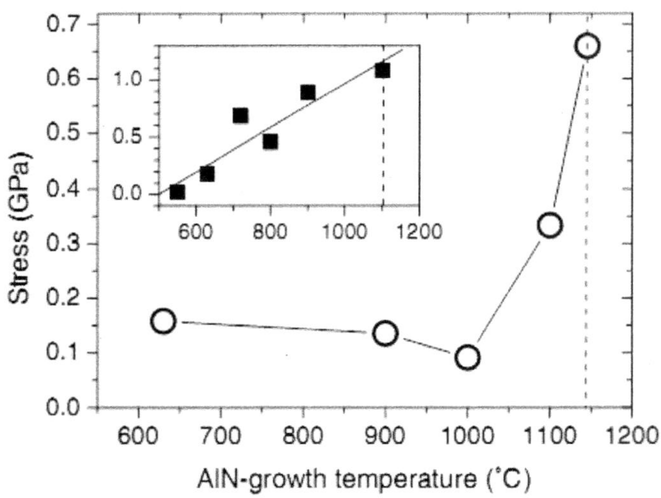

Figure 1.10 Tensile stress of a ~1.3 μm thick GaN layer with one AlN interlayer after 400 nm GaN growth, as determined by curvature measurements. The inset shows the stress as determined by XRD of the series, where the AlN interlayer and the GaN layers were doped with Si. Reproduced from Ref. [13] with permission from John Wiley and Sons.

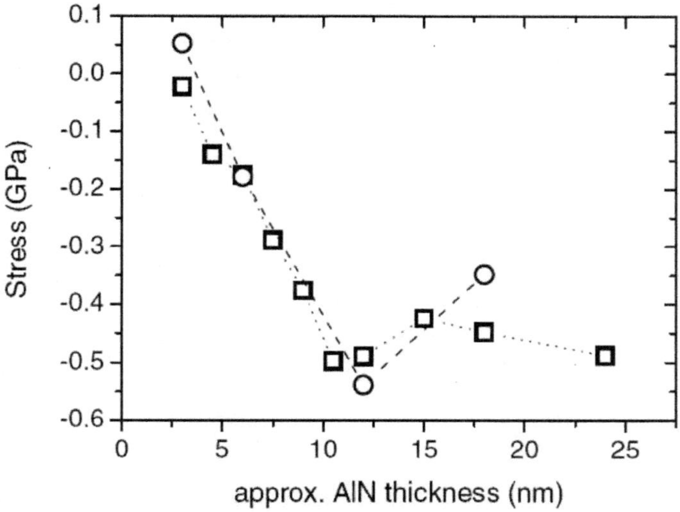

Figure 1.11 In situ–measured stress of a 1.3 μm thick GaN layer on Si(111) at 1145°C. The samples were grown with LT-AlN interlayers of different thickness, deposited at 630°C (□) and 1000°C (○) after ~450 nm of GaN growth. Reproduced from Ref. [13] with permission from John Wiley and Sons.

The thickness of the AlN interlayer is a critical parameter. It influences the stress state of the AlN and subsequent GaN. Figure 1.11 shows the in situ-measured stress of GaN during growth with LT-AlN interlayers of different thickness, which was reported by Krost and coworkers [13]. The results show that AlN interlayers deposited at 630°C and 1000°C had a similar tendency between stress and thickness. When the thickness of AlN rose from 3 nm to 12 nm, the compressive stress in the GaN film increased rapidly, which is beneficial in compensating the tensile thermal stress. When the AlN interlayer thickness was further increased to 25 nm, no significant changes happened. This can be explained by the strain-relaxing process of the AlN layer grown on GaN, as indicated by the TEM image of the AlN layer on GaN with varied thicknesses (see Fig. 1.12 [68]). When the AlN layer on GaN was very thin, the AlN was fully strained, which had the same in-plane lattice constant as GaN. Thus, negligible compressive stress was introduced into the subsequent GaN layer. With an increase of thickness, AlN began to relax because of the generation of dislocations, and the in-plane lattice constant of AlN became smaller, which led to an increase of compressive stress in the subsequent GaN layer. After the AlN layer fully relaxed, the lattice constant would not change with an increase in the AlN layer thickness, resulting in an invariable value of the compressive stress.

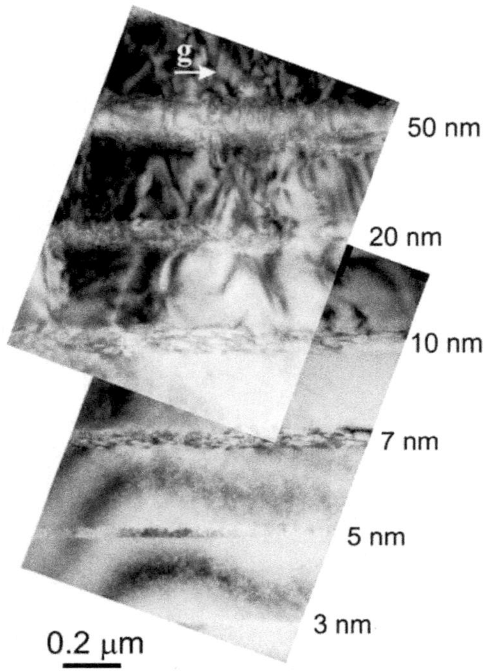

Figure 1.12 Cross-sectional bright-field TEM image of AlN interlayers with various thicknesses grown sequentially with 150 nm GaN spacers. The image has been recorded under two beam conditions off the $[10\bar{1}0]$ zone axis, with **g** = $[\bar{1}2\bar{1}0]$. The specimen has been tilted about the $[\bar{1}2\bar{1}0]$ tilt axis (i.e., the axis parallel to **g**) by ~6° in order to broaden the projection of the interfaces. Misfit dislocation arrays are observed only after the 5 nm interlayer. Reproduced from Ref. [68] with permission from John Wiley and Sons.

The AlGaN interlayer can also introduce compressive stress since its lattice constants are smaller than that of GaN. By inserting three LT-AlGaN interlayers, 4.4 μm thick crack-free GaN grown on an Si substrate was obtained by Fritze et al. in 2012 [69], as shown in Fig. 1.13. Although thick GaN on Si was obtained by LT-Al(Ga)N interlayers, He et al. suggested that high leakage current appeared in the epilayer due to the two-dimensional electron gas (2DEG) in the LT-AlN/GaN interface [70].

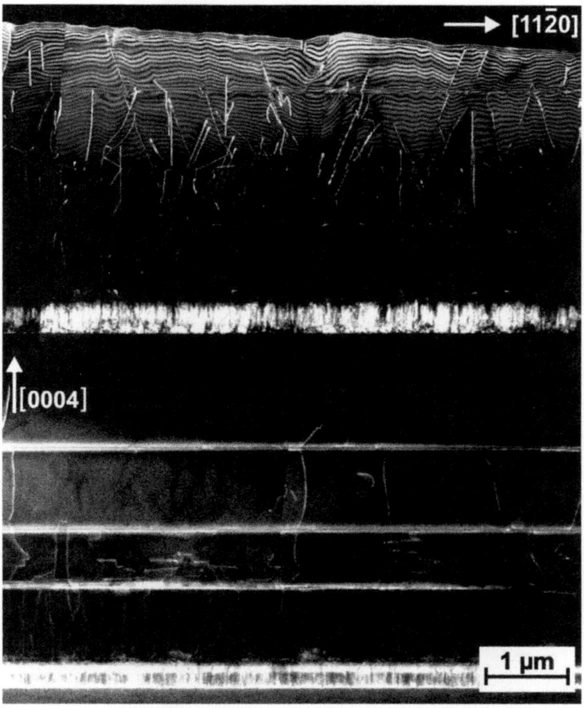

Figure 1.13 Weak-beam, dark-field cross-sectional TEM images of GaN on Si with AlGaN interlayers. In the upper TEM image, all dislocations with an *a* component are visible; dislocations with a *c* component are visible in the lower TEM image. Reprinted from Ref. [69], with the permission of AIP Publishing.

1.3.2 The AlGaN Buffer Layer

Although compressive stress was introduced by the AlN nucleation layer because of the lattice mismatch between GaN and AlN, it relaxed very quickly as dislocations are generated. As we know, the critical thickness of dislocation generation is inversely related to the lattice mismatch. Inserting AlGaN buffer layers between the AlN nucleation layer and GaN reduces the lattice mismatch between the GaN layer and the AlN nucleation layer; thus it increases the critical thickness and postpones the relaxation of the compressive stress. AlGaN buffer layers with fixed [14, 15], continuously graded [16–20], and step-graded [21–24] Al composition were inserted between the AlN nucleation layer and GaN to grow crack-free GaN on Si.

In 1999, Ishikawa et al. [14] adopted a 250 nm AlGaN buffer layer on an AlN seeding layer and achieved 1 μm thick crack-free GaN on a Si substrate for the first time. In 2004, a GaN transistor on 4-inch Si for linear power applications was achieved by Hanson and coworkers in Nitronex Corporation [71], and AlGaN transition layers were used to obtain flat and crack-free epilayers [72]. Able and coworkers used a graded AlGaN buffer with a continuously decreasing Al composition to introduce compressive stresses, and crack-free GaN layers exceeding 2 μm in thickness were grown on a Si(111) substrate [16]. Step-graded AlGaN buffer layers were introduced to grow crack-free GaN on Si by Cheng et al. in 2006 [21], which can keep the compressive stress in the GaN layer and prevent the generation of cracks in the GaN layer. And the dislocation density was also enormously reduced, indicated by the TEM images of GaN on Si with and without an AlGaN buffer [21], as shown in Fig. 1.14.

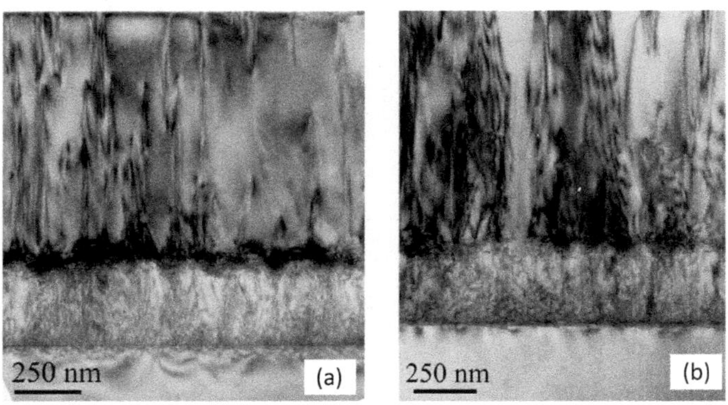

Figure 1.14 Cross-sectional TEM images of the sample with 1 mm GaN on a 320 nm AlN template. Image (a) is taken under the (0002) two-beam diffraction condition, and image (b) is taken under the (11-20) two-beam diffraction condition. Reproduced from Ref. [21]. With permission of Springer.

In 2008, a 3.7 μm thick crack-free (Al)GaN epilayer was grown on a silicon substrate using a graded AlGaN buffer by Arslan et al. [48]. Raghavan and coworkers [17, 18] studied the influence of the thickness of the graded AlGaN buffer layer. A thicker GaN layer could grow under compressive stress with a thicker graded AlGaN buffer layer (Fig. 1.16), which increased the critical thickness of the crack-free epilayer. Using the optimized growth condition of the AlGaN buffer layer, 2 μm thick GaN was kept under compressive stress.

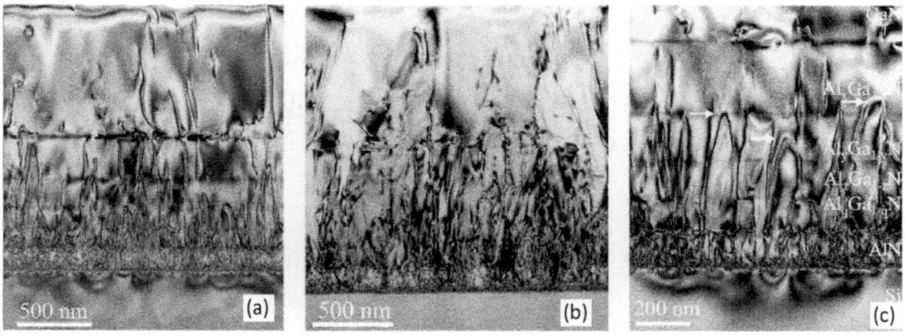

Figure 1.15 Cross-sectional TEM images of the sample with 1 mm GaN grown on a 160 nm AlN template with 750 nm step-graded AlGaN intermediate layers. Images (a) and (c) are taken under the (0002) two-beam diffraction condition, and image (b) is taken under the (11-20) two-beam diffraction condition. In image (c), the arrows point out the formation of dislocation loops or annihilation of dislocations. Reproduced from Ref. [21]. With permission of Springer.

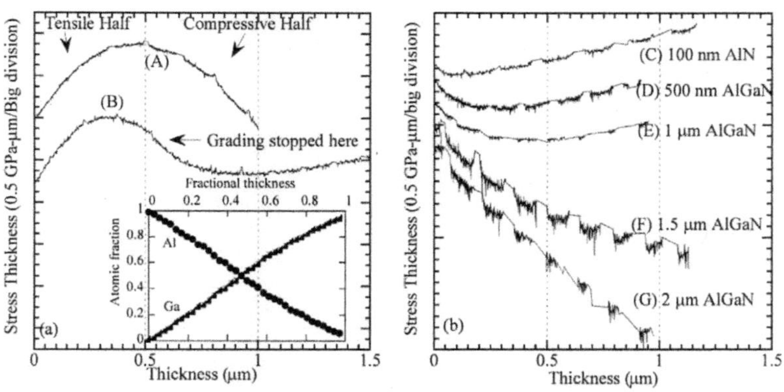

Figure 1.16 Stress thickness versus thickness plots for the linearly graded AlGaN layer and the GaN layer, deposited on different buffer layers, as indicated. Inset in (a) shows the linear change in composition in the graded AlGaN buffer layer. Reprinted from Ref. [18], with the permission of AIP Publishing.

By using graded AlGaN buffer layers, high-quality AlGaN/GaN high-electron-mobility transistors (HEMTs) grown on 150 mm and 200 mm Si substrates were successively achieved by Cheng et al. [73, 74]. A high buffer breakdown voltage of 1380 V was achieved for a sample grown on 200 mm Si substrates with a total buffer thickness of 4.6 μm. The wafer bow was controlled to less than 20 μm for GaN on 200 mm Si substrates, indicating an effective stress engineering of AlGaN buffer layers. In addition, an AlGaN buffer was also utilized to demonstrate AlGaN/GaN/AlGaN double heterostructures (Fig. 1.17), which offers better confinement of the 2DEG and could avoid electron spillover to the barrier layers and buffer layers at high-power conditions.

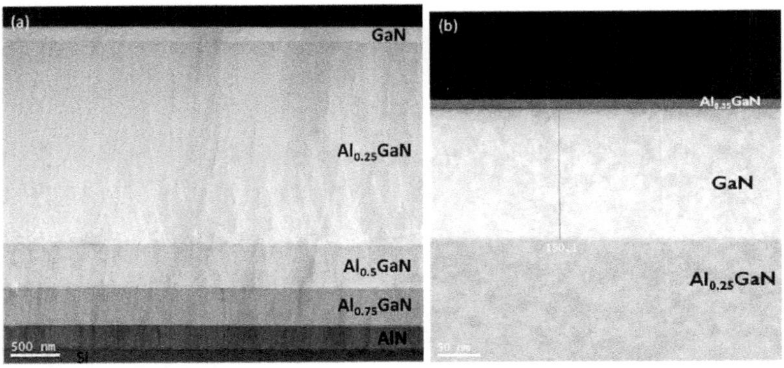

Figure 1.17 Cross-sectional scanning TEM of (a) the whole layer stack and (b) the active region of the AlGaN/GaN HEMT grown on 200 mm Si substrates [74]. Copyright (2012) the Japan Society of Applied Physics.

1.3.3 Al(Ga)N/GaN SLs

An SL structure is an effective way to reduce dislocations in epitaxy growth of Ge on Si and conventional III-As material ascribed to the dislocation bending in the strain field. In 1999, Nikishin et al. introduced AlN/GaN SLs to GaN on Si by molecular beam epitaxy (MBE). In 2001, Feltin et al. demonstrated high-quality 3.8 μm thick GaN on Si by inserting AlN/GaN SLs [25]. The dislocation density was reduced to 2×10^9 cm^{-2}, and the thicknesses of AlN and GaN in SLs were 3 nm and 4 nm, respectively. Although the relaxation of 3 nm AlN on GaN was small, it became considerable in tens of AlN/GaN stacks. Similar to the LT-Al(Ga)N interlayer, compression can be introduced into the subsequent GaN epilayer because of the lattice mismatch.

Sugahara et al. studied the influence AlN/GaN thickness and pairs of AlN/GaN SLs on the stress engineering effect [26]. The optimized thicknesses of AlN and GaN in SLs were 5 nm and 20 nm, respectively. The wafer bending decreased when more pairs of AlN/GaN SLs were inserted. The dislocation filter effect of AlN/GaN SLs was investigated by Ubukata et al. [27] and Wang et al. [28]. The cross-sectional TEM images (Fig. 1.18 and Fig. 1.20) shows that the AlN/GaN SLs are effective in eliminating dislocations because the strain field introduced dislocation bending. The AlN and GaN layers in the AlN/GaN SLs are strained, indicated by the asymmetrical reciprocal-lattice space mapping (RSM) for GaN on a Si substrate with AlN/GaN SLs, as shown in Fig. 1.19.

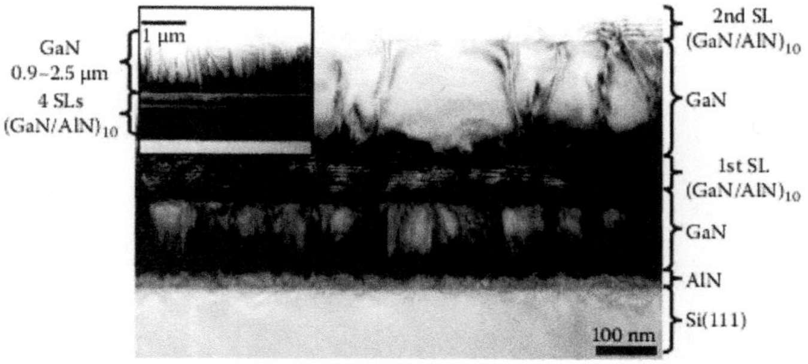

Figure 1.18 Cross-sectional TEM image of an AlN/GaN superlattice deposited on a 160 nm thick GaN layer and a 40 nm thick AlN buffer on Si(111). Reprinted from Ref. [25], with the permission of AIP Publishing.

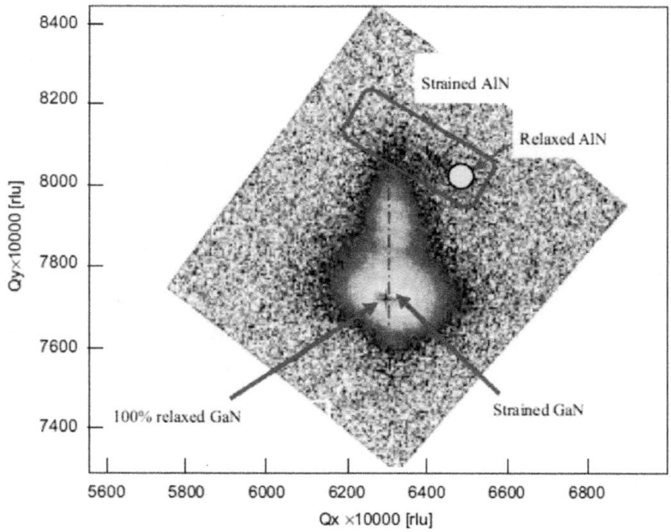

Figure 1.19 Asymmetrical reciprocal-lattice space mapping (RSM) for GaN on a Si substrate with AlN/GaN SLs. Reprinted from Ref. [27], Copyright (2007), with permission from Elsevier.

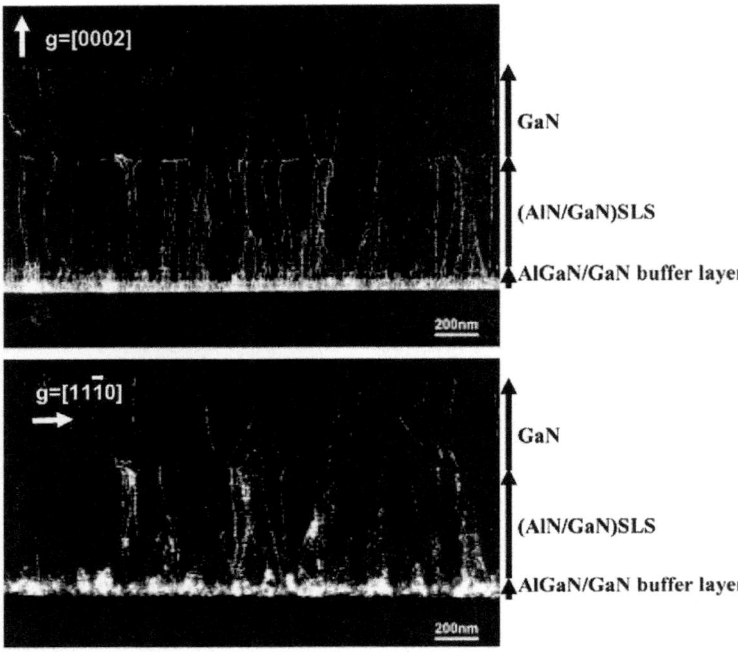

Figure 1.20 Cross-sectional TEM image of GaN on Si. Reprinted from Ref. [27], Copyright (2007), with permission from Elsevier.

In 2009, Selvaraj et al. demonstrated high-quality AlGaN/GaN HEMT structures with an AlN/GaN SLs buffer layer, which shows a high breakdown voltage of 1813 V with the maximum epilayer thickness of 9 μm on a 4-inch Si substrate [29].

Similar to the AlN/GaN SLs, AlGaN/GaN SLs [75] and AlN/AlGaN SLs [76, 77] were also introduced to achieve stress engineering and to reduce the threading dislocation density of GaN on Si. Other methods were also proposed to reduce the tensile stress, such as the patterned silicon substrate [61], the GaN/AlN sequence [64], and the Ga-rich GaN buffer [65]. Corder et al. inserted a GaN (250 nm)/AlN (250 nm) sequence to grow an AlGaN/GaN HEMT on Si with MBE, of which the total crack-free epilayer thickness was about 2 μm [64, 78]. A patterned silicon substrate can also partly release the tensile stress and reduce the cracks, and light-emitting diodes (LEDs) based on GaN-on-Si technology were obtained with this method [61]. However, it is not effective to reduce the wafer bow when enlarging the chip size; thus it is not suitable for GaN-on-Si power devices.

1.4 Leakage Reduction and Breakdown Voltage Enhancement

Although crack-free GaN films with low wafer bow have been obtained on Si substrates by using stress engineering methods, there still are various issues to overcome, such as the high buffer leakage and the current collapse. For GaN on Si, the reduction of buffer leakage and the enlargement of the breakdown voltage are critical issues for power applications. Firstly, the Si substrate and the AlN/Si interface are conducting [79], which causes a prebreakdown and high leakage from the Si substrate when the electric field is still much lower than the critical field of GaN [80]. Secondly, the density of the background donor in unintentionally doped GaN is 10^{16}–10^{17}/cm^3 due to the existence of oxygen and silicon impurities and nitrogen vacancies, which increases the leakage current of the AlGaN/GaN HEMT. Compensational doping, thick buffer layers, the AlGaN/GaN/AlGaN double heterostucture, and Si substrate removal are effective in increasing the breakdown voltage.

1.4.1 Compensational Doping

Deep-level acceptors were introduced to compensate the background electrons and increase the resistivity of GaN since the conductivity of unintentionally doped GaN is caused by donors such as oxygen and silicon impurities. A high-resistance GaN film was achieved by Fe [32–35], Mg [81], and carbon dopants [36–39]. Fe and Mg doping have a strong memory effect, which led to residual dopants in the subsequent layers and in the chambers. Carbon doping can be performed by adjusting the growth conditions without introducing extra sources, since there is carbon in the metalorganic (MO) precursors. Unlike Fe and Mg dopants, a C dopant shows no segregation or memory effect, which is recognized as a favorite agent.

In 1979 Monemar and Lagerstedt found that GaN contaminated by Fe and Cr from the reactor hardware and sources was electrically insulating, which was grown by hydride vapor-phase epitaxy [32]. Photoluminescence spectroscopy and other studies proposed that the Fe^{3+} and Fe^{2+} acceptor level is located 2.6 and 3.17 eV, respectively, above the valence band maximum [82, 83]. In 2002,

Heikman and coworkers grew Fe-doped GaN layers on a sapphire substrate by metal-organic chemical vapor deposition (MOCVD) using ferrocene as the Fe precursor. High resistivity of $7 \times 10^9\ \Omega$/sq. of a GaN film was attained combined with high crystalline quality, very similar to an undoped film [33]. Both carrier concentration and Hall mobility decreased with increasing Fe concentration in GaN films codoped with Si and Fe, as shown in Fig. 1.21. The memory effect of Fe doping was studied [34], which is found to be related to the segregation of Fe from the GaN surface, rather than to the reactor environment (Fig. 1.22), which is different from the memory effect of the Mg dopant. Compared to carbon doping, the advantage of Fe doping is that it allows for growth of semi-insulating GaN without the high threading dislocation densities or high carbon levels. Young Chul and coworkers employed a Fe-doped GaN buffer layer in the growth of AlGaN/GaN HEMTs on Si substrates [35]. The Fe-doped GaN buffer layer successfully suppressed the premature failure caused by substrate-induced breakdown, resulting in enhanced and much more consistent breakdown voltages.

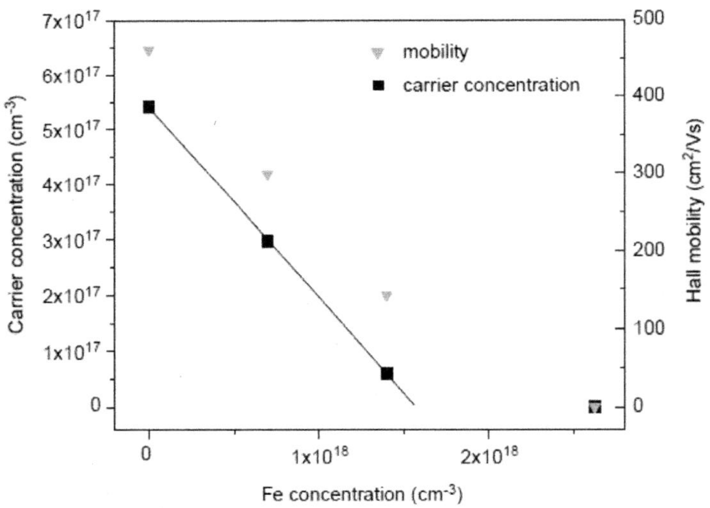

Figure 1.21 Carrier concentration and Hall mobility versus Fe concentration in GaN films codoped with Si and Fe. The Si concentration was kept constant at 6.4×10^{16} cm^{-3}. Reprinted from Ref. [34], Copyright (2003), with permission from Elsevier.

Carbon was reported to act as a shallow acceptor in GaN when substituted for nitrogen (C_N) [84], which trapped the background electrons. In 1999, Webb et al. reported that semi-insulating GaN epilayers were grown with intentional carbon doping by ammonia MBE through introducing methane gas [36]. High-quality C-doped GaN layers with resistivity greater than 10^6 Ω·cm had been obtained to demonstrate a high-quality HEMT, and resistivity greater than 10^8 Ω·cm was also achieved by Tang et al. [37]. For GaN grown by MOCVD, the carbon can be doped by adjusting the growth condition of GaN. Usually, lower growth pressure, temperature, and V/III ratio result in high carbon concentration in the GaN epilayer. Wickenden et al. reported that unintentionally doped GaN films grown by MOCVD at varying pressures exhibit reduced carbon and oxygen impurity incorporation and reduced resistivity with increasing growth pressure [38], as shown in Fig. 1.23. When the growth pressures varied from 65 to 500 torr, as shown in Fig. 1.24, the carbon doping level was reduced from 4×10^{17} to 5×10^{16} cm^{-3}, resulting in the resistivity of GaN to vary from $>10^{10}$ Ω·cm at low pressure to 10^3 Ω·cm at 500 torr. Ikeda et al. reported the enhancement of the breakdown voltage of the GaN HEMT on Si with carbon doping, which increased from 100 V with 10^{16}–10^{17} carbon doping level to 800 V with near-10^{19} carbon doping level [39].

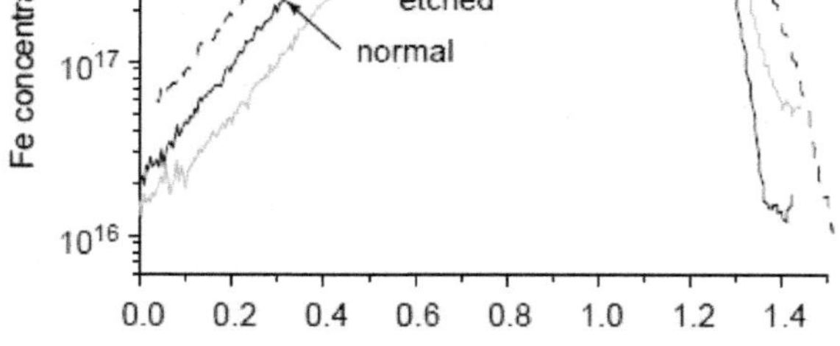

Figure 1.22 Secondary ion mass spectrometry (SIMS) profiles of Fe-doped GaN layers followed by nominally undoped layers. The slow Fe concentration turn-off was improved by treating the sample in acids between the growth of the GaN:Fe layer and the nominally undoped layer. The broadening of the "interrupted" profile is believed to be an artifact of the SIMS measurement. Reprinted from Ref. [34], Copyright (2003), with permission from Elsevier.

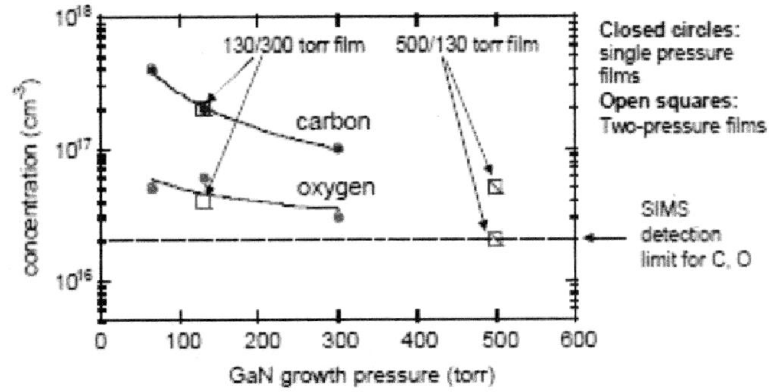

Figure 1.23 SIMS analysis of carbon and oxygen for UID-GaN films grown at MOCVD pressures of 65, 130, and 300 torr (closed circles) demonstrates the trend of reduced carbon and carbon:oxygen ratio with increasing pressure. Reprinted from Ref. [38], Copyright (2004), with permission from Elsevier.

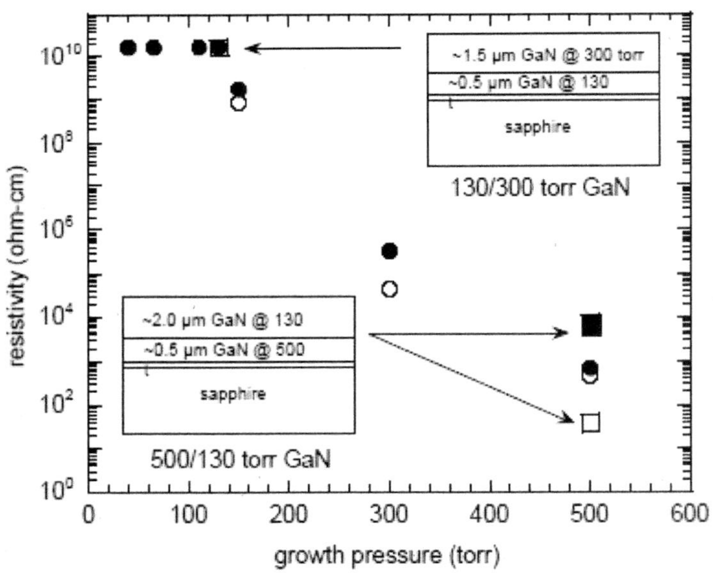

Figure 1.24 Resistivity of UID-GaN films as a function of epitaxial growth pressure measured in the dark (closed symbols) and in ambient room light (open symbols). The values were derived from *I*/*V* characterization using indium stripe contacts on squares patterned in the GaN films. The experimental limit of the resistivity analysis is reached at a level of 10^{10} Ω·cm. Data for the two-pressure films are again plotted relative to the pressure of the underlying film. Inserts for the two-pressure films illustrate the respective film structures. Reprinted from Ref. [38], Copyright (2004), with permission from Elsevier.

1.4.2 Other Methods

Increasing of the thickness of the buffer layer will enlarge the distance from the electrode to the Si substrate, which thus can raise the breakdown voltage. The results reported by Ikeda et al. showed a proportional relation between the breakdown voltage and the total thickness of the epilayer, and a 2000 V breakdown voltage was achieved with a 7 μm epilayer [85]. Dadgar et al. also reported that the breakdown voltage increased from ~1200 V to ~2000 V with the increase of epilayer thickness from 4.5 μm to 7.8 μm [31], as shown in Fig. 1.25. But a thick buffer layer often leads to large wafer bow and long growth time. An effective method is to increase the critical field of the buffer layer, which can achieve a high voltage with a limited buffer thickness. The critical field of AlGaN is higher than that of GaN; thus a high breakdown voltage can be achieved with an AlGaN buffer layer. An AlGaN/GaN HEMT grown on 150 mm and 200 mm Si substrates with a high breakdown voltage and low wafer bow was achieved using an AlGaN buffer layer [73, 74]. Due to the low critical field and high conductivity of Si, the breakdown voltage of the AlGaN/GaN HEMT is greatly reduced. One solution is to remove the Si substrate. A breakdown voltage as high as 2200 V was achieved by Srivastava et al. with removal of the Si substrate [40]. However, this method will increase the complexity and cost of the process.

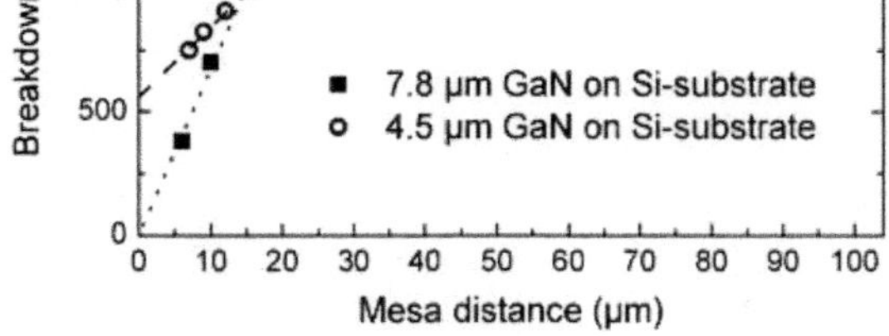

Figure 1.25 High-voltage breakdown measurements of field-effect transistor (FET)-type layer structures processed with different contact distances and with a removed AlGaN/GaN interface between the contacts, all measured with a floating substrate potential. The thicker the structure, the higher the voltage that can be achieved. Reproduced from Ref. [31] with permission from John Wiley and Sons.

1.5 Conclusions

Various ways of stress engineering have paved the way for achieving crack-free GaN-on-Si materials. However, plenty of defects, including threading dislocations, microcracks, micropits, and point defects, are still present in the buffer layers and even active regions of the epistructures on Si. The influence of these defects on device operation is still under investigation. Efforts from both epitaxy and device processing are required to improve the epilayer quality further.

References

1. Liu, L., and Edgar, J. H. (2002). Substrates for gallium nitride epitaxy, *Mater. Sci. Eng., R*, **37**, 61.
2. Yim, W. M., and Paff, R. J. (1974). Thermal expansion of AlN, sapphire, and silicon, *J. Appl. Phys.*, **45**, 1456.
3. Krost, A., and Dadgar, A. (2002). GaN-based optoelectronics on silicon substrates, *Mater. Sci. Eng., B*, **93**, 77–84.
4. Zhang, B., and Liu, Y. (2014). A review of GaN-based optoelectronic devices on silicon substrate, *Chin. Sci. Bull.*, **59**, 1251–1275.
5. Komiyama, J., Abe, Y., Suzuki, S., Nakanishi, H., and Koukitu, A. (2009). MOVPE of AlN-free hexagonal GaN/cubic SiC/Si heterostructures for vertical devices, *J. Cryst. Growth*, **311**, 2840–2843.
6. Strittmatter, A., Bimberg, D., Krost, A., Bläsing, J., and Veit, P. (2000). Structural investigation of GaN layers grown on Si(1 1 1) substrates using a nitridated AlAs buffer layer, *J. Cryst. Growth*, **221**, 293–296.
7. Fenwick, W. E., Li, N., Xu, T., Melton, A., Wang, S., Yu, H., Summers, C., Jamil, M., and Ferguson, I. T. (2009). MOCVD growth of GaN on Si(111) substrates using an ALD-grown Al2O3 interlayer, *J. Cryst. Growth*, **311**, 4306–4310.
8. Dadgar, A., Poschenrieder, M., Bläsing, J., Contreras, O., Bertram, F., Riemann, T., Reiher, A., Kunze, M., Daumiller, I., Krtschil, A., Diez, A., Kaluza, A., Modlich, A., Kamp, M., Christen, J., Ponce, F. A., Kohn, E., and Krost, A. (2003). MOVPE growth of GaN on Si(1 1 1) substrates, *J. Cryst. Growth*, **248**, 556–562.
9. Dadgar, A., Bläsing, J., Diez, A., Alam, A., Heuken, M., and Krost, A. (2000). Metalorganic chemical vapor phase epitaxy of crack-free GaN on Si (111) exceeding 1 μm in thickness, *Jpn. J. Appl. Phys.*, **39**, L1183.

10. Dadgar, A., Hums, C., Diez, A., Bläsing, J., and Krost, A. (2006). Growth of blue GaN LED structures on 150-mm Si(111), *J. Cryst. Growth*, **297**, 279–282.

11. Bläsing, J., Reiher, A., Dadgar, A., Diez, A., and Krost, A. (2002). The origin of stress reduction by low-temperature AlN interlayers, *Appl. Phys. Lett.*, **81**, 2722.

12. Reiher, A., Blasing, J., Dadgar, A., Diez, A., and Krost, A. (2003). Efficient stress relief in GaNheteroepitaxy on Si(111) using low-temperature AlNinterlayers, *J. Cryst. Growth*, **248**, 563–567.

13. Krost, A., Dadgar, A., Strassburger, G., and Clos, R. (2003). GaN-based epitaxy on silicon: stress measurements, *Phys. Status Solidi A*, **200**, 26.

14. Ishikawa, H., Zhao, G. Y., Nakada, N., Egawa, T., Soga, T., Jimbo, T., and Umeno, M. (1999). High-quality GaN on Si substrate using AlGaN/AlN intermediate layer, *Phys. Status Solidi A*, **176**, 599.

15. Zhang, B., Egawa, T., Ishikawa, H., Nishikawa, N., Jimbo, T., and Umeno, M. (2001). InGaN multiple-quantum-well light emitting diodes on Si(111) substrates, *Phys. Status Solidi A*, **188**, 151–154.

16. Able, A., Wegscheider, W., Engl, K., and Zweck, J. (2005). Growth of crack-free GaN on Si(111) with graded AlGaN buffer layers, *J. Cryst. Growth*, **276**, 415–418.

17. Raghavan, S., and Redwing, J. (2005). Growth stresses and cracking in GaN films on (111) Si grown by metalorganic chemical vapor deposition. II. Graded AlGaN buffer layers, *J. Appl. Phys.*, **98**, 023515.

18. Raghavan, S., Weng, X., Dickey, E., and Redwing, J. M. (2006). Correlation of growth stress and structural evolution during metalorganic chemical vapor deposition of GaN on (111) Si, *Appl. Phys. Lett.*, **88**, 041904.

19. Zhu, D., McAleese, C., McLaughlin, K. K., Haberlen, M., Salcianu, C. O., Thrush, E. J., Kappers, M. J., Phillips, W. A., Lane, P., Wallis, D. J., Martin, T., Astles, M., Thomas, S., Pakes, A., Heuken, M., and Humphreys, C. J. (2009). GaN-based LEDs grown on 6-inch diameter Si (111) substrates by MOVPE, *Proc. SPIE*, **7231**, 723118.

20. Li, J., Lin, J. Y., and Jiang, H. X. (2006). Growth of III-nitride photonic structures on large area silicon substrates, *Appl. Phys. Lett.*, **88**, 171909.

21. Cheng, K., Leys, M., Degroote, S., Van Daele, B., Boeykens, S., Derluyn, J., Germain, M., Van Tendeloo, G., Engelen, J., and Borghs, G. (2006). Flat GaN epitaxial layers grown on Si(111) by metalorganic vapor phase epitaxy using step-graded AlGaN intermediate layers, *J. Electron. Mater.*, **35**, 592–598.

22. Kim, M.-H., Do, Y.-G., Kang, H. C., Noh, D. Y., and Park, S.-J. (2001). Effects of step-graded Al $_{x}Ga_{1-x}N$ interlayer on properties of GaN grown on Si(111) using ultrahigh vacuum chemical vapor deposition, *Appl. Phys. Lett.*, **79**, 2713.

23. Huang, C. C., Chang, S. J., Chuang, R. W., Lin, J. C., Cheng, Y. C., and Lin, W. J. (2010). GaN grown on Si(111) with step-graded AlGaN intermediate layers, *Appl. Surf. Sci.*, **256**, 6367–6370.

24. Zhu, D., McAleese, C., Häberlen, M., Salcianu, C., Thrush, T., Kappers, M., Phillips, A., Lane, P., Kane, M., Wallis, D., Martin, T., Astles, M., and Humphreys, C. (2010). InGaN/GaN LEDs grown on Si(111): dependence of device performance on threading dislocation density and emission wavelength, *Phys. Status Solidi C*, **7**, 2168–2170.

25. Feltin, E., Beaumont, B., Laügt, M., de Mierry, P., Vennégués, P., Lahrèche, H., Leroux, M., and Gibart, P. (2001). Stress control in GaN grown on silicon (111) by metalorganic vapor phase epitaxy, *Appl. Phys. Lett.*, **79**, 3230.

26. Tomoya, S., Jeong-Sik, L., and Kohji, O. (2004). Role of AlN/GaN multilayer in crack-free GaN layer growth on 5″φ Si (111) substrate, *Jpn. J. Appl. Phys.*, **43**, L1595.

27. Ubukata, A., Ikenaga, K., Akutsu, N., Yamaguchi, A., Matsumoto, K., Yamazaki, T., and Egawa, T. (2007). GaN growth on 150-mm-diameter (111) Si substrates, *J. Cryst. Growth*, **298**, 198–201.

28. Wang, H.-M., Zhang, J.-P., Chen, C.-Q., Fareed, Q., Yang, J.-W., and Khan, M. A. (2002). AlN/AlGaN superlattices as dislocation filter for low-threading-dislocation thick AlGaN layers on sapphire, *Appl. Phys. Lett.*, **81**, 604.

29. Selvaraj, S. L., Suzue, T., and Egawa, T. (2009). Breakdown enhancement of AlGaN/GaN HEMTs on 4-in silicon by improving the GaN quality on thick buffer layers, *IEEE Electron Device Lett.*, **30**, 587–589.

30. Dadgar, A., Fritze, S., Schulz, O., Hennig, J., Bläsing, J., Witte, H., Diez, A., Heinle, U., Kunze, M., Daumiller, I., Haberland, K., and Krost, A. (2013). Anisotropic bow and plastic deformation of GaN on silicon, *J. Cryst. Growth*, **370**, 278–281.

31. Dadgar, A., Hempel, T., Bläsing, J., Schulz, O., Fritze, S., Christen, J., and Krost, A. (2011). Improving GaN-on-silicon properties for GaN device epitaxy, *Phys. Status Solidi C*, **8**, 1503–1508.

32. Monemar, B., and Lagerstedt, O. (1979). Properties of VPE-grown GaN doped with Al and some iron-group metals, *J. Appl. Phys.*, **50**, 6480.

33. Heikman, S., Keller, S., DenBaars, S. P., and Mishra, U. K. (2002). Growth of Fe doped semi-insulating GaN by metalorganic chemical vapor deposition, *Appl. Phys. Lett.*, **81**, 439–441.

34. Heikman, S., Keller, S., Mates, T., DenBaars, S. P., and Mishra, U. K. (2003). Growth and characteristics of Fe-doped GaN, *J. Cryst. Growth*, **248**, 513–517.

35. Young Chul, C., Pophristic, M., Ho-Young, C., Peres, B., Spencer, M. G., and Eastman, L. F. (2006). The effect of an Fe-doped GaN buffer on OFF-state breakdown characteristics in AlGaN/GaN HEMTs on Si substrate, *IEEE Trans. Electron Devices*, **53**, 2926–2931.

36. Webb, J. B., Tang, H., Rolfe, S., and Bardwell, J. A. (1999). Semi-insulating C-doped GaN and high-mobility AlGaN/GaN heterostructures grown by ammonia molecular beam epitaxy, *Appl. Phys. Lett.*, **75**, 953.

37. Tang, H., Webb, J. B., Bardwell, J. A., Raymond, S., Salzman, J., and Uzan-Saguy, C. (2001). Properties of carbon-doped GaN, *Appl. Phys. Lett.*, **78**, 757.

38. Wickenden, A. E., Koleske, D. D., Henry, R. L., Twigg, M. E., and Fatemi, M. (2004). Resistivity control in unintentionally doped GaN films grown by MOCVD, *J. Cryst. Growth*, **260**, 54–62.

39. Ikeda, N., Niiyama, Y., Kambayashi, H., Sato, Y., Nomura, T., Kato, S., and Yoshida, S. (2010). GaN power transistors on Si substrates for switching applications, *Proc. IEEE*, **98**, 1151–1161.

40. Srivastava, P., Das, J., Visalli, D., Van Hove, M., Malinowski, P. E., Marcon, D., Lenci, S., Geens, K., Cheng, K., and Leys, M. (2011). Record breakdown voltage (2200 V) of GaN DHFETs on Si with 2-μm buffer thickness by local substrate removal, *IEEE Electron Device Lett.*, **32**, 30–32.

41. Zamir, S., Meyler, B., Zolotoyabko, E., and Salzman, J. (2000). The effect of AlN buffer layer on GaN grown on (1 1 1)-oriented Si substrates by MOCVD, *J. Cryst. Growth*, **218**, 181–190.

42. Meng, W. J., Sell, J. A., Perry, T. A., Rehn, L. E., and Baldo, P. M. (1994). Growth of aluminum nitride thin films on Si(111) and Si(001): structural characteristics and development of intrinsic stresses, *J. Appl. Phys.*, **75**, 3446.

43. Marchand, H., Zhao, L., Zhang, N., Moran, B., Coffie, R., Mishra, U. K., Speck, J. S., DenBaars, S. P., and Freitas, J. A. (2001). Metalorganic chemical vapor deposition of GaN on Si(111): stress control and application to field-effect transistors, *J. Appl. Phys.*, **89**, 7846.

44. Poschenrieder, M., Schulze, F., Bläsing, J., Dadgar, A., Diez, A., Christen, J., and Krost, A. (2002). Bright blue to orange photoluminescence

emission from high-quality InGaN/GaN multiple-quantum-wells on Si(111) substrates, *Appl. Phys. Lett.*, **81**, 1591.

45. Zhang, B. S., Wu, M., Shen, X. M., Chen, J., Zhu, J. J., Liu, J. P., Feng, G., Zhao, D. G., Wang, Y. T., and Yang, H. (2003). Influence of high-temperature AlN buffer thickness on the properties of GaN grown on Si(111), *J. Cryst. Growth*, **258**, 34–40.

46. Raghavan, S., and Redwing, J. M. (2004). In situ stress measurements during the MOCVD growth of AlN buffer layers on (111) Si substrates, *J. Cryst. Growth*, **261**, 294–300.

47. Lin, K.-L., Chang, E.-Y., Huang, J.-C., Huang, W.-C., Hsiao, Y.-L., Chiang, C.-H., Li, T., Tweet, D., Maa, J.-S., and Hsu, S.-T. (2008). MOVPE high quality GaN film grown on Si (111) substrates using a multilayer AlN buffer, *Phys. Status Solidi C*, **5**, 1536–1538.

48. Arslan, E., Ozturk, M. K., Teke, A., Ozcelik, S., and Ozbay, E. (2008). Buffer optimization for crack-free GaN epitaxial layers grown on Si(1 1 1) substrate by MOCVD, *J. Phys. D: Appl. Phys.*, **41**, 155317.

49. Yang, J. H., Kang, S. M., Dinh, D. V., and Yoon, D. H. (2009). Influence of AlN buffer layer thickness and deposition methods on GaN epitaxial growth, *Thin Solid Films*, **517**, 5057–5060.

50. Yang, J. W., Lunev, A., Simin, G., Chitnis, A., Shatalov, M., Khan, M. A., Van Nostrand, J. E., and Gaska, R. (2000). Selective area deposited blue GaN–InGaN multiple-quantum well light emitting diodes over silicon substrates, *Appl. Phys. Lett.*, **76**, 273.

51. Chen, P., Zhang, R., Zhao, Z. M., Xi, D. J., Shen, B., Chen, Z. Z., Zhou, Y. G., Xie, S. Y., Lu, W. F., and Zheng, Y. D. (2001). Growth of high quality GaN layers with AlN buffer on Si(1 1 1) substrates, *J. Cryst. Growth*, **225**, 150.

52. Fanning, D., Witkowski, L., Lee, C., Dumka, D., Tserng, H., Saunier, P., Gaiewski, W., Piner, E., Linthicum, K., and Johnson, J. (2005). 25W X-band GaN on Si MMIC, *GaAs MANTECH Digest* (unpublished).

53. Chen, P., Zhang, R., Zhao, Z. M., Xi, D. J., Shen, B., Chen, Z. Z., Zhou, Y. G., Xie, S. Y., Lu, W. F., and Zheng, Y. D. (2001). Growth of high quality GaN layers with AlN buffer on Si(1 1 1) substrates, *J. Cryst. Growth*, **225**, 150–154.

54. Lahreche, H., Vennegués, P., Tottereau, O., Laugt, M., Lorenzini, P., Leroux, M., Beaumont, B., and Gibart, P. (2000). Optimisation of AlN and GaN growth by metalorganic vapour-phase epitaxy (MOVPE) on Si(1 1 1), *J. Cryst. Growth*, **217**, 13–25.

55. Xiang, P., Liu, M., Yang, Y., Chen, W., He, Z., Leung, K. K., Surya, C., Han, X., Wu, Z., Liu, Y., and Zhang, B. (2013). Improving the quality of GaN

on Si(111) substrate with a medium-temperature/high-temperature bilayer AlN buffer, *Jpn. J. Appl. Phys.*, **52**, 08JB18.

56. Lobanova, A. V., Mazaev, K. M., Talalaev, R. A., Leys, M., Boeykens, S., Cheng, K., and Degroote, S. (2006). Effect of V/III ratio in AlN and AlGaN MOVPE, *J. Cryst. Growth*, **287**, 601–604.

57. Krishnan, B., Lee, S., Li, H., Su, J., Lee, D., and Paranjpe, A. (2013). Growth of $Al_xGa_{1-x}N$ structures on 8 in. Si (111) substrates, *Sens. Mater.*, **25**, 205–217.

58. *Kelly*, M. K., Vaudo, R. P., Phanse, V. M., Gorgens, L., Ambacher, O., and Stutzmann, M. (1999). Large free-standing GaN substrate by hydride vapor phase epitaxy and laser-induced liffoff, *Jpn. J. Appl. Phys.*, **38**, L217–L219.

59. Creighton, J. R., Wang, G. T., Breiland, W. G., and Coltrin, M. E. (2004). Nature of the parasitic chemistry during AlGaInN OMVPE, *J. Cryst. Growth*, **261**, 204–213.

60. Cheng, K. (2015). Trimming the thickness of high-voltage GaN-on-silicon HEMTs, *Compd. Semicond.*, **21**, 36.

61. Dadgar, A., Alam, A., Riemann, T., Blasing, J., Diez, A., Poschenrieder, M., Strassburg, M., Heuken, M., Christen, J., and Krost, A. (2001). Crack-free InGaN/GaN light emitters on Si(111), *Phys. Status Solidi A*, **188**, 155.

62. Tungare, M., Leathersich, J. M., Tripathi, N., and Suvarna, P. (2011). Crack-free iii-nitride structures (> 3.5 μm) on silicon, *MRS Proc.*, **1324**, 9–15.

63. Dadgar, A., Alam, A., Riemann, T., Blasing, J., Diez, A., Poschenrieder, M., Strassburg, M., Heuken, M., Christen, J., and Krost, A. (2001). Crack-free InGaN/GaN light emitters on Si(111), *Phys. Status Solidi A*, **188**, 155–158.

64. Cordier, Y., Semond, F., Lorenzini, P., Grandjean, N., Natali, F., Damilano, B., Massies, J., Hoël, V., Minko, A., Vellas, N., Gaquière, C., DeJaeger, J. C., Dessertene, B., Cassette, S., Surrugue, M., Adam, D., Grattepain, J. C., Aubry, R., and Delage, S. L. (2003). MBE growth of ALGaN/GaN HEMTs on resistive Si(111) substrate with RF small signal and power performances, *J. Cryst. Growth*, **251**, 811–815.

65. Mo, C., Fang, W., Pu, Y., Liu, H., and Jiang, F. (2005). Growth and characterization of InGaN blue LED structure on Si(111) by MOCVD, *J. Cryst. Growth*, **285**, 312.

66. Amano, H., Iwaya, M., Kashima, T., Katsuragawa, M., Akasaki, I., Han, J., Hearne, S., Floro, J. A., Chason, E., and Figiel, J. (1998). Stress and defect control in GaN using low temperature interlayers, *Jpn. J. Appl. Phys.*, **37**, 1540–1542.

67. Li, T., Dadgar, A., and Mastro, M. (2011). III-V compound semiconductors integration with silicon based microelectronics, (CRC Press, Boca Raton).

68. Dimitrakopulos, G. P., Kalesaki, E., Komninou, P., Kehagias, T., Kioseoglou, J., and Karakostas, T. (2009). Strain accommodation and interfacial structure of AlN interlayers in GaN, *Cryst. Res. Technol.*, **44**, 1170–1180.

69. Fritze, S., Drechsel, P., Stauss, P., Rode, P., Markurt, T., Schulz, T., Albrecht, M., Bläsing, J., Dadgar, A., and Krost, A. (2012). Role of low-temperature AlGaN interlayers in thick GaN on silicon by metalorganic vapor phase epitaxy, *J. Appl. Phys.*, **111**, 124505.

70. Zhiyuan, H., Yiqiang, N., Fan, Y., Jin, W., Yao, Y., Zhen, S., Peng, X., Minggang, L., Shuo, W., Jincheng, Z., Zhisheng, W., Baijun, Z., and Yang, L. (2014). Investigations of leakage current properties in semi-insulating GaN grown on Si(1 1 1) substrate with low-temperature AlN interlayers, *J. Phys. D: Appl. Phys.*, **47**, 045103.

71. Hanson, A., Borges, R., Brown, J., Cook Jr, J., Gehrke, T., Johnson, J., Linthicum, K., Peters, S., Piner, E., and Rajagopal, P. (2004). Development of a GaN transistor process for linear power applications, in *2004 International Conference on Compound Semiconductor Manufacturing Technology* (GaAs MANTECH), Miami, FL.

72. Rajagopal, P., Gehrke, T., Roberts, J. C., Brown, J., Weeks, T. W., Piner, E. L., and Linthicum, K. J. (2003). Large-area, device quality GaN on Si using a novel transition layer scheme, *Materials Research Society Symposium Proceedings*, Warrendale, PA, pp. 3–8.

73. Cheng, K., Leys, M., Degroote, S., Derluyn, J., Sijmus, B., Favia, P., Richard, O., Bender, H., Germain, M., and Borghs, G. (2008). AlGaN/GaN high electron mobility transistors grown on 150 mm Si(111) substrates with high uniformity, *Jpn. J. Appl. Phys.*, **47**, 1553–1555.

74. Cheng, K., Liang, H., Van Hove, M., Geens, K., De Jaeger, B., Srivastava, P., Kang, X., Favia, P., Bender, H., and Decoutere, S. (2012). AlGaN/GaN/AlGaN double heterostructures grown on 200 mm silicon (111) substrates with high electron mobility, *Appl. Phys. Express*, **5**, 011002.

75. Einfeldt, S., Heinke, H., Kirchner, V., and Hommel, D. (2001). Strain relaxation in AlGaN/GaN superlattices grown on GaN, *J. Appl. Phys.*, **89**, 2160.

76. Wang, H.-M., Zhang, J.-P., Chen, C.-Q., Fareed, Q., Yang, J.-W., and Khan, M. A. (2002). AlN/AlGaN superlattices as dislocation filter for low-

threading-dislocation thick AlGaN layers on sapphire, *Appl. Phys. Lett.*, **81**, 604.

77. Zhang, J. P., Wang, H. M., Gaevski, M. E., Chen, C. Q., Fareed, Q., Yang, J. W., Simin, G., and Khan, M. A. (2002). Crack-free thick AlGaN grown on sapphire using AlN/AlGaN superlattices for strain management, *Appl. Phys. Lett.*, **80**, 3542.

78. Cordier, Y., Baron, N., Semond, F., Massies, J., Binetti, M., Henninger, B., Besendahl, M., and Zettler, T. (2007). In situ measurements of wafer bending curvature during growth of group-III-nitride layers on silicon by molecular beam epitaxy, *J. Cryst. Growth*, **301–302**, 71–74.

79. Visalli, D., Van Hove, M., Derluyn, J., Srivastava, P., Marcon, D., Das, J., Leys, M. R., Degroote, S., Cheng, K., and Vandenplas, E. (2010). Limitations of field plate effect due to the silicon substrate in AlGaN/GaN/AlGaN DHFETs, *IEEE Trans. Electron Devices*, **57**, 3333–3339.

80. Lu, B., and Palacios, T. (2010). High breakdown (>1500 V) AlGaN/GaN HEMTs by substrate-transfer technology, *IEEE Electron Device Lett.*, **31**, 951.

81. Wang, Y., Yu, N., Deng, D., Li, M., Sun, F., and Lau, K. (2010). Improved breakdown voltage of AlGaN/GaN HEMTs grown on Si substrates using partially Mg-doped GaN buffer layer by MOCVD, *Science China Physics, Mechanics and Astronomy*, **53**, 1578–1581.

82. Baur, J., Maier, K., Kunzer, M., Kaufmann, U., Schneider, J., Amano, H., Akasaki, I., Detchprohm, T., and Hiramatsu, K. (1994). Infrared luminescence of residual iron deep level acceptors in gallium nitride (GaN) epitaxial layers, *Appl. Phys. Lett.*, **64**, 857.

83. Heitz, R., Maxim, P., Eckey, L., Thurian, P., Hoffmann, A., Broser, I., Pressel, K., and Meyer, B. K. (1997). Excited states of Fe^{3+} in GaN, *Phys. Rev. B*, **55**, 4382.

84. Seager, C. H., Wright, A. F., Yu, J., and Götz, W. (2002). Role of carbon in GaN, *J. Appl. Phys.*, **92**, 6553.

85. Ikeda, N., Kaya, S., Li, J., Sato, Y., Kato, S., and Yoshida, S. (2008). High power AlGaN/GaN HFET with a high breakdown voltage of over 1.8 kV on 4 inch Si substrates and the suppression of current collapse, in *20th International Symposium on Power Semiconductor Devices and IC's*, Orlando, FL, pp. 287–290.

Chapter 2

The Characteristics of Polarization Effects in GaN Heterostructures

Zhigang Wang
The School of Information Science and Technology, Southwest Jiaotong University, Chengdu, China
zhigangwang@swjtu.edu.cn

In this chapter, a comprehensive insight of polarization charges and 2DEG origin is presented with respect to material properties. The origin of 2DEG induced by polarization in AlGaN/GaN heterostructures is demonstrated. Specially, the ab initio theory with eliminating discrepancies of experimental results is also introduced for calculating polarization effects in III–V heterostructures. When the nonlinear polarization effects are taken into account, the ab initio theory can predict an explicit calculation of polarization in simulations or experiments.

2.1 Introduction

Gallium nitride (GaN) and its alloys with III nitride materials such as AlN and InN became the third-generation semiconductors

---

Gallium Nitride Power Devices
Edited by Hongyu Yu and Tianli Duan

ISBN 978-981-4774-09-3 (Hardcover), 978-1-315-19662-6 (eBook)
www.panstanford.com

for power applications. The main reason is attributed to their outstanding characteristics in the III–V family of semiconductors: wide bandgap, high-field electron velocity, large breakdown voltage (BV), large thermal conductivity, and robustness. The wide-bandgap materials of GaN, AlN, InN, and the ternary alloy are direct-bandgap semiconductors with a wurtzite structure. Figure 2.1 shows the major intrinsic limits of BV versus the specific on-resistance ($R_{on,sp}$) of Si, 4H-SiC, 6H-SiC, and GaN material [1]. In this limit chart, we can obviously find that wide-bandgap materials can be used to obtain a higher performance for fabricating power devices in comparison with conventional Si material. The trade-off between BV and the specific $R_{on,sp}$ as the main issue can be used to evaluate the merit of these operation characteristics in power devices. Therefore, from the limit graph of BV versus $R_{on,sp}$, the GaN material regarded as the third-generation semiconductor is an attractive candidate for replacing Si power devices in some power applications. Especially, due to the high density of a two-dimensional electron gas (2DEG) and high carrier mobility in the AlGaN/GaN heterostructure field-effect transistors (HFETs), a wider frequency range at a certain BV can be achieved. Figure 2.2 shows the limits of the cutoff frequency and BV [2]. Given the paramount interest in high-speed and high-voltage devices, the AlGaN/GaN HFETs have been mostly fabricated by a combination of a high electron velocity and a high breakdown field for modern terahertz and power electronics.

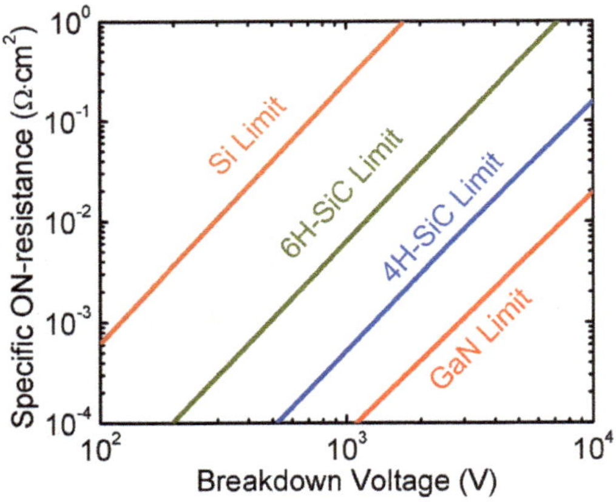

Figure 2.1 The theoretical limit diagram of BV and $R_{on,sp}$.

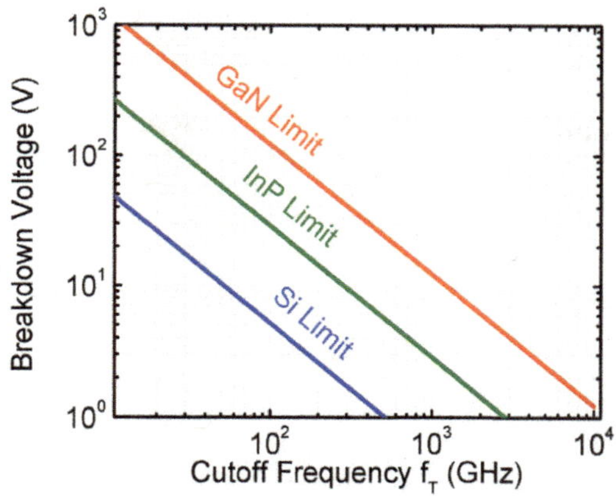

Figure 2.2 The theoretical limit diagram of cutoff frequency and BV.

Owing to most traditional semiconductors based on the diamond or zinc-blende structure, the polar properties are limited at a low level, such as polarization effects. The III nitride material with a wurtzite structure can generate strong polarization. Especially, in the heterostructure, which is formed by the two kinds of wide-bandgap material, the piezoelectric and spontaneous polarization can create a strong built-in potential and electric field for accumulating free carriers at quantum exceeding 10^{13} cm^{-2}. The polarization effect should be explored for device design. The ab initio theory is introduced in this chapter, which takes into account the macroscopic polarization in group III nitride alloys. In Al(In)GaN/GaN heterostructures, nanostructures, and devices, the generation of 2DEG introduced by nonlinear polarization should be modeled in an explicit expression for future simulation and fabrication.

2.2 The ab initio Theory in III–V Semiconductors

As demonstrated in the previous paragraphs, the polarization effect as a main issue needs to be understood for developing polar heterostructure devices. We should develop a model for formulizing the following:

(i) How to calculate the spontaneous polarization depends upon material parameters

(ii) How to calculate the piezoelectric polarization depends upon the strain and stress in the heterostructure

(iii) How to obtain an approximation of the 2DEG density in Ga-face AlGaN/GaN or AlInN/GaN HFET structures

The piezoelectric polarization effects are exhibited by stress or strain induced by a lattice mismatch, thermal relaxation, or electric effects. We first illustrate a qualitative description of spontaneous polarization linked to their charge symmetry, typical of bulk III nitride materials, before introducing the piezoelectric polarization.

2.2.1 Spontaneous Polarization

Usually, the III-V nitride semiconductors manifest two main structures, namely the wurtzite structure and the sphalerite structure, as shown in Fig. 2.3. The sphalerite structure, with a charge central-symmetrical structure, shows no characteristics of spontaneous polarization. But in the wurtzite structure, as shown in Fig. 2.3a, the noncoincident center of the positive charge and negative charge leads to spontaneous polarization. And its polarization direction depends on its material growth direction. For information on the model of spontaneous polarization, the reader can refer to Refs. [3, 4]. The spontaneous polarization of the group III nitride semiconductors can be calculated by the component fraction x as

$$P_{\text{ABN,sp}} = x \cdot P_{\text{AN,sp}} + (1 - x) \cdot P_{\text{AN,sp}} + bx \cdot (1-x), \tag{2.1}$$

where the index of ABN is a III–N material and b is the coefficient of nonlinearity in ab initio theory. The first two terms in Eq. 2.1 represent the standard linear interpolation between the binary compounds, such as GaN and AlN. The nonlinear term is described in the last term by parabolic fits. In this term, b as a modulated parameter is used to reduce the discrepancy between experimental data and theoretical calculation. The conventional parameters of a ternary compound can be approximated by the quadratic equation of their binary compounds.

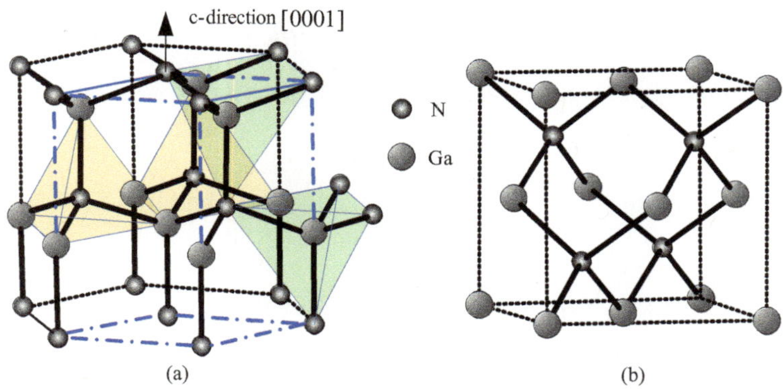

Figure 2.3 III nitride lattice structure diagram: (a) wurtzite structure and (b) sphalerite structure.

2.2.2 Piezoelectric Polarization

Owing to the lattice misfit between constituents in the heterojunction, lattice stress or strain is generated at the heterointerface [5]. Corresponding to the stress or strain, piezoelectric polarizations are gained. The detailed piezoelectric polarization model based on ab initio calculations [6, 7] is introduced in this section by tensor calculation.

The piezoelectric moduli (e_{ijk} and d_{ijk}) in the III–V semiconductors are all tensors [8]. Especially, in the hexagonal system, as shown in Fig. 2.3, the tensors along the *i* axis are associated with strain tensor (ε_{jk}) and stress tensor (σ_{jk}). Thus, the piezoelectric polarization in this direction is conventionally modeled by strain and stress as the following formula [6]:

$$P_{\mathrm{pz},i} = \sum_{l} e_{ij}\varepsilon_{j} = \sum_{l} d_{ij}\sigma_{j}, i = 1, 2, 3; j, k = 1, ..., 6 \tag{2.2}$$

According to Hook's law, the internal or external stress of the material can be also calculated by

$$\sigma_{ij} = \sum_{k,l} c_{ijkl}\varepsilon_{kl}, \tag{2.3}$$

where c_{ijkl} is the elastic tensor with a 6 × 6 matrix. And tensor elements can use Voigt's notation $xx \rightarrow 1$, $yy \rightarrow 2$, $zz \rightarrow 3$, $yz \rightarrow 4$, $zy \rightarrow 5$, and $yx \rightarrow 6$. According to the basic elements definition of elastic tensor, c_{ijkl} can be rewritten as c_{mn}, where $c_1 = c_{xx}$, $c_2 = c_{yy}$, $c_3 = c_{zz}$, $c_4 = c_{yz}$, $c_5 = c_{zy}$, and $c_6 = c_{yx}$. Meanwhile, the strain tensor (ε_{jk}) and stress tensor (σ_{jk}) can also be expressed in concise matrix notation with $\varepsilon_1 = \varepsilon_{xx}$, $\varepsilon_2 = \varepsilon_{yy}$, $\varepsilon_3 = \varepsilon_{zz}$, $\varepsilon_4 = \varepsilon_{yz}$, $\varepsilon_5 = \varepsilon_{zy}$, and $\varepsilon_6 = \varepsilon_{zx}$ and $\sigma_1 = \sigma_{xx}$, $\sigma_2 = \sigma_{yy}$, $\sigma_3 = \sigma_{zz}$, $\sigma_4 = \sigma_{yz}$, $\sigma_5 = \sigma_{zy}$, and $\sigma_6 = \sigma_{yx}$, where $i, j, k, l = x, y, z$ and $m, n = 1,..., 6$. Thus, c_{ijkl} can be described by

$$c_{mn} = \begin{pmatrix} c_{11} & c_{12} & c_{13} & 0 & 0 & 0 \\ c_{12} & c_{11} & c_{13} & 0 & 0 & 0 \\ c_{13} & c_{13} & c_{11} & 0 & 0 & 0 \\ 0 & 0 & 0 & c_{44} & 0 & 0 \\ 0 & 0 & 0 & 0 & c_{44} & 0 \\ 0 & 0 & 0 & 0 & 0 & 0.5(c_{11} - c_{12}) \end{pmatrix}, \tag{2.4}$$

and then from Eq. 2.3, the piezoelectric polarization can be given by

$$\begin{pmatrix} \sigma_{xx} \\ \sigma_{yy} \\ \sigma_{zz} \\ \sigma_{yz} \\ \sigma_{zy} \\ \sigma_{yx} \end{pmatrix} = \begin{pmatrix} \sigma_1 \\ \sigma_2 \\ \sigma_3 \\ \sigma_4 \\ \sigma_5 \\ \sigma_6 \end{pmatrix} = \begin{pmatrix} c_{11} & c_{12} & c_{13} & 0 & 0 & 0 \\ c_{12} & c_{11} & c_{13} & 0 & 0 & 0 \\ c_{13} & c_{13} & c_{11} & 0 & 0 & 0 \\ 0 & 0 & 0 & c_{44} & 0 & 0 \\ 0 & 0 & 0 & 0 & c_{44} & 0 \\ 0 & 0 & 0 & 0 & 0 & 0.5(c_{11} - c_{12}) \end{pmatrix} \begin{pmatrix} \varepsilon_1 \\ \varepsilon_2 \\ \varepsilon_3 \\ \varepsilon_4 \\ \varepsilon_5 \\ \varepsilon_6 \end{pmatrix}. \tag{2.5}$$

In the III–V nitride system, semiconductors with a wurtzite structure have the isotropic characteristic along the base plane. The piezoelectric polarization along the c axis is mainly caused by the lattice mismatch of the heterojunction, so it is necessary to obtain the strain and stress to calculate the tensors.

Due to the structure symmetry perpendicular to the c axis, $\varepsilon_1 = \varepsilon_2$ is obtained and then $\sigma_1 = \sigma_2$ and $\sigma_3 = 0$ are achieved from Eq. 2.5. On the basis of Eq. 2.5, the relationship between ε_3 and ε_1 along the x direction is simplified by

$$\varepsilon_3 = -2\frac{c_{13}}{c_{33}}\varepsilon_1. \tag{2.6}$$

Along the [0001] direction, as shown in Fig. 2.3, ε_3 and ε_1 can be given by

$$\varepsilon_1 = \frac{a - a_0}{a_0} \tag{2.7}$$

and

$$\varepsilon_3 = \frac{c - c_0}{c_0}. \tag{2.8}$$

In the III–V nitride semiconductors with the wurtzite structure, due to the symmetry of the crystal structure, the piezoelectric tensor d_{ij}, composed of 27 elements formerly, can be reduced to the tensor with 3 × 6 elements, which is simplified as

$$d_{ij} = \begin{pmatrix} 0 & 0 & 0 & 0 & d_{15} & 0 \\ 0 & 0 & 0 & d_{24} & 0 & 0 \\ d_{31} & d_{32} & d_{33} & 0 & 0 & 0 \end{pmatrix}, \tag{2.9}$$

where $d_{31} = d_{32}$ and $d_{15} = d_{24}$. Associating Eq. 2.2 with Eq. 2.9, piezoelectric polarization can be expressed by

$$\begin{pmatrix} P_{\mathrm{pz},1} \\ P_{\mathrm{pz},2} \\ P_{\mathrm{pz},3} \end{pmatrix} = \begin{pmatrix} 0 & 0 & 0 & 0 & d_{15} & 0 \\ 0 & 0 & 0 & d_{24} & 0 & 0 \\ d_{31} & d_{31} & d_{33} & 0 & 0 & 0 \end{pmatrix} \begin{pmatrix} \sigma_1 \\ \sigma_2 \\ \sigma_3 \\ \sigma_4 \\ \sigma_5 \\ \sigma_6 \end{pmatrix}. \tag{2.10}$$

Then, from Eq. 2.10, the matrix is rewritten as

$$\begin{pmatrix} P_{\mathrm{pz},1} \\ P_{\mathrm{pz},2} \\ P_{\mathrm{pz},3} \end{pmatrix} = \begin{pmatrix} \frac{1}{2} d_{15}\sigma_5 \\ \frac{1}{2} d_{15}\sigma_4 \\ d_{31}(\sigma_1 + \sigma_2) + d_{33}\sigma_3 \end{pmatrix}. \tag{2.11}$$

Because of the feature of symmetry in biaxial stress, $\sigma_1 = \sigma_2$ and $\sigma_3 = 0$, the strain perpendicular to the axis can be ignored, that is, $\sigma_4 = \sigma_5 = 0$. Hence, the last term of polarization by the stress is along the growth direction and its expression can be described by

$$P_{\mathrm{pz},3} = 2d_{31}\sigma_1 = 2d_{31}\varepsilon_1\left(c_{11} + c_{12} - 2\frac{c_{13}^2}{c_{33}}\right). \tag{2.12}$$

The elastic coefficient c_{13} and c_{33} in Eq. 2.12 should be given to calculate the corresponding strain $P_{PZ,3}$.

From Eq. 2.2, we can obviously find the stress $e_{jk} = c_{ij}d_{ik}$.

In III–V nitride semiconductors with the wurtzite structure, the piezoelectric tensor d_{ij}, composed of 27 elements formerly, can be reduced to the tensor with 18 elements as the symmetry of the crystal structure, which can be described as

$$e=\begin{pmatrix} 0 & 0 & 0 & 0 & e_{14} & 0 \\ 0 & 0 & 0 & e_{15} & 0 & 0 \\ e_{31} & e_{31} & e_{33} & 0 & 0 & 0 \end{pmatrix}. \tag{2.13}$$

Along the [0001] direction (i = 3), as the symmetry of the wurtzite structure ($\varepsilon_{15} = \varepsilon_{24}$, $\varepsilon_{31} = \varepsilon_{32}$), and $\sigma_1 = \sigma_2$, $\sigma_3 = 0$, the biaxial polarization can be given by

$$P_{pz,3} = \varepsilon_1 \frac{c_{11} - c_{12}}{c_{11} - \dfrac{c_{13}^2}{c_{33}}}\left(e_{31} - e_{33}\frac{c_{13}}{c_{33}}\right). \tag{2.14}$$

In analogy to the determination of piezoelectric polarization along the c axis in dependence of biaxial strain, equations can be derived in the case of uniaxial ($\sigma_1 = \sigma_2 = 0, \sigma_3 \neq 0$) as well as hydrostatic strain ($\sigma_1 = \sigma_2 = \sigma_3$).

(a) Uniaxial strain:

$$P_{pz,3} = e_{33}\varepsilon_3 + e_{31}(\varepsilon_1 + \varepsilon_2) = 2\varepsilon_1\left(e_{31} - e_{33}\frac{c_{13}}{c_{33}}\right) \tag{2.15}$$

(b) Hydrostatic strain:

$$P_{pz,3} = \varepsilon_1\left(2e_{31} + e_{33}\frac{c_{11} + c_{12} - c_{13}}{c_{33} - c_{13}}\right) \tag{2.16}$$

2.2.3 The Analytical Model of a 2DEG at the AlGaN/GaN Interface

For the Ga-face AlGaN/GaN heterostructure along the [0001] plane, the potential from the surface to the substrate is zero. From Fig. 2.4, we can obtain that

$$q\phi_{\mathrm{b}}(x)+\left(\frac{\varepsilon_1\varepsilon_1}{d_{\mathrm{AlGaN}}}\right)\left(n_{\mathrm{2DEG}}-\frac{\sigma(x)}{q}\right)-\Delta E_{\mathrm{c}}+E_{\mathrm{F}}(x)=0. \tag{2.17}$$

The 2DEG density (n_{2DEG}) at the heterointerface can be calculated by Eq. 2.17 as

$$n_{\mathrm{2DEG}}=\frac{\sigma(x)}{q}-\left(\frac{\varepsilon_1\varepsilon_1}{d_{\mathrm{AlGaN}}}\right)[q\phi_{\mathrm{b}}(x)-\Delta E_{\mathrm{c}}+E_{\mathrm{F}}(x)], \tag{2.18}$$

where ΔE_c is the conduction band offset between the $Al_xGa_{1-x}N$ and GaN layers, E_F is the Fermi level with respect to the GaN conduction-band-edge energy, q is the electronic charge, and $\phi_b(x)$ is the $Al_xGa_{1-x}N$ barrier height. The polarization density $\sigma(x)$ can be obtained by Eq. 2.2 and Eq. 2.3.

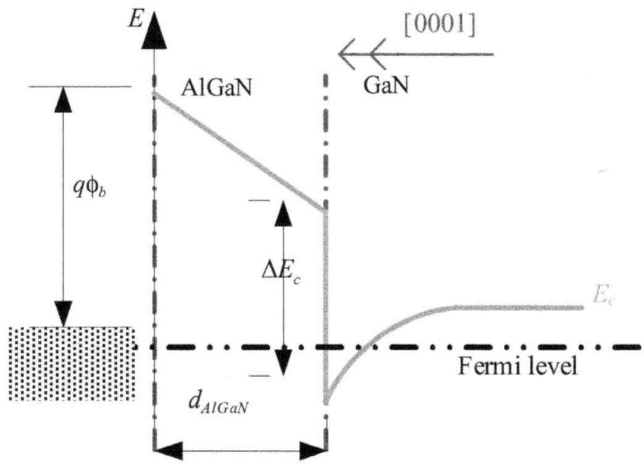

Figure 2.4 Schematic diagram of the energy band at the AlGaN/GaN heterostructure for potential conservation. The high 2DEG density is accumulated in the quantum well at the AlGaN/GaN interface.

2.3 Polarization Effects Discussion

A general overview of the material bandgap of III–N materials is given in Fig. 2.5. From AlN, GaN, and InN, we can observe that with the bandgap of InN of ~0.8 eV and the bandgap of AlN of 6.2 eV at room temperature, these compound semiconductors can be synthesized to be a broad range of bandgaps covering a very broad range of energies and thus emission wavelengths, from the infrared to the deep ultraviolet. As the bandgap of various semiconductor materials is related to lattice constants, a broad range of lattice constants can result in different strain and stress at the heterointerface and a huge bandgap can result in high BV values. Because in the bulk III–N semiconductor, the BV as an intrinsic parameter is strongly limited by available intrinsic carrier concentrations, in a high-voltage application, the wide-bandgap semiconductors have the state-of-the-art characteristic for obtaining high performance as compared to narrow-bandgap materials.

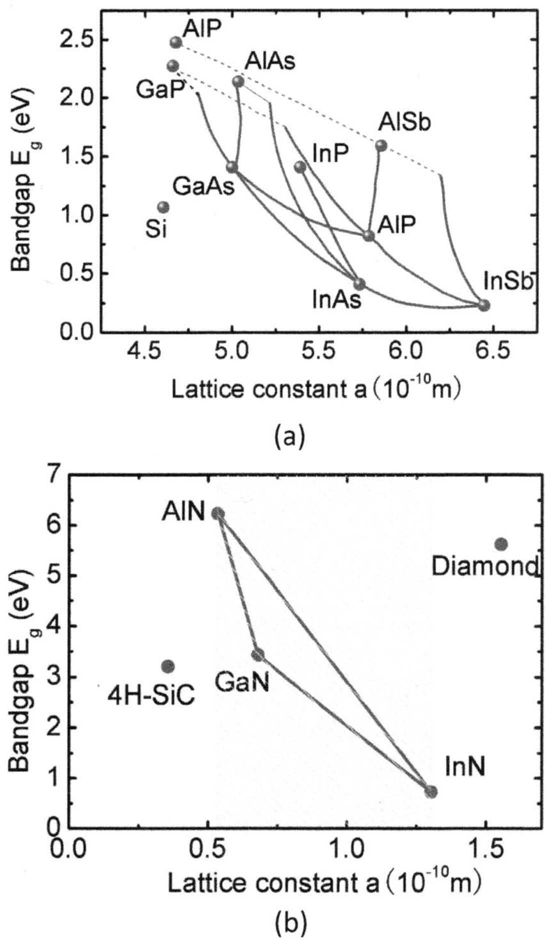

Figure 2.5 Bandgap with respect to lattice constant for compound semiconductors [9].

Meanwhile, in the case of III–N heterostructures, the formation of piezoelectric polarization in wurtzite III nitride materials is intrinsically provided by a lattice mismatch, which is modeled by elastic constants as the correspondent parameters given in Table 2.1. In lattice-matched heterostructures, the extremely small difference between two kinds of semiconductor interfaces along the growth direction of epitaxy can't yield strong polarization. In the AlGaN/GaN heterostructure, the lattice mismatch is unavoidable to be generated at the AlGaN/GaN heterointerface. In these mismatch systems, if the lattice in the AlGaN barrier is relaxed, cracks, defects, and piezoelectric polarization degeneration are generated. From ab initio theory, the lattice constant versus bandgap can be modulated by a fraction of the compound. Hence, the corresponding parameters of bandgap and lattice constant can be adjusted by changing the component in compound semiconductors.

Table 2.1 Elastic compliance and piezoelectric constants in group III nitrides of the wurtzite crystal structure [7, 10]

Crystal property	AlN	GaN	InN
c_{11}(GPa)	396	367	223
c_{12}(GPa)	137	135	115
c_{13}(GPa)	108	103	92
c_{33}(GPa)	373	405	224
c_{44}(GPa)	116	95	48
e_{33}(cm^{-2})	1.55	0.73	0.73
e_{31}(cm^{-2})	-0.58	-0.47	-0.49
e_{15}(cm^{-2})	-0.48	-0.40	-0.40
P_{sp}(C/m^2)	-0.081	-0.029	-0.032

The spontaneous polarization and piezoelectric polarization in that polarization charge density are varied with respect to the rotated angle. The typical angle of the plane is given in Table 2.2. According to these rotated angles, the spontaneous polarization charge density of AlGaN and GaN material is calculated as a function of the rotated angle of the crystal plane, as given in Fig. 2.6a–d. But in the AlGaN/GaN heterostructure, the GaN and AlGaN spontaneous polarization is strongly affected by the rotated angle. It is observed that the maximum spontaneous polarization charge density is obtained along the *c* axis. Because the rotated angle is corresponding with the spontaneous polarization direction along the *c* axis, with an increase of the rotated angle along the *c* axis, the spontaneous polarization is reduced. At 90°, the spontaneous polarization is decreased to zero. The piezoelectric polarization is enhanced with the increase of the Al mole in AlGaN. The change trend of charge density is the same as spontaneous polarization. For example, when its rotated angle is perpendicular to the *c* axis, the piezoelectric polarization is zero. The increase of the Al mole in AlGaN can enhance the piezoelectric polarization, as shown in Fig. 2.6.

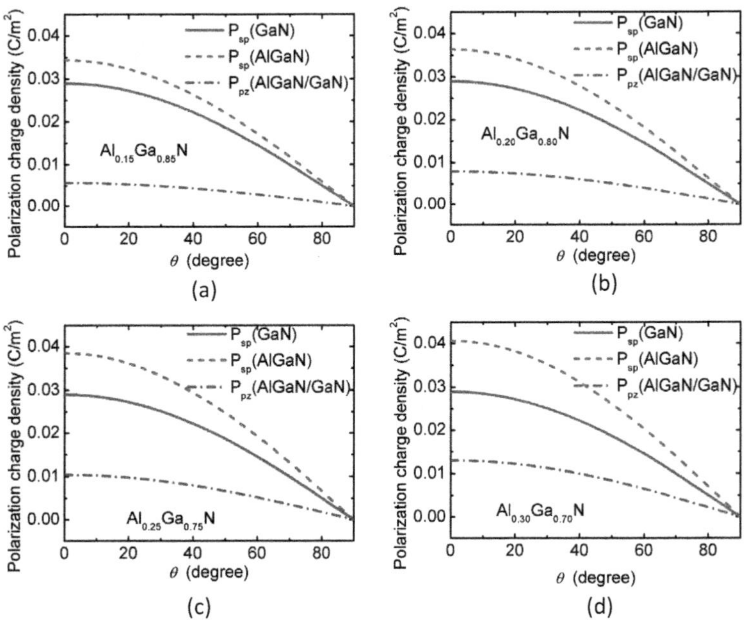

Figure 2.6 Polarization charge density and inclination angle between the *c* axis: (a) $Al_{0.15}Ga_{0.85}N$, (b) $Al_{0.20}Ga_{0.80}N$, (c) $Al_{0.25}Ga_{0.75}N$, and (d) $Al_{0.30}Ga_{0.70}N$.

In Eq. 2.18, the calculation of the 2DEG density is modeled by material and structure parameters. Figure 2.7 shows a theoretical 2DEG density at a different Al mole fraction in AlGaN. With the increment of the Al mole fraction to $x = 0.8$ in a 6 nm AlGaN barrier, the 2DEG density is obtained as 8.2×10^{12} cm^{-2} piezoelectric polarization and spontaneous polarization is increased.

Table 2.2 Angle (θ) with respect to the crystal plane of group III nitrides in the wurtzite crystal structure [8]

Plane constant	AlN θ	GaN θ	InN θ
$(10\bar{1}\bar{1})$	1.0749	1.0814	1.0777
$(11\bar{2}2)$	1.0125	1.0194	1.0154
$(10\bar{1}\bar{3})$	0.5522	0.5592	0.5552

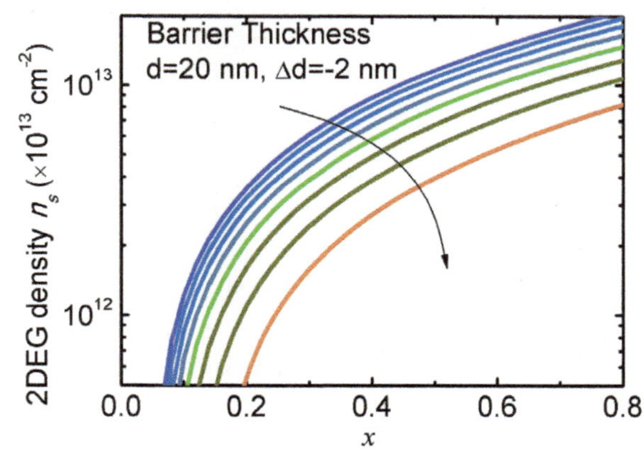

Figure 2.7 Calculation of 2DEG density in AlGaN/GaN as a function of the Al mole in the $Al_xGa_{1-x}N$ barrier. With the increase of the $Al_xGa_{1-x}N$ barrier thickness from 4 nm to 20 nm, the 2DEG density can be increased to more than 10^{13} cm^{-2} as x from 0.2 to 1.

Acknowledgments

The project is supported by the National Natural Science Foundation of China (61404110) and the National Higher-Education Institution General Research and Development Project (2682014CX097).

References

1. Ambacher, O., Foutz, B., Smart, J., Shealy, J. R., Weimann, N. G., Chu, K., et al. (2000). Two dimensional electron gases induced by spontaneous

and piezoelectric polarization in undoped and doped AlGaN/GaN heterostructures, *J. Appl. Phys.*, **87**, 334–344.

2. Ambacher, O., Smart, J., Shealy, J. R., Weimann, N. G., Chu, K., Murphy, M., et al. (1999). Two-dimensional electron gases induced by spontaneous and piezoelectric polarization charges in N- and Ga-face AlGaN/GaN heterostructures, *J. Appl. Phys.*, **85**, 3222–3233.
3. Bernardini, F., Fiorentini, V., and Vanderbilt, D. (1997). Spontaneous polarization and piezoelectric comstants of III-V nitrides, *Phys. Rev. B*, **56**, 10024–10025.
4. Deen, D. A. (2011). *Advanced Design of Ultra-Thin Barrier AlN/GaN HEMTs* (University of Notre Dame, IN).
5. Gualtieri, J. G., Kosinski, J. A., and Ballato, A. (1994). Piezoelectric materials for acoustic wave applications, *IEEE Trans. Ultrason. Ferroelectr. Freq. Control*, **41**(1), 53–59.
6. Ikeda, N., Kaya, S., Jiang, L., Sato, Y., Kato, S., and Yoshida, S. (2008). *High Power AlGaN/GaN HFET with a High Breakdown Voltage of over 1.8 kV on 4 inch Si Substrates and the Suppression of Current Collapse.* Paper presented at the 20th International Symposium on Power Semiconductor Devices and IC's, 287–290 (in Florida).
7. Romanov, A. E., Baker, T. J., Nakamura, S., Speck, J. S., and Grp, E. J. U. (2006). Strain-induced polarization in wurtzite III-nitride semipolar layers, *J. Appl. Phys.*, **100**(2), 023522.
8. Rosker, M. J., Albrecht, J. D., Cohen, E., Hodiak, J., and Chang, T. H. (2010). *DARPA's GaN Technology Thrust.* Paper presented at the 2010 IEEE MTT-S International Microwave Symposium Digest (MTT), 1214–1217 (in Canada).
9. Wood, C., and Jena, D. (2007) *Polarization Effects in Semiconductors: From ab initio Theory to Device Applications* (Springer, Boston).
10. Wright, A. F. (1997). Elastic properties of zinc-blende and wurtzite AlN, GaN, and InN, *J. Appl. Phys.*, **82**(6), 2833–2839.

Chapter 3

GaN Transistor Fabrication Process

Liu Zhihong
Singapore-MIT Alliance for Research and Technology Center, 1 CREATE Way, Innovation Wing, #05-09, 138602, Singapore
liuzhihong@ieee.org

In this chapter a few key process steps to fabricate GaN high-electron-mobility transistors (HEMTs) for power electronics applications will be described. The whole process flow to fabricate a HEMT mainly includes five steps, namely mesa isolation, ohmic contact formation, gate formation, pad and interconnection metallization, and surface passivation. Sometimes, to reduce the gate leakage current, an additional step of gate dielectric deposition is included before the gate metal deposition in order to form a structure called metal-insulator-semiconductor HEMT (MIS-HEMT) or metal-oxide-semiconductor HEMT (MOS-HEMT). Usually a field plate is also required to improve the breakdown voltage and suppress the surface trapping effect. A typical fabrication process flow for GaN HEMTs is shown in Fig. 3.1.

This chapter will be divided into five sections. In the five sections a few key processes will be discussed sequentially, namely device isolation in the first section, ohmic contacts in the second section,

Gallium Nitride Power Devices
Edited by Hongyu Yu and Tianli Duan

ISBN 978-981-4774-09-3 (Hardcover), 978-1-315-19662-6 (eBook)
www.panstanford.com

gate fabrication in the third section, surface passivation in the fourth section, and field plates in the last section.

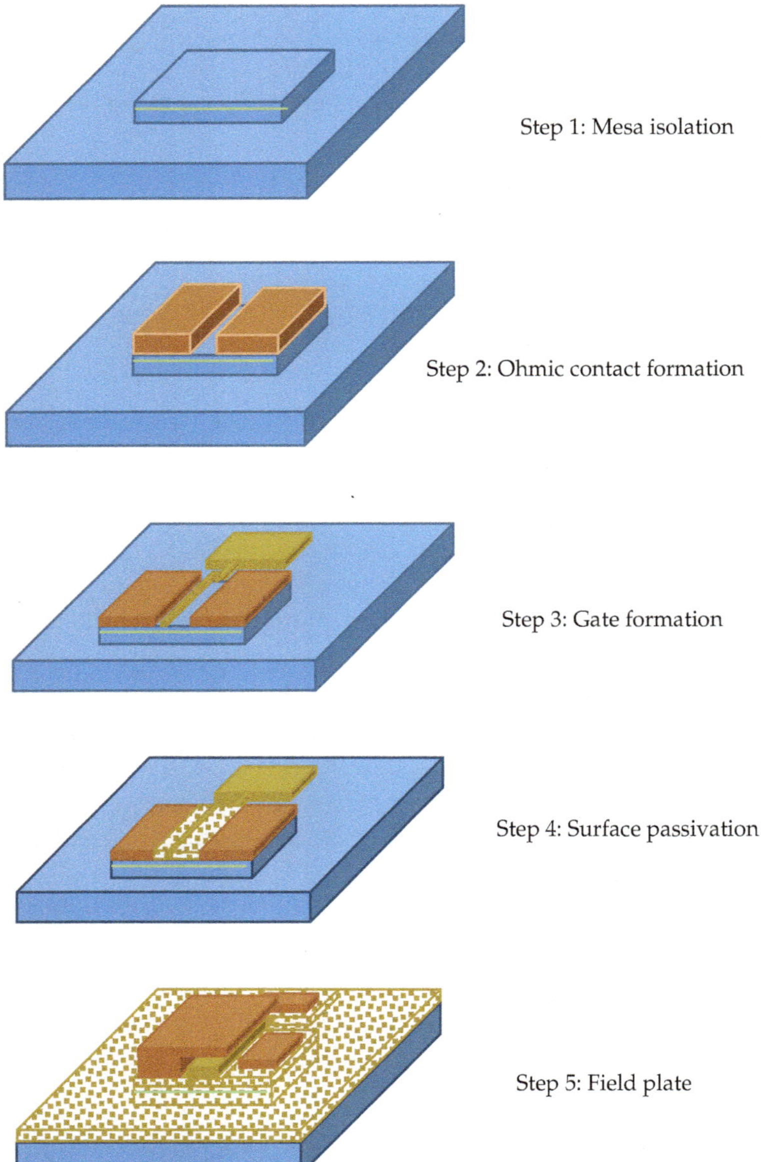

Figure 3.1 Fabrication process flow for GaN HEMTs.

3.1 Device Isolation

The separate devices, interconnects, and pads must be isolated from each other to prevent a short circuit. Device isolation can be realized by mesa (trench) etching or ion implantation. Mesa etching is a simple and cost-effective process technology, and it can usually be accomplished by wet etch (using chemicals) or dry etch (using plasmas). Wet etching has the advantages of a simple and cheap process, a smooth sidewall and surface with slight damage, possible high etch selectivity between different doped materials and between different composition materials, a possible isotropic etch profile, etc. Dry etch has the advantages of a vertical etch profile, better uniformity across a large wafer area, higher reproducibility, etc.

3.1.1 Wet Etch

GaN-based semiconductor materials are inert to commonly used chemicals under etch normal conditions, such as HCl, H_2SO_4, HNO_3, and HF [1]. Table 3.1 lists the etch rates of commonly used chemicals for Ga-face GaN at room temperature or elevated temperatures. The wet-etching rates of GaN by chemicals highly depend on the quality of the material. If there is a higher density of defects in the material, GaN could be attacked by some chemicals, such as potassium hydroxide (KOH) solution. Some early reports showed that GaN can be etched in an aqueous NaOH solution; however, the etching ceased soon after the formation of an insoluble coating of presumably $Ga(OH)_3$. Further etching requires to remove this coating by continual shaking or vibration. Carosella et al. [2] found that the AZ400k photoresist developer can etch the GaN grown by molecular beam epitaxy (MBE) at 800°C but cannot etch the GaN grown by metal-organic chemical vapor deposition (MOCVD) or hydride vapor-phase epitaxy (HVPE). The discrepancy in the etch rates reported by the different groups using the same chemical and at the same temperature is believed to be related to the different qualities of the GaN material, mainly due to the different defect densities.

Table 3.1 Etch rates of GaN by various chemicals

Chemical	Temperature (°C)	Etch rate (μm/min)	Etching planes
Acetic acid (Ch_3COOH)	30	<0.001	None
Citric acid ($C_6H_8O_7$)	75	0	
Succinic acid (HOOC-$(CH_2)_2$-COOH)	75	0	
Oxalic acid (HOOCCOOH)	75	0	
Hydrochloric acid (HCl)	80	<0.001	None
Hydrofluoric acid (HF)	25	0	
Hydroiodic acid (HI)	25	0	
Nitric acid (HNO_3)	85	<0.001	None
Phosphoric acid (H_3PO_4)	82	0	
Phosphoric acid (H_3PO_4)	108–195	0.013–3.2	{10--}, {10-3}
Sulfuric acid (H_2SO_4)	93	<0.001	None
Boric acid (HBO_3)	25	0	
Hydrogen peroxide (H_2O_2)	25	0	
HCl + H_2O_2 + H_2O	25	0	
HI + H_2O_2 + H_2O	25	0	
HCl + HNO_3 + H_2O_2	25	0	
Potassium hydroxide (K_3PO_4)	25	0	
Sodium hydroxide (NaOH)	25	0	
Potassium hydroxide (KOH)	25	0	
AZ400K photoresist developer	25	0	
Potassium hydroxide (KOH), molten	150–247	0.003–23	{10-0}, {10--}
50% KOH in H_2O	83	<0.001	None
10%–50% KOH in ethylene (CH_2OHCH_2OH)	90–182	0.015–1.3	{10-0}
50% NaOH in H_2O	100	<0.001	None
20% NaOH in ethylene glycol	178	0.67–1.0	None
Potassium iodide	25	0	
2% Bromine/methanol	25	0	
n-Methyl-2-pyrrolidonone	25	0	

Source: Reprinted from Ref. [1], with the permission of AIP Publishing.

The wet-etch rate also depends of the polarity of GaN. In [3] Palacios et al. observed that only N-face MBE-grown GaN epilayers were etched in aqueous KOH from room temperature to 80°C, while Ga-face crystals were not etched under the same etching conditions. Shimuzu et al. in Ref. [4] also observed that the presumably N-face MBE-grown GaN was severely degraded in morphology after etching in phosphoric acid at temperatures higher than 220°C. Li et al. proposed a mechanism to interpret the polarity selectivity of chemical etching for GaN [5, 6]. For N-face GaN, if the surface termination layer is nitrogen, each nitrogen atom at the terminated layer has one negatively charged dangling bond. When the sample contacts the KOH solution, the hydroxide ions (OH^-) are first absorbed on the sample surface and subsequently react with the Ga atom. The reaction is as below:

$$2GaN + 3H_2O \xrightarrow{KOH} Ga_2O_3 + 2NH_3 \uparrow \quad (3.1)$$

GaN and H_2O react and form Ga_2O_3 and NH_3. KOH works as a catalyst, and the product of Ga_2O_3 is dissolved in the KOH solution. The sequence is repeated and hence the N-face is etched in the KOH solution. If Ga is the surface termination layer, Ga atoms react with OH^- first and the above-mentioned sequence is repeated. As for Ga-face GaN, each nitrogen atom has three negatively charged dangling bonds, and the repulsion between these three negative charges and OH^- hinders the hydroxide ions from approaching Ga atoms and suppresses their reaction. So Ga-face GaN is difficult to be etched in a chemical.

Exposure to photos will enhance the wet-etch rates for GaN. Photoelectrochemical (PEC) etching can realize wet etch for GaN at room temperature. And it is worth noting that there is little difference between the etch rates for Ga-face GaN and N-face GaN, or in other words, PEC etching has almost no polarity selectivity. The disadvantage of PEC etching over conventional wet etching is that it needs a light source with photon energy larger than the bandgap of GaN, that is, 3.42 eV, or a corresponding wavelength of smaller than 360 nm. Table 3.2 summarizes some results of the etch rates and surface roughness after PEC etching of GaN.

Table 3.2 Etch rates of GaN by photoelectrochemical (PEC) etching [7–10]

Chemical	Etch rate (μm/min)	Roughness
Potassium hydroxide (KOH)	50 for 0.02 M >300 for 0.04 M	1.5 nm
Sodium hydroxide (NaOH)	80	NA
HCl	1.5–40	NA
Tartaric acid/Ethylene glycol	160	81.5 nm
AZ400K developer	110	None
$K_2S_2O_8$ + KOH	50	7 nm

A typical PEC etching apparatus is shown in Fig. 3.2 [8]. A Pt wire was used as the cathode, and a thin layer of Ti (100 nm) was used as a mask and electrical contact to the samples. No external bias was applied during etching. The UV illumination could be a He-Cd laser with 325 nm or a Hg arc lamp. The etch rates were proportional to the light intensities and can be up to 300 nm/min with a light intensity of 50 mW/cm^2.

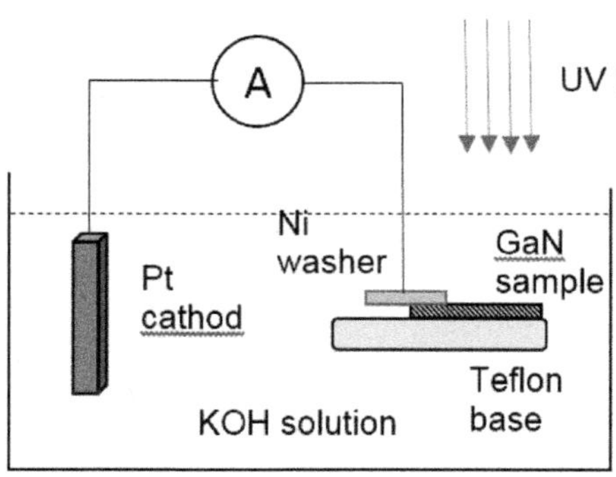

Figure 3.2 Schematic of a typical apparatus for photoelectrochemical (PEC) etching for GaN.

The mechanism responsible for the PEC etching of GaN is similar to that of PEC etching for other semiconductor materials, such as GaAs. Under UV illumination, electrons and holes are generated and the holes assist in the oxidization of GaN. The resulting oxides are dissolved in the solution, and the excess electrons are consumed by the cathode through a reduction reaction. The oxidization process is based on the chemical reaction equation below:

$$2\text{GaN} + 6\text{h}^{+} \longrightarrow 2\text{Ga}^{3} + \text{N}_2 \uparrow, \tag{3.2}$$

where h^+ denotes a hole. Besides the KOH solution, HCl and H_3PO_4 have also been found to be able to PEC-etch GaN, and HNO_3 and CH_3COOH can etch GaN with an external bias [11].

Although a high etch rate can be obtained by PEC etching, it has a drawback, that is, the substrate of the GaN device needs to be conductive. This limits a lot of practical applications for PEC etching as many devices require the substrate and buffer highly resistive in order to have a high breakdown voltage and reduce the radio frequency (RF) power loss. And the electrode design also makes the etching apparatus complicated. A few groups have investigated the photoassisted contactless etching for GaN using $K_2S_2O_8$ and KOH [12]. This technique does not require an electrical contact to the sample or a cathode, so the wet-etch apparatus is simple and suitable for wet etching of a GaN sample with a high-resistivity substrate. The relevant reactions in the UV light–assisted wet etching for GaN in K_2SO_8 and KOH solution include

$$\text{GaN} + \lambda \xrightarrow{\text{UV}} \text{GaN} + \text{e}^{-} + \text{p}^{+} \tag{3.3}$$

$$\text{GaN} + 3\text{p}^{+} \longrightarrow \text{Ga}^{3} + \tfrac{1}{2}\text{N}^{2} \uparrow \tag{3.4}$$

$$\text{S}_2\text{O}_8{}^{2-} + 2\text{e}^{-} \longrightarrow \text{SO}_4{}^{2-} \tag{3.5}$$

$$\text{SO}_4^{-*} + \text{OH}^{-} \longrightarrow \text{SO}_4^{2-} + \text{OH}^{*} \tag{3.6}$$

$$\text{SO}_4^{-*} + \text{e}^{-} \longrightarrow \text{SO}_4^{2-} \tag{3.7}$$

$$\text{OH}^{*} + \text{e}^{-} \longrightarrow \text{OH}^{-} \tag{3.8}$$

$$\text{OH}^{*} + \text{O}_2 \longrightarrow \text{O}_3^{-} + \text{H}^{+} \tag{3.9}$$

Compared to PEC etching, the difference of the mechanisms for the contactless photoassisted etching is that the excess electrons are not consumed by the cathode but consumed by the reduction of peroxydisulfate ions ($S_2O_8{}^{2-}$) and the sulfate ion radicals ($SO_4{}^{-*}$) and/or hydroxyl radicals (OH^*). The sulfate ion radicals ($SO_4{}^{-*}$) and/or hydroxyl radicals (OH^*) are formed by the peroxydisulfate ions ($S_2O_8{}^{2-}$) under UV illumination. The etch rate will be faster if a Pt mask is used instead of a SiO_2 mask, as Pt can work as a catalyst to produce sulfate ion radicals under UV illumination [10]. Besides the lower etch rate than PEC etching, another drawback of the contactless photoassisted etching is that the etch rate is sensitive to the pH value of the etchant. The etch depth will saturate with time when the pH value of the etchant is lower than 12.0, resulting from the formation of monopersulfuric (H_2SO_5) and sulfuric acids due to the photolysis of the perxoydisulfate ions ($S_2O_8{}^{2-}$) under UV exposure [13, 14].

3.1.2 Dry Etch

Dry etching by plasma is widely used in the fabrication of various GaN devices, including PN diodes and Schottky barrier diodes (SBDs), HEMTs, light-emitting diodes (LEDs), and lasers. Compared to wet etch, dry etch is much easier to realize a fast etch rate, vertical-profile smooth etch surface, high uniformity and reproducibility, etc. The plasma energy and the chemistry need to be controlled to reduce the damage to the material surface. Due to the inert chemical nature and high bond energy of GaN-based materials, the development of dry etch is complicated. Table 3.3 lists the bond energies of various semiconductors, including III nitrides, GaAs, and Si.

Table 3.3 Bond energies (eV/atom) for various semiconductor materials

Material	GaN	AlN	InN	GaAs	Si
bonding energy	8.92	11.52	7.72	6.52	4.42

Plasma etching is completed through a reactive ion etcher (RIE) system, whose schematic diagram is shown in Fig. 3.3a. RIE plasmas are typically generated by applying RF power at 13.56 MHz between two parallel electrodes in a reactive gas. The ion energies are typically a few hundred electron volts. The sample is put on a wafer chuck that is biased at a negative voltage. RIE is operated at low pressures from several tens of millitorr (mT) to hundreds of millitorr, which promotes anisotropic etching benefitting from the lengthened mean free paths, which suppresses the collisional scattering of ions during the acceleration in the sheath.

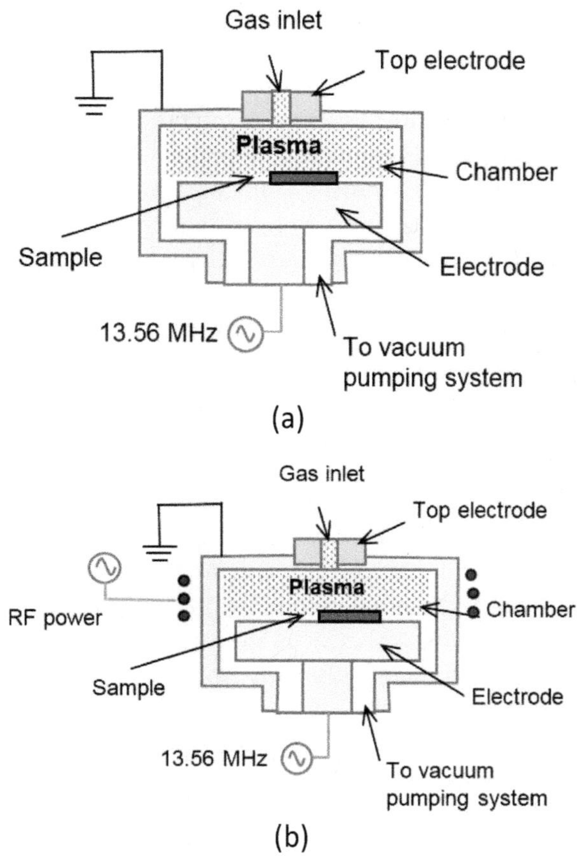

Figure 3.3 Schematic diagrams of a dry-etcher system for (a) an RIE system and (b) an ICP system.

To obtain a high dry-etch rate, a high RF power is required to increase the energies of ions so as to break the III–N bonds and desorb the etch products from the surface more efficiently by sputtering. However, these high-energy ions will bombard the III nitride atoms away from their original crystalline locations and form serious damage. The surface damages, such as Ga vacancies and N vacancies created by the high energy bombarding, will form carrier traps and cause serious trapping effects, such as current collapse and dynamic on-resistance degradation in electronic devices. Increasing the chemical activity in the plasma could minimize the damage, while keeping a fast etch rate, but it often results in a less anisotropic etched profile, which could increase the isolation leakage current and reduce the reverse blocking voltage of the devices. Therefore, it is necessary to seek an alternative tool that combines a high etch rate and low damage. An inductively coupled plasma (ICP) etching system meets such requirements. Its schematic diagram is shown in Fig. 3.3b. Similar to an RIE system, an RF power at 13.56 MHz is applied to the sample chuck or power electrode to accelerate the ions and bombard the wafer surface. Different from the RIE system, the plasmas of an ICP are formed in a dielectric vessel encircled by an inductive coil into which RF power is applied at a different frequency, such as 2.54 GHz or 2 MHz. The alternating electrical field between the coils induces a strong alternating magnetic field, which traps the electrons in the center of the chamber and generates high-density plasmas. Since ion energy and plasma density can be effectively tuned by two separate RF power sources, ICP etching can have a high etch rate, while keeping damage low.

Several different plasma chemistries have been investigated to etch GaN and its base materials. The etch rates and profiles are strongly affected by the volatility of the etch products formed. Table 3.4 lists the boiling points of the etch products of GaN and its based materials by halogen or CH_4-/H_2-based plasmas [15]. Due to the poor volatility of the products from the reactions between the III nitrides and F-based plasmas, which can be seen from Table 3.4, F-based plasmas are difficult to etch GaN. Cl-based gases such as Cl_2, BCl_3, or $SiCl_4$ are generally used for dry etching of GaN. The reactions in plasma etching of GaN by Cl_2 plasmas involve

$$Cl_2 \xrightarrow{\text{plasma}} Cl + Cl \tag{3.10}$$

$$Cl + GaN \xrightarrow{\text{plasma}} GaCl_3 + N_2 \uparrow \tag{3.11}$$

$$Cl + AlN \xrightarrow{\text{plasma}} AlCl_3 + N_2 \uparrow \tag{3.12}$$

Generally the plasma etching proceeds with the combination of both chemical reaction and physical sputtering. Physical sputtering can help to desorb the reduction products and realize a smooth etched sidewall and surface. That's why although Cl_2 can effectively etch GaN, a second gas is usually involved in GaN plasma etching. The second gas, such as Ar, BCl_3, H_2, or N_2, is to enhance the physical sputtering besides Cl_2. Cl_2/BCl_3 plasmas are most widely used to etch GaN films. The addition of BCl_3 to a Cl_2 plasma can improve the sputter desorption due to the higher mass ions and reduce the surface oxidation.

Table 3.4 Boiling points of various possible etch products of GaN and its based materials etched by halogen or CH_4/H_2-based plasmas [15]

Etch products	Boiling points (°C)
$AlCl_3$	183
AlF_3	NA
AlI_3	360
$AlBr_3$	263
$(CH_3)_3Al$	126
$GaCl_3$	201
GaF_3	1000
GaI_3	345
$GaBr_3$	279
$(CH_3)_3Ga$	55.7
$InCl_3$	600
InF_3	>1200
InI_3	NA
$InBr_3$	Sublimes
$(CH_3)_3In$	134
NCl_3	<71
NF_3	−129
NBr_3	NA
NI_3	Explodes
NH_3	−33
N_2	−196
$(CH_3)_3N$	−33

Figure 3.4 shows the etch rate for GaN using an ICP system as a function of the Cl_2 proportion in Cl_2/BCl_3 plasmas [15]. As the proportion of Cl_2 increases from zero, the GaN etch rates increase till 80% due to the higher reactivity of Cl to GaN. Slower GaN etch rates in a pure Cl_2 plasma are attributed to less efficient sputter

desorption of the etch products in the absence of BCl_3. Similar trends have been observed for the etch rates of GaN in Cl_2/H_2 plasmas and Cl_2/N_2 plasmas, as shown in Fig. 3.5 and Fig. 3.6, respectively [15].

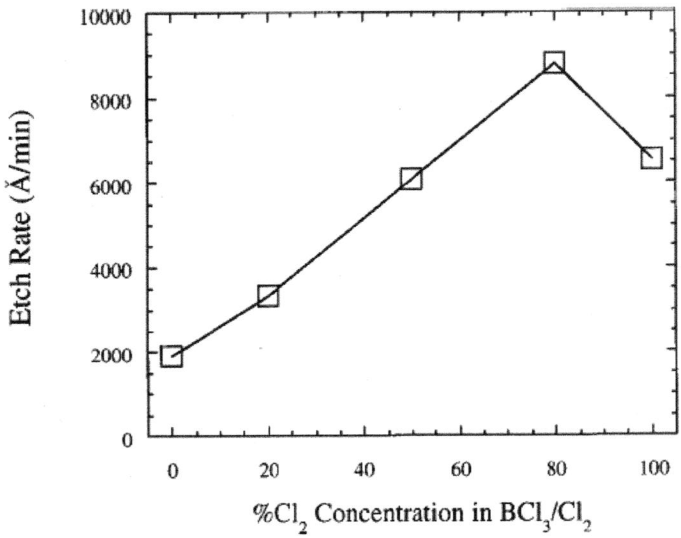

Figure 3.4 GaN etch rates by an ICP system using BCl_3/Cl_2 plasmas as a function of the Cl_2 proportion. Reprinted from Ref. [15], with the permission of AIP Publishing.

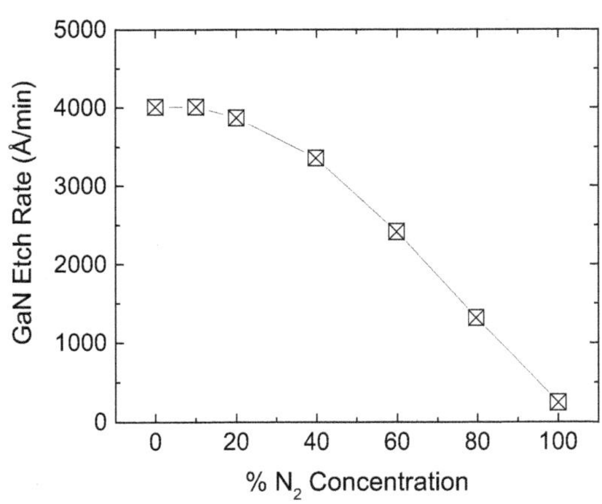

Figure 3.5 GaN etch rags by an ICP system using $Cl_2/N_2/Ar$ plasmas as a function of the N_2 proportion. Reprinted from Ref. [15], with the permission of AIP Publishing.

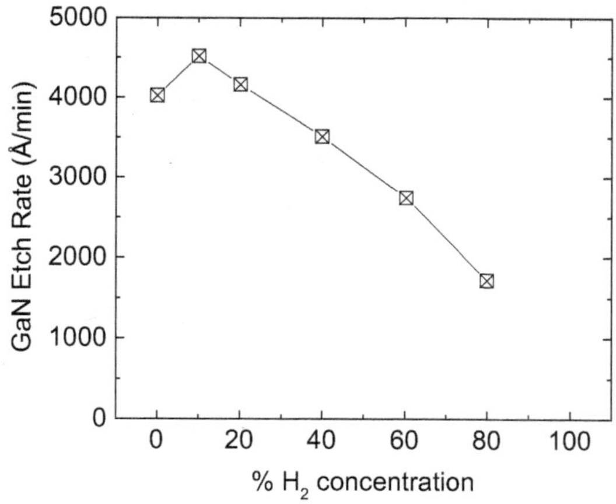

Figure 3.6 GaN etch rags by an ICP system using Cl_2/H_2 plasmas as a function of the H_2 proportion. Reprinted from Ref. [15], with the permission of AIP Publishing.

It was found that the etch rate of GaN by Cl_2 plasma can only be stable after some time from the start of etching. At the start the etch of GaN is very slow, and there exists a delay time for the GaN to be etched linearly with time. The surface native oxide was proved to be the root cause, as the native oxide GaO_x is difficult to be etched by Cl_2. Chemical pretreatment, for example, $HCl:H_2O$ or $HF:H_2O$, before Cl_2 plasma etch can suppress this phenomenon. Low-energy BCl_3 plasma pretreatment before the Cl_2 actual etch is an effective way to solve this issue. It is much slower for BCl_3 plasma to etch GaN than Cl_2 but much faster to etch GaO_x or AlO_x due to the formation of BClO. And BCl_3 also helps to remove the residual oxidizing agents from the etching chamber. So lower-energy BCl_3 treatment can successfully remove the native oxide at the surface of the GaN or AlGaN and thereby improve the controllability of the following Cl_2 etching. Figure 3.7 shows an example of the etch rate versus the etch time for Cl_2 plasma to etch GaN with and without BCl_3 pretreatment [16]. It is clearly shown that a low-power BCl_3 pretreatment can effectively eliminate the delay time of the GaN etch by Cl_2 and improve the etch rate linearity.

Figure 3.8 shows an image of the etched mesa for device isolation during our GaN HEM fabrication. Before etching, a layer of photoresist AZ1518 was coated and patterned using a UV light (320 nm) contact mask aligner. The photoresist residue after development was descumed using O_2 plasma followed by a hard bake. Then the wafer was loaded into the ICP chamber. The etching was carried out with 10/20 sccm Cl_2/BCl_3 gases and 50/100 W RF/ICP power and at 10 mTorr chamber pressure to give a low etch rate of around 13 nm/min. The successful mesa isolation was ensured by etching down to the GaN buffer (~150 nm).

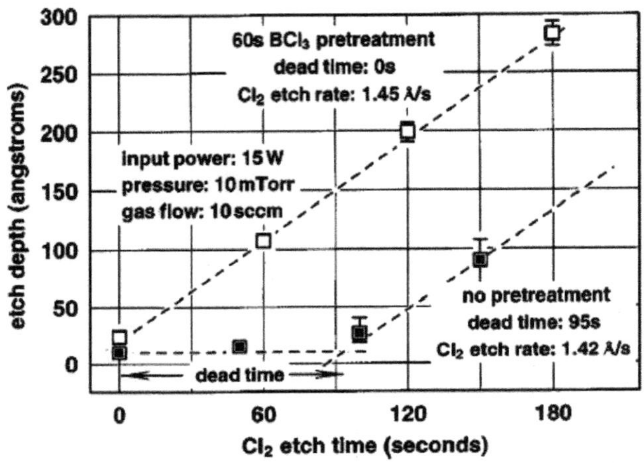

Figure 3.7 The etch depth versus the etch time for Cl_2 plasma to etch GaN with and without BCl_3 pretreatment. Reprinted from Ref. [16], with the permission of AIP Publishing.

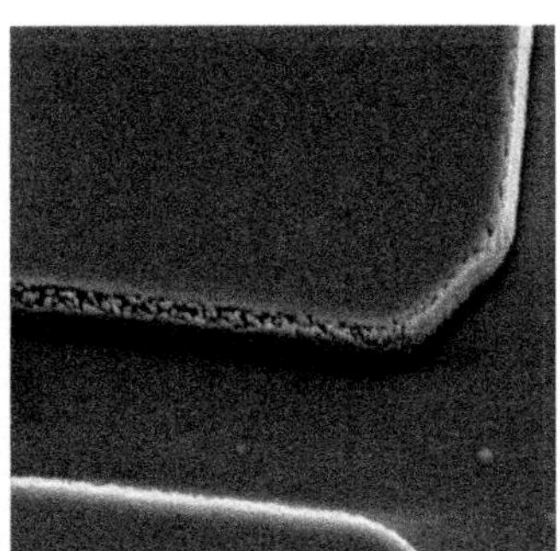

Figure 3.8 Scanning electron microscope (SEM) image of a mesa etched by Cl_2 and BCl_3 plasmas in an ICP system for GaN HEMT fabrication.

The isolation current is one of the key parameters to monitor used to evaluate the quality of mesa etching. Poor isolation not only results in a high off-state leakage current and high power consumption but also limits the reverse blocking voltage of the devices. Ideally the isolation should be enough just after etching down beyond the two-dimensional electron gas (2DEG) channel, but some researchers found the device breakdown voltage can be affected by the etch depth. The isolation current also depends on the GaN buffer quality, such as the threading dislocation density (TDD) and background doping determined by the material growth. A typical current–voltage (*I*–*V*) curve for isolation leakage measurement is shown as Fig. 3.9. Typically the isolation leakage current is less than 10^{-6} mA/mm at 20 V in a GaN HEMT.

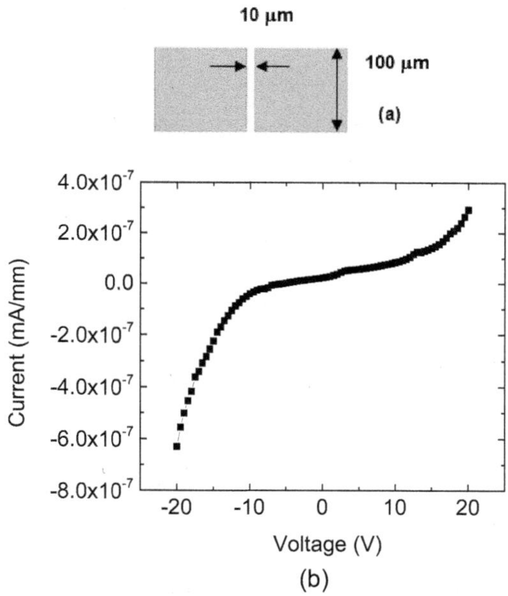

Figure 3.9 (a) Layout of the pattern for isolation leakage measurement. (b) Typical current–voltage characteristics for mesa isolation in a GaN HEMT.

3.1.3 Implantation Isolation

Besides mesa etching, device isolation can also be realized by ion implantation. Compared to mesa isolation, implantation isolation has the advantage that it is a planar process and can ease the subsequent process. As illustrated by Fig. 3.10, planar isolation by ion implantation can avoid the sidewall leakage between the conductive channel (2DEG) and the Schottky gate. Normally the metal or dielectric on a vertical sidewall is thinner than that on a flat surface. Such thin metal or dielectric will cause the device to break down prematurely at a high voltage or current level. A device with implantation isolation prevents such premature breakdown and improves the device's reverse blocking voltage and reliability. In addition, the planar process of implantation isolation will also improve the yield of the device on a large-size wafer.

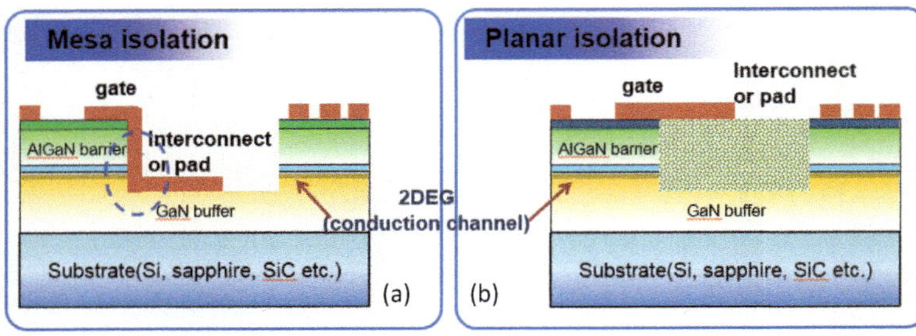

Figure 3.10 Schematic of the device isolation by (a) mesa etch and (b) ion implantation for a GaN HEMT.

Different ions have been tried for implantation isolation in GaN HEMTs, including He, P, Ar, N, O, and As. Table 3.5 lists the reported results of implantation isolation using various ions and implantation conditions for GaN HEMTs. Most of the ion implantation was done with a 7° tilt to suppress the channel effect. The mechanism of isolation by ion implantation in a GaN HEMT is dominated by the damage created by the high-energy ion bombardment. A lot of point defects or simple defect complexes were generated by the collision cascade of energetic ions implanted into the nitride materials. The deep-level traps associated with these defects, like Ga or N vacancies or interstitials, compensate the background doping carriers and lead to high resistivity. This mechanism of damage compensate is validated by the annealing behavior of the ion-implanted samples. After relatively low-temperature annealing for an ion-implanted GaN sample, the resistivity becomes higher than the as-implanted one, as part of the defects are annealed out and the hopping sites are reduced so that the hopping current is lower. However, if the annealing temperature is higher than a certain value, the sample's resistivity decreases because the hopping density is lower than the carrier density and the leakage current is dominated by the increased carrier density.

Table 3.5 Summary of various ions and implantation conditions for implantation isolation in GaN HEMTs

Implant ions	Ion energy (keV)/ Dosage(cm^{-2})	Resulted R_{sh} ($\Omega/\square$)	Research group	Ref.
Al	$520/5 \times 10^{13} + 250/4 \times 10^{13}$	1.8×10^{11}	IET	[19]
Ar	$40/4 \times 10^{13}$	3×10^{12}	Panasonic	[17]
As	$80/5 \times 10^{14}$	3×10^{11}	SMART	Unpub.
As, He	NA	NA	UCLA	[18]
B	$10/4.5 \times 10^{13}$	4×10^{12}	Panasonic	[17]
C	$800/1.5 \times 10^{13} + 300/1 \times 10^{13}$	5.8×10^{11}	IET	[19]
Fe	$50/3 \times 10^{13}$	4×10^{12}	Panasonic	[17]
He	20–190	NA	IEMN	[20]
Kr	$5/7.4 \times 10^{12} + 25/7.1 \times 10^{12} + 45/1.11 \times 10^{13}$	NA	NTU	[21]
N	$30/6 \times 10^{12} + 160/1.8 \times 10^{13} + 400/2.5 \times 10^{13}$	1.1×10^{11}	U. Florida	[22, 23]
N	$25/2 \times 10^{14} + 120/6.5 \times 10^{14}$	NA	MIT	[24]
N	$15/5 \times 10^{13}$	4×10^{12}	Panasonic	[17]
O	$25/5 \times 10^{14} + 50/5 \times 10^{14} + 75/5 \times 10^{14}$	10^{12}	NCTU	[25, 26]
P/He	$(P)75/5 \times 10^{11} + (P)180/2 \times 10^{12} + (He)75/6 \times 10^{13}$	10^{12}	UCSD	[27]
Zn	NA	NA	Mitsubishi	[28]

Abbreviations in the table:
IET: Institute of Electron Technology, Poland
U. Florida: University of Florida
UCSD: University of California, San Diego
IEMN: Institut d'électronique de microélectronique et de nanotechnologie
UCLA: University of California, Los Angeles
SMART: Singapore-MIT Alliance for Research and Technology
NTU: Nanyang Technological University
MIT: Massachusetts Institute of Technology
NCTU: National Chiao Tung University
Unpub.: Unpublished

Monto Carlo–based simulation can be carried out using the Stopping and Range of Ions in Matter (SRIM) tool to estimate the damage distribution created by the ion implantation with a certain energy and an angle to tilt. Figure 3.11 shows the simulated damage profile after As implantation with 80 keV energy, a 5×10^{14} cm^{-2} dosage, and a 7° tilt angle. A wafer structure including a 3 nm GaN cap, a 20 nm AlGaN barrier, a 1 nm AlN spacer, and a 1.5 um GaN buffer on a Si substrate was used for the simulation and experiment. The schematic of the wafer structure and the band diagram are also shown in the figure, aligned with the damage profile. As seen from the simulation, the 2DEG electrons fully disappear after the As implantation.

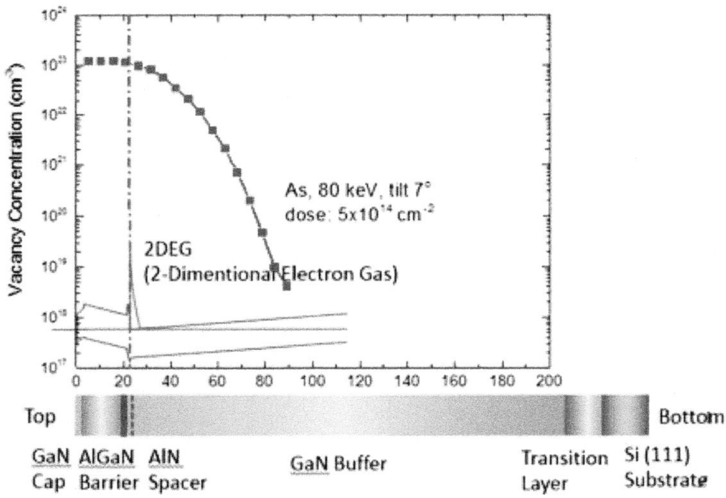

Figure 3.11 Simulation results of As implantation with 80 keV energy, a 5×10^{14} cm^{-2} dosage, and a 7°C tilt. The wafer structure and band diagram are also included in the figure.

Figure 3.12 shows the leakage current between two adjacent mesas with a distance of 10 μm. On the mesa a Ti/Al/Ni/Au ohmic contact with a contact resistance (R_C) of around 0.5 Ω·mm was formed after 850°C annealing. The isolation characteristic of mesa isolation with the same pattern dimension is also included in the figure. The implanted area shows excellent electrical insulation properties. The leakage current of a 10 μm gap is lower than 3.3

× 10^{-5} mA/mm over a voltage range of 100 V. The sheet resistance (R_{sh}) of the implanted region is as high as 3 × 10^{11} Ω/□, even higher than the plasma-etched trench region. This value is also comparable to those reported in the literature (10^9–10^{12} Ω/□).

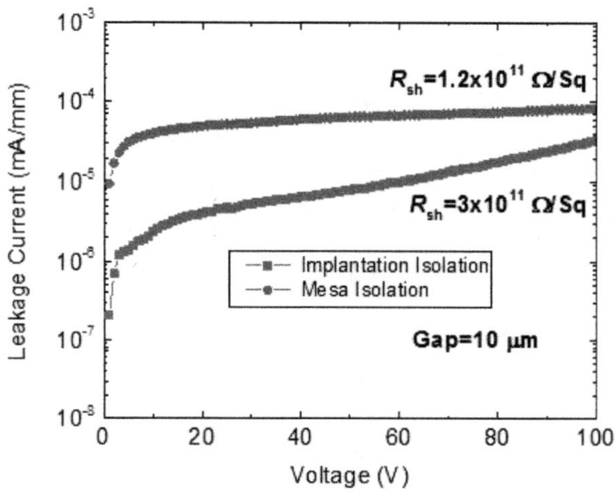

Figure 3.12 Leakage current between two mesas with a distance of 10 μm isolated by ion implantation and mesa etching.

3.2 Ohmic Contacts

3.2.1 The Ti/Al/X/Au Metal Scheme

Ti/Al/X/Au multilayer metals are commonly used for ohmic contacts in GaN HEMTs, where X can be Ti , Ni, Pt, Mo, etc. [29–32]. Ti/Al/Ni/Au is the most widely used ohmic contact metal scheme. The thickness of each layer in the metal scheme is important to get low contact resistance and a smooth surface. Although the contact properties are varied for different sources of wafers, there are some similar trends for the influence of each metal thickness on the contact.

It was found that the Ti/Al ratio plays the most important role in determining the final contact resistance. Figure 3.13 shows the ohmic contact resistance (R_c) and specific contact resistivity (ρ_C) as a function of the Al/Ti thickness ratio in the Ti/Al/Ni/Au metal scheme for a GaN HEMT, where Ni/Au = 40/150 nm and annealed at 900°C in N_2 ambient [33]. It can be seen that the contact resistance reaches a minimum when the Al/Ti ratio is around 6. The influence of the Ti/Al thickness on the contact resistance is shown in Fig. 3.14, in which are plotted the results of R_c and ρ_C as a function of the Ti thickness when the Al/Ti ratio is fixed at 6 and Ni/Au ratio is fixed at 40/150 nm and annealed at 900°C in N_2 ambient [33]. If the Al/Ti ratio is fixed at 6, the contact resistance will reach a minimum when the thickness of Ti is around 30 nm. The thickness of Ni also influences the contact resistance, and there exists an optimum value of 40 nm, as shown in Fig. 3.15 [33].

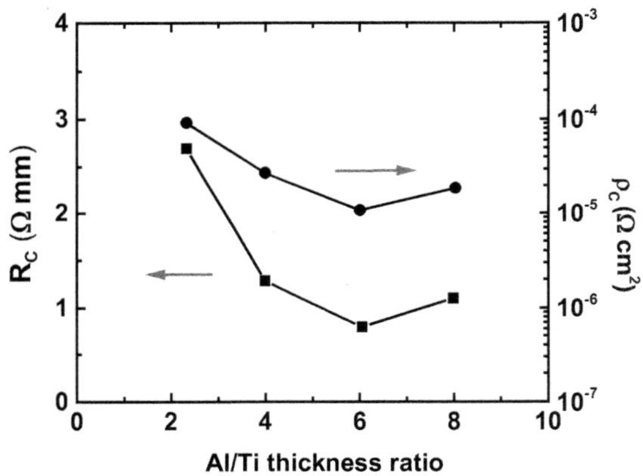

Figure 3.13 Ohmic contact resistance (R_c) and specific contact resistivity (ρ_C) as a function of the Al/Ti thickness ratio in the Ti/Al/Ni/Au metal scheme for a GaN HEMT. Ni/Au = 40/150 nm and annealed at 900°C in N_2 ambient. Reprinted from Ref. [33], Copyright (2002), with permission from Elsevier.

Compared to the top layer of Au, the thickness of Ni is more important for the contact resistance, which is also reported by H. P. Xin from Northrop Grumman Corporation [34]. Figure 3.16 shows the contact results with different Au thicknesses and Ni thicknesses when Ti/Al is fixed [34]. Generally, the thicker the Ni, the lower the contact resistance. And increasing the Au thickness also gives a lower contact resistance when Ni is thin. If Ni is thick, the change of the Au thickness will not bring much change to the contact resistance. On the other side, the Au thickness obviously affects the ohmic metal surface after annealing. The thicker the Au is, the rougher the metal will be after annealing at the same temperature. Figure 3.17 plots

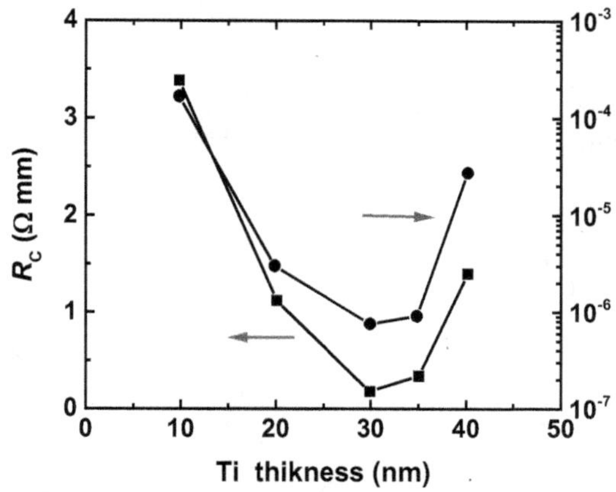

Figure 3.14 Ohmic contact resistance (R_c) and specific contact resistivity (ρ_C) as a function of the Ti thickness in the Ti/Al/Ni/Au metal scheme for a GaN HEMT. The Al/Ti ratio is 6 and Ni/Au is 40/150 nm and annealed at 900°C in N_2 ambient. Reprinted from Ref. [33], Copyright (2002), with permission from Elsevier.

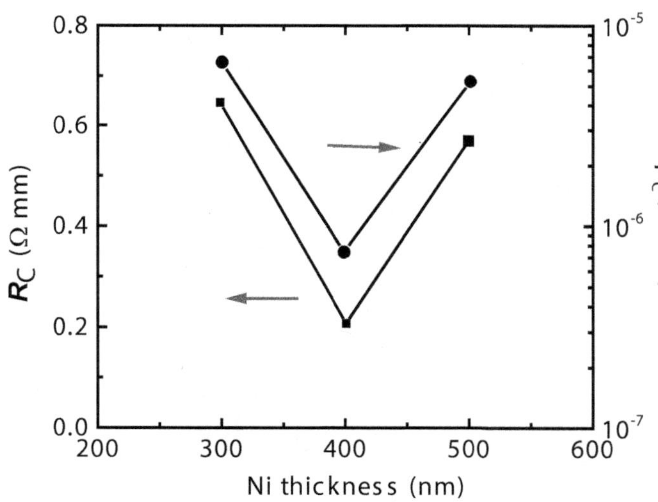

Figure 3.15 Ohmic contact resistance (R_c) and specific contact resistivity (ρ_C) as a function of the Ni thickness in the Ti/Al/Ni/Au metal scheme for a GaN HEMT. Ti/Al = 30/180 nm; Ni + Au = 190 nm annealed at 900°C in N_2 ambient [. Reprinted from Ref. [33], Copyright (2002), with permission from Elsevier.

the values of roughness as a function of the Au thickness at different Ni thickness levels [28]. It can be clearly seen that compared to the Ni thickness, the thickness of the top Au greatly affects the annealed surface roughness.

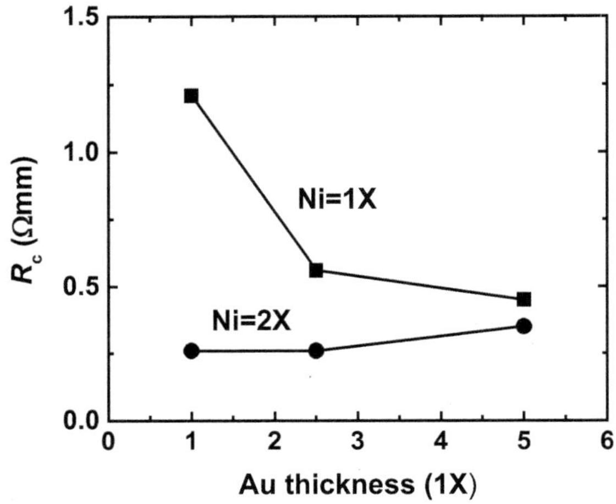

Figure 3.16 Contact resistance (R_c) with various Ni and Au resistance values annealed at 830°C for 30 s [34].

Postmetallization annealing is usually required to form a good ohmic contact for Ti/Al/X/Au metal schemes. In most cases a good ohmic contact of $R_c < 0.5$ Ω·mm can be achieved for a Ti/Al/Ni/Au ohmic contact in AlGaN/GaN or InAlN/GaN HEMTs after annealing at 800°C–900°C in N_2 ambient for 30 s to 1 min.

Figure 3.19 shows a typical plot of the contact resistance (R_C) and specific contact resistivity (ρ_C) as a function of the post-metal-annealing temperature for a Ti/Al/Ni/Au ohmic contact in a GaN HEMT fabricated in our lab. The ohmic contact formation started from the pattern definition using the image reversal process of photoresist AZ5214. AZ 5214 was used to eliminate the chlorobenzene, which can cause rough edges and nonuniform critical dimensions (here mainly source-to-drain distances) over the whole wafer. At first a layer of AZ5214 with a thickness of ~1.4 μm was coated on the wafer. After a dehydration bake, the wafer was exposed by 320 nm UV light using a contact mask aligner. Then a second hard bake was used and followed by flood exposure. The patterns were developed using CD 26 developer. After appropriate surface treatment, a four-layer metal

Ti/Al/Ni/Au was deposited using an electron beam evaporator and then followed by lift-off using acetone and an AZ 300T photoresist stripper. Finally, the wafer with ohmic metals was annealed to form an ohmic contact between the source/drain metal electrode and the AlGaN/GaN heterostructure.

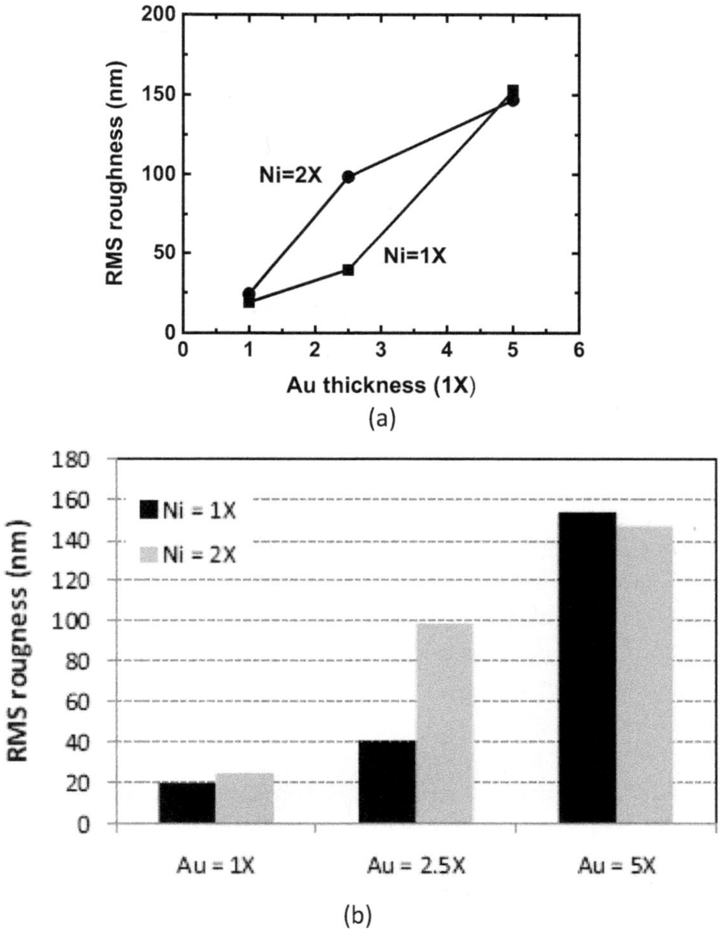

Figure 3.17 (a) Microscope photographs of the annealing metal surface for Ti/Al/Ni/Au in GaN HEMTs with different Au thicknesses. (b) The values of roughness as a function of the Au thickness at different Ni thickness levels [34].

Appropriate surface treatment before metal deposition can reduce the ohmic contact resistance and improve long-term reliability through removal of the residual photoresist, native oxide, and foreign contamination. A clean, oxide-free surface is critical for metal adhesion and low specific contact resistance. Before metallization, short-time oxygen plasma RIE was used to remove the residual photoresist brought during the previous lithography processes. This was followed by acid treatment using buffered oxide etch (BOE) solution for 30 s to remove the native oxide and contaminations.

Figure 3.18 shows the R_t versus L line measured from a transmission line method (TLM) pattern with L = 5, 10, 20, 40, and 80 μm and W = 100 μm. The ohmic contact was formed by Ti/Al/Ni/Au (20/120/40/50 nm) deposited on the AlGaN/GaN surface and followed by rapid thermal annealing (RTA) at 850°C for 30 s in a N_2 atmosphere. The calculated values of R_{sh}, R_c, and ρ_c are 493 Ω/□, 0.31 Ω·mm, and 2 × 10^{-6} Ω·cm^2, respectively. Figure 3.19 shows the measured values of R_c and ρ_c annealed at different temperatures. Good ohmic contact resistance R_c < 0.3 Ω·mm and specific contact resistivity ρ_C < 2 × 10^{-6} Ω·cm^2 are achieved after annealing at 850°C for 30 s. After annealing at 850°C and 875°C, ρ_c reaches the minimum but the metal surface is obviously rougher at 875°C than that at 850°C. A rough surface can blur the pattern edge definition and results in inaccurate critical dimension values, as well as increase the risk of a short between the gate and source electrodes, since the gate–source distance is generally very small in the GaN HEMT in order to obtain a low source access resistance. Hence the annealing temperature of the ohmic contact with Ti/Al/Ni/Au (20/120/40/50 nm) at 850°C for 30 s in a N_2 atmosphere using RTA was used considering the requirements for both low contact resistance and a smooth surface.

Traditional interpretation for the contact mechanisms between Ti/Al/Ni/Au and AlGaN/GaN (or InAlN/GaN and AlN/GaN) heterostructures is described as follows. After postmetallization annealing, the Ti and Al attract the nitrogen from the AlGaN layer below (including the GaN cap layer, if available) and form TiAlN [35]. Then a large amount of N vacancies are formed in the barrier. These N vacancies will work as high concentration n-dopants and pull down the conduction band of the barrier below the Fermi level, which enhances the electron tunneling between the ohmic metal and the highly n-doped barrier/2DEG. The TiAlN formed is very thin, and electrons can easily tunnel through. Ti is necessary since it works as the catalyst for the reaction between Al and GaN/AlGaN. And Ti itself also reactivates with GaN/AlGaN to form TiN, which has a low work function and thus a small barrier height with the GaN/AlGaN. At the same time, Ti works as a thin "wetting agent" layer to promote the adhesion of Al to the semiconductor. At the top of the ohmic metal, Au is included to avoid the oxidation of Ti/Al and to work as the bonding and interconnecting metal owing to its low electric resistivity and good chemical stability. To prevent the interdiffusion between the Au and Ti/Al during the high-temperature annealing and/or long time working, Ni (or Ti, Pt, Mo, etc.) is inserted between the Au and Ti/Al as a blocking layer. Compared to Ti used as the blocking layer, Ni (or Pt, Mo, etc.) can also improve the smoothness of the metal surface after annealing.

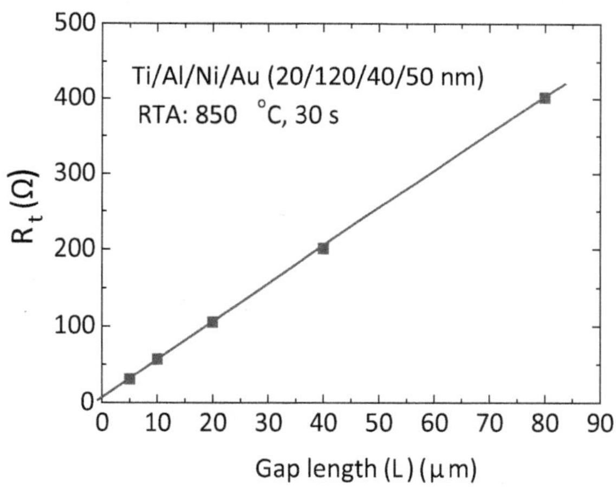

Figure 3.18 R_t versus L measured for the TLM pattern with Ti/Al/Ni/Au (20/120/40/50 nm) on an AlGaN/GaN heterostructure after RTA at 850°C for 30 s in a N_2 atmosphere.

However, the above-mentioned interpretation of the ohmic contact realization between the annealed Ti/Al/X/Au and AlGaN/GaN began to be questioned recently. A fact was observed that although good contact can be achieved with a Au layer on the top of Ti/Al/X, an ohmic contact is very difficult to form after the top Au layer is removed or replaced with other metals. Figure 3.20 compares the ohmic contact resistance (R_C) of Ti/Al/Ni/Au (20/120/40/50 nm), Ti/Al/Ni (20/120/40 nm), and Ti/Al/Ni/Pt (20/120/40/50 nm) fabricated on one AlGaN/GaN HEMT wafer after annealing at 850°C for 30 s. This means that the top Au layer plays a more important role than just decreasing the metal resistance of the whole electrode.

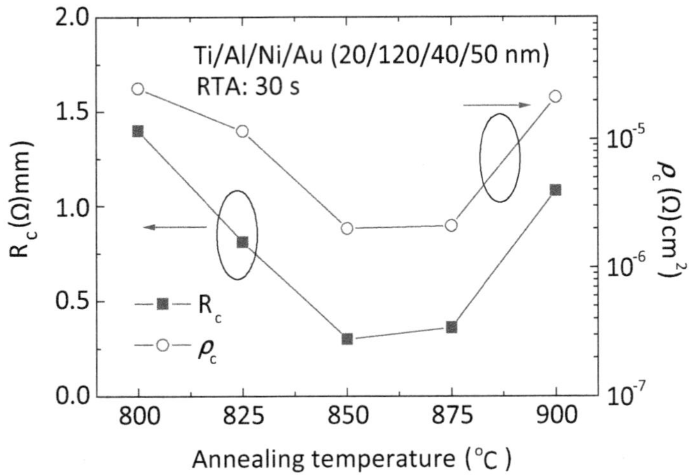

Figure 3.19 Measured R_c and ρ_c for the Ti/Al/Ni/Au (20/120/40/50 nm) on an AlGaN/GaN heterostructure annealed at different temperatures.

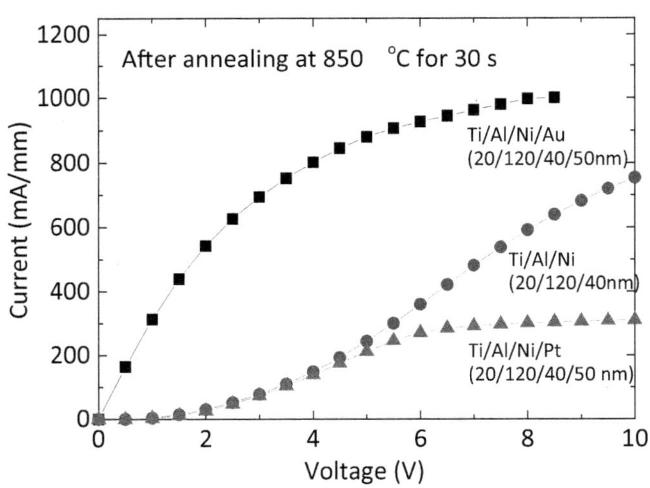

Figure 3.20 Comparison of the ohmic contact resistance (R_C) of Ti/Al/Ni/Au (20/120/40/50 nm), Ti/Al/Ni (20/120/40 nm), and Ti/Al/Ni/Pt (20/120/40/50 nm) on one AlGaN/GaN HEMT wafer after annealing at 850°C for 30 s.

A second theory argues that the key to forming an ohmic contact between Ti/Al/Ni/Au and AlGaN/GaN is that the Ti(Al)N penetrates through the AlGaN barrier and forms a contact directly to the electron channel [36–38]. Ti(Al)N metal inclusions that have been observed to penetrate into the AlGaN barrier and the GaN channel are contained by a very thin layer of Au (or AuAl). This Au (or AuAl) diffusion front is believed to play an important role in the Ti(Al)N penetration.

We have tried to replace the top Au in the ohmic metal scheme Ti/Al/Ni/Au with Pt and annealed at the same temperature [38]. The wafers and processes were the same. The ohmic contact resistance of Ti/Al/Ni/Au is good, with $R_c = \sim 0.5\ \Omega\cdot\text{mm}$, but that of Ti/Al/Ni/Pt is very bad and even Shottky like, as shown in Fig. 3.20. Figure 3.21 shows two transmission electron microscopy (TEM) images of Ti/Al/Ni/Au and Ti/Al/Ni/Pt with the same thickness (20/100/40/50 nm) after annealing. It is obvious that in the Ti/Al/Ni/Au ohmic contact, the TiAlN penetrates through the AlGaN barrier and forms a contact directly to the electron channel. However, in the case of the Ti/Al/Ni/Pt ohmic contact, even though it was annealed at a higher temperature, much less TiAlN penetrates through the AlGaN barrier. This confirms that Au plays a critical role in assisting the TiAlN in penetrating the AlGaN/AlN/GaN structure. This is possibly due to the higher diffusion coefficient for Au in Al and GaN(AlGaN) than Pt and other metals. However, the details of the diffusion mechanism are still unclear and need to be investigated further.

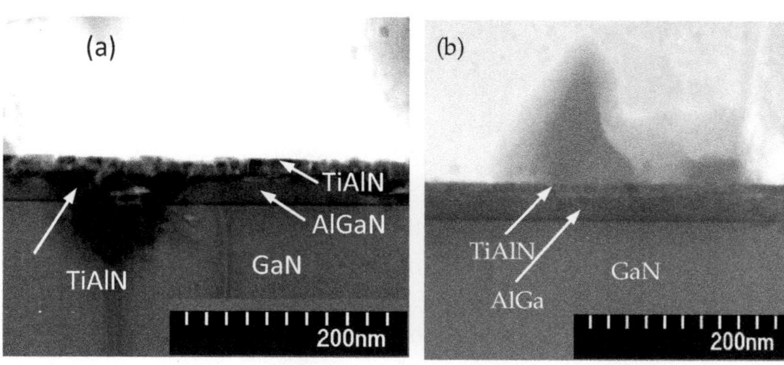

Figure 3.21 TEM images of the ohmic metal surface for AlGaN/AlN/GaN HEMTs. (a) Ti/Al/Ni/Au (20/100/40/50 nm) after RTA at 825°C for 30 s; (b) Ti/Al/Ni/Pt (20/100/40/50 nm) after RTA at 950°C for 30 s.

3.2.2 CMOS-Compatible Ohmic Contacts

Although standard GaN devices fabricated using Au-bearing technologies are very successful, the Au-contained process is not compatible with the complementary metal-oxide-semiconductor (CMOS) foundries as Au can form deep trap levels in Si and cause serious issues of threshold voltage shift and instability. To enable the GaN HEMTs to be fabricated in standard Si CMOS foundries in order to utilize the mature Si process technologies and to reduce the device cost, Si CMOS–compatible fabrication technologies need to be developed.

It is found that it is not easy to achieve good ohmic contacts for GaN HEMTs unless Au is used in the metallization. Lots of efforts have been made, and a few different Au-free technologies have been reported in literature. So far, low ohmic contact resistance values (R_C < 1.0 Ω·mm) comparable to the value of a Au-bearing ohmic contact have been achieved by the approaches below:

- Using different metal schemes. After replacing the top metal from Au with Pt [38, 39] or W [40] or replacing the Ni/Au with Ni [41], W [42], TiN [43], Ta [44], etc., and with optimized metal thickness and annealing temperature, good ohmic contacts can be achieved. Some reports show that Ta/Al-based [44–46], Ti/TiN-based [47], or Hf/Al-based metal schemes [48] could work to achieve good ohmic contacts for AlGaN/GaN HEMTs. The drawback of such methods is that the obtained ohmic contact resistance strongly depends on the wafer structures and the wafer material quality. For example, a Hf-based metal scheme has been reported to form a good ohmic contact for an InAlN/GaN HEMT structure but cannot achieve low contact resistance for mostly used AlGaN/AlN/GaN HEMT structures. We have developed Ti/Al/Ni/Pt metallization with optimized metal thickness and annealing at 975°C to achieve R_C = ~0.5 Ω·mm and a smooth metal surface [38].
- Doing recess before the metallization of the ohmic contacts [49, 50]. With such a technology and a Ti/Al or Ti/Al/W ohmic stack, our group reported a contact resistance of R_C < 0.3 Ω·mm at as low an annealing temperature as 450°C [50]. The downside of this technology is that the need for ohmic recess increases the process complexity and sensitivity to the recess etch nonuniformity.

- Doing ion implantation in the AlGaN/GaN materials underneath the ohmic metals and making the III nitride highly doped with n-dopants [51–53]. Although as low as $R_C < 0.1\ \Omega\cdot$mm has been achieved using an ion implantation technology, the dopant activation needs a very high temperature (>1100°C), which could create defects and degrade the device performance.
- Using an AlGaN/GaN HEMT structure with a very thin AlGaN barrier. R_C of 0.65 $\Omega\cdot$mm was achieved with Ti/Al/W ohmic metallization in the samples with a ~10 nm AlGaN barrier [54]. In these samples, an in situ metal-organic chemical vapor deposition (MOCVD) SiN cap layer is required to reduce the depletion of the 2DEG by the AlGaN surface and thereby keep a high 2DEG sheet carrier density in the access regions.
- Using an n^+-GaN capped n-AlGaN/GaN HEMT structure. Ultralow contact resistance $R_C < 0.1\ \Omega\cdot$mm can be achieved using Ti/Al metals and low annealing temperatures for an n^+-GaN/n-AlGaN/GaN structure [55]. However, this approach needs an etch recipe with high selectivity between GaN and AlGaN to remove the n^+-GaN cap at the active region; and also the n-AlGaN barrier increases the gate leakage current, especially when the gate is with a Schottky contact.
- Regrowth of n^+-GaN in the ohmic contact regions [56–58]. The AlGaN or InAlN barrier is removed before the regrowth, and the regrown n^+-GaN layer directly contacts the 2DEG. The ohmic metal, typically Ti/Al, fabricated on the n^+-GaN layer, can form a good ohmic contact with the n^+-GaN without annealing or just with low-temperature annealing. The regrowth can be done by MBE or MOCVD, but MBE is better as the growth temperature can be much lower (~700°C versus ~1000°C) and the surface decomposition at high temperature can be avoided. The shortcoming of the regrowth approach is its high cost.

3.3 Gate Fabrication

Two types of gate contact structures are generally used in GaN HEMTs, Schottky contact and metal-insulator-semiconductor (MIS) structure. For the former, the gate metal comes directly in contact with the nitride and forms a Schottky contact. For the latter, a thin layer of dielectric is inserted between the gate metal and the III nitride surface in order to reduce the leakage current and increase the off-state breakdown voltage.

3.3.1 Schottky Gate

To ensure a low gate leakage current in the GaN device, a gate metal with a high Schottky barrier height and high thermal stability is desirable. The work function of the gate metal directly influences the Schottky barrier height between the gate and the nitride semiconductor. Various metals, such as Ni [50, 63, 64], Ti [65], Pt [68], Pd [68], Ir [72, 73], and Re [73], have been tried as the Schottky gate in a GaN metal-semiconductor field-effect transistor (MESFET) or HEMTs. Ni/Au is the most widely used Schottky gate metal scheme in GaN HEMTs. Ni is chosen due to its high work function, good adhesion to the semiconductor surface, good thermal stability, etc.

Other metals have also been investigated to be adopted as the gate contact, and their Schottky barrier heights are listed in Table 3.6. The measured values of the Schottky barrier height are varying with reports, as many other factors, such as the surface states and native oxide, also influence the actually measured barrier height.

Many researchers have found that annealing after gate metallization can improve the Schottky contact characteristics, such as the ideality factor decrease toward unity, the Schottky barrier increase, and the gate leakage current reduction [93–95]. The physical mechanism exists in the fact that after annealing, the bottom metal of the gate contact, for example, Ni, will contact the GaN or AlGaN surface more closely. Generally O impurities exist on the nitride surface in the forms of native oxide GaO_x or $AlGaO_x$ before the metal deposition. O impurities also exist in the AlGaN or InAlN barrier introduced during epitaxy, which work as carrier trap centers and enhance the trap-assisted current, which decreases the effective barrier height. After annealing, Ni reacts with O existing in the native oxide on the surface or in the barrier and NiO_x is formed, which decreases the gate leakage current and increases the effective Schottky barrier height.

Table 3.6 Schottky barrier heights of various metals as the gate contact to GaN and AlGaN/GaN HEMT structures

Metal	Schottky barrier height to GaN	Schottky barrier height (eV) to AlGaN/GaN; Al: 0.2–0.3
Ni	1.12 [59]; 0.84 [60]; 0.96 [61]; 0.96 [62]	1.01 [50]; 1.13 [63]; 0.98 [64]
Ti	0.63 [53]	0.95 [65]
Pt	0.93 [66]; 1.08 [67]	0.98 [68]
Pd	0.93 [52]; 1.11 [69]; 0.96 [60]; 0.91 [70]	1.08 [59]
Ir	1.13 [63]; 0.69 [71]; 1.12 [64]	1.45 [72]; 1.47 [73]
Re	0.82 [74]	1.68 [64]
ITO	0.68 [75]	0.62 [76]
TaN	0.68 [77]	0.54 [67]
Cr	0.53 [78]	~0.9 [79]
W	0.8 [80]	NA
Cu	0.85 [81]	1.15 [82]
Graphene	1.24 [83]	0.4 [84]
WN_x	0.67 [85]	~1.07–1.21 [86]
RuO_x	0.92 [87]	~0.82 [88]
TiN	0.75–0.79 [89]; 0.61 [90]	0.67 [91]; 0.69 [92]

3.3.2 Metal-Insulator-Semiconductor Gate

Typically the GaN HEMTs with the Schottky metal gate suffer an issue of high gate leakage current [96–98]. The high gate leakage current limits the maximum positive voltage, which can be biased on the gate, and then limits the maximum drain current and available output power. The structure of metal-insulator-semiconductor high-electron-mobility transistors (MIS-HEMTs), like a Si metal-oxide-semiconductor field-effect transistor (MOSFET), with a thin dielectric layer between the metal gate and the nitride epilayer, is a good solution to the issue of high gate leakage current. A detailed discussion on the MIS-HEMT will be given in another chapter.

3.4 Surface Passivation

The large number of polarization-induced charges need to be compensated, and the composition occurs externally; therefore, surface states cannot be avoided. These surface states degrade the gate leakage current and current collapse. Recently, it is found that using an insulate layer could passivate the surface states and improve the device performance. The detail of the surface passivation is elaborated in Chapter 5.

3.5 Field Plates

For a field-effect transistor (FET) structure, when drain is biased at a high voltage, the electrical field usually does not distribute evenly along the channel. The electrical peak occurs at the edge of the gate close to the drain, as shown in Fig. 3.23 [99]. Thus the device will prematurely break down once the electrical field peak at the edge of the gate reaches the critical electrical field of the GaN material before the whole channel is punched through. If there is a field plate on top of the gate, as illustrated in Fig. 3.22, there will be two peaks at high drain bias, one of which is located at the gate edge close to drain and the other of which is located at the field plate edge close to drain, as shown in Fig. 3.23. Therefore, the electrical field is distributed more evenly and the breakdown voltage will be increased.

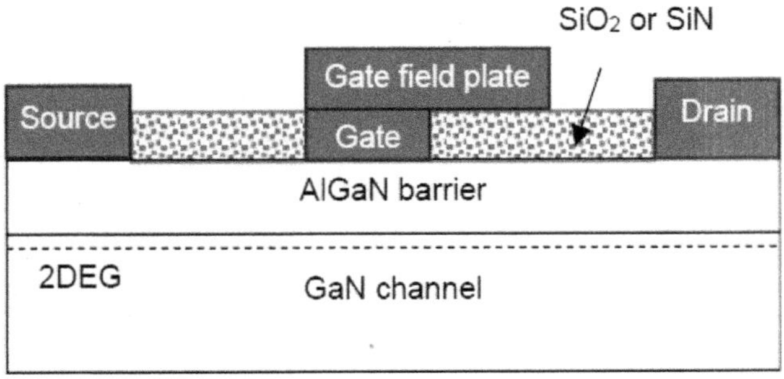

Figure 3.22 The schematic of a GaN HEMT with a gate field plate [99].

The factors to be designed in a field plate HEMT include the field plate length *l*, the insulator thickness under the field plate *t*, and the insulator's dielectric constant ε. The breakdown voltage will increase with the increase of the insulator's thickness t till saturation and also increase with the increase of the field plate length l till saturation. The adverse effect of the long field plate length is the increased feedback capacitance, which will increase the power loss of the device.

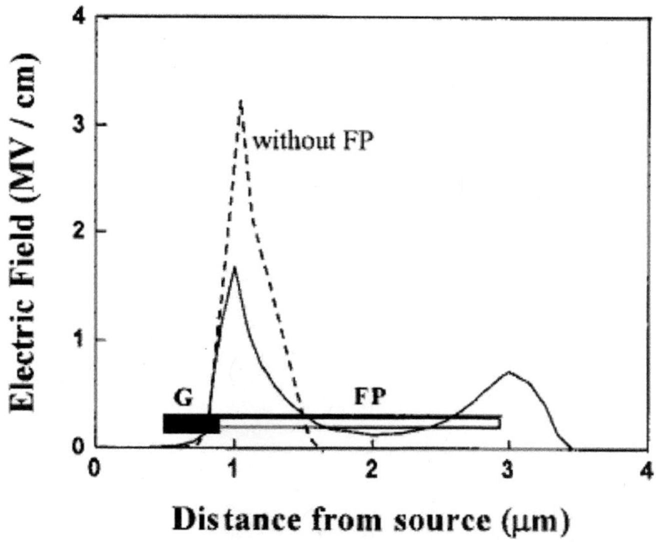

Figure 3.23 Electrical field distribution along the channel for a GaN HEMT at high drain bias with and without a field plate structure From Ref. [103]. Copyright @ 2001 IEEE.

The breakdown voltage of a field plate HEMT also depends on the insulator's dielectric constant ε. The larger the dielectric constant is, the higher breakdown voltage will be, as more electric displacement field lines will be in the insulator and fewer in the III nitride. From simulation [99], the breakdown voltage can be twice for the device with a field plate above SiN than that of a device with a field plate above SiO_2. However, a higher dielectric constant will also increase the feedback capacitance formed by the field plate and the 2DEG and increase the power loss of the device. So there exists a trade-off in designing the field plate material and dimensions considering both high breakdown voltage and low feedback capacitance for the device.

To obtain a high breakdown voltage while keeping a low field plate capacitance, more field plate shapes have been proposed and applied into the GaN power HEMTs, such as the slant field plate, multistep field plate, multiple-field plate, and source field plate. Figure 3.24 shows various field plate designs to balance the breakdown voltage and gate feedback capacitance: (a) slant field plate [100], (b) multiple-gate-field plate [101], (c) multiple-field plate [102], (d) multistep field plate [103], and (e) source field plate [104].

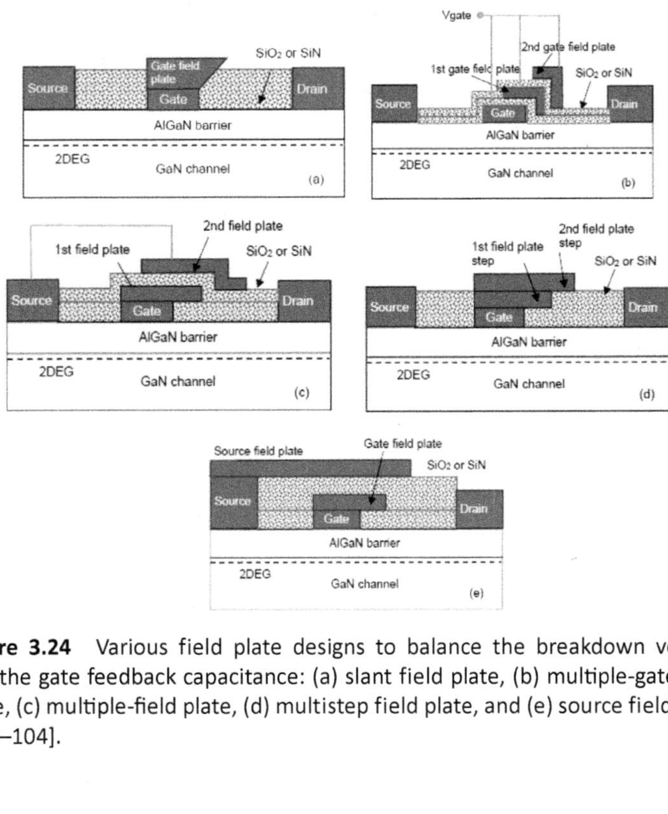

Figure 3.24 Various field plate designs to balance the breakdown voltage and the gate feedback capacitance: (a) slant field plate, (b) multiple-gate-field plate, (c) multiple-field plate, (d) multistep field plate, and (e) source field plate [100–104].

References

1. Stocker, D. A., Schubert, E. F., and Redwing, J. M. (1998). Crystallographic wet chemical etching of GaN, *Appl. Phys. Lett.*, **73**, 2654–2656.
2. Carosella, C. A., Molnar, B., Schiestel, S., and Sprague, J. A. (2000). Wet etching of ion-implanted GaN crystals by AZ-400K photoresist, *MRS Internet J. Nitride Semicond. Res.*, **5S1**, W11.70.

3. Palacios, T., Calle, F., Varela, M., Ballesteros, C., Monroy, E., Naranjo, F. B., Sánchez-García, M. A., Calleja, E., and Muñoz, E. (2000). Wet etching of GaN grown by molecular beam epitaxy on Si (111), *Semicond. Sci. Technol.*, **15**, 996–1000.

4. Shimizu, M., Suzuki, A., Watanabe, M., Shirakashi, J., Balakrishan, K., and Okumura, H. (eds.) (1998). Kentaro Onabe, Blue Laser and Light Emitting Diodes II, in *International Symposium on Blue Laser and Light Emitting Diodes II*, Chiba, Japan, p. 723.

5. Li, D., Sumiya, M., Yoshimura, K., Suzuki, Y., Fukuda, Y., and Fuke, S. (2000). Effect of passivation on AlGaN/GaN HEMT device performance, *Phys. Status Solidi A*, **180**, 357–363.

6. Li, D., Sumiya, M., Fuke, S., Yang, D., Que, D., Suzuki, Y., and Fukuda, Y. (2001). Selective etching of GaN polar surface in potassium hydroxide solution studied by x-ray photoelectron spectroscopy, *J. Appl. Phys.*, **90**, 4219–4223.

7. Cho, H., Donovan, S. M., Abernathy, C. R., Pearton, S. J., Ren, F., and Han, J. (1999). Photo-electrochemical etching of $In_xGa_{1-x}N$, *MRS Internet J. Nitride Semicond. Res.*, **4S1**, G6.40.

8. Youtsey, C., Adesida, I., and Bulman, G. (1997). Highly anisotropic photoenhanced wet etching of n-type GaN, *Appl. Phys. Lett.*, **71**, 2151–2153.

9. Minsky, M. S., White, M., and Hu, E. L. (1996). Room-temperature photoenhanced wet etching of GaN, *Appl. Phys. Lett.*, **68**, 1531–1533.

10. Bardwell, A., Webb, J. B., Tang, H., Fraser, J., and Moisa, S. (2001). Ultraviolet photoenhanced wet etching of GaN in $K_2S_2O_8$ solution, *J. Appl. Phys.*, **89**, 4142–4149.

11. Seo, J. W., Oh, C. S., Yang, J. W., Yang, G. M., Lim, K. Y., Yoon, C. J., and Lee, H. J. (2001). Allowable substrate bias for the etching of n-GaN in photo-enhanced electrochemical etching, *Phys. Status Solidi A*, **188**, 403–406.

12. Bardwell, J. A., Foulds, I. G., Webb, J. B., Tang, H., Fraser, J., Moisa, S., and Rolfe, S. J. (1999). A simple wet etch for GaN, *J. Electron. Mater.*, **28**, L24–L26.

13. Dogliotti, L., and Hayon, E. (2511). Flash photolysis of per[oxydi]sulfate ions in aqueous solutions. The sulfate and ozonide radical anions, *J. Phys. Chem.*, **71**, 2511–2516.

14. Kolthoff, I. M., and Miller, I. K. (1951). The chemistry of persulfate. I. The kinetics and mechanism of the decomposition of the persulfate ion in aqueous medium, *J. Am. Chem. Soc.*, **73**, 3055–3059.

15. Pearton, S. J., Zolper, J. C., Shur, R. J., and Ren, F. (1999). GaN: processing, defects, and devices, *J. Appl. Phys.*, **86**, 1–78.

16. Buttari, D., Chini, A., Palacios, T., Coffie, R., Shen, L., Xing, H., Heikman, S., McCarthy, L., Chakraborty, A., Keller, S., and Mishra, U. K. (2003). Origin of etch delay time in Cl_2 dry etching of AlGaN/GaN structures, *Appl. Phys. Lett.*, **83**, 4779–4781.

17. Umeda, H., Takizawa, T., Anda, Y., Ueda, T., and Tanaka, T. (2013). High-voltage isolation technique using Fe ion implantation for monolithic integration of AlGaN/GaN Transistors, *IEEE Trans. Electron Devices*, **60**, 771–775.

18. Chung, Y., Hang, C. Y., Cai, S., Chen, Y., Lee, W., Wen, C. P., Wang, K. L., and Itoh, T. (421). Effects of output harmonic termination on PAE and output power of AlGaN/GaN HEMT power amplifier, *IEEE Microwave Compon. Lett.*, **12**, 421–423.

19. Taube, A., Kamińska, E., Kozubal, M., Kaczmarski, J., Wojtasiak, W., Jasiński, J., Borysiewicz, M. A., Ekielski, M., Juchniewicz, M., Grochowski, J., Myśliwiec, M., Dynowska, E., Barcz, A., Prystawko, P., Zając, M., Kucharski, R., and Piotrowska, A. (2015). Ion implantation for isolation of AlGaN/GaN HEMTs using C or Al, *Phys. Status Solidi A*, **212**, 1162–1169.

20. Ducatteau, D., Minko, A., Hoël, V., Morvan, E., Delos, E., Grimbert, B., Lahreche, H., Bove, P., Gaquière, C., De Jaeger, J. C., and Delage, S. (2006). Output power density of 5.1/mm at 18 GHz with an AlGaN/GaN HEMT on Si substrate, *IEEE Electron Device Lett.*, **27**, 7–9.

21. Arulkumaran, S., Ng, G. I., Ranjan, K., Saw, G. Z., Murmu, P. P., and Kennedy, J. (2014). Improved device isolation in AlGaN/GaN HEMTs on Si by heavy Kr+ Ion implantation, in *Proceedings of the Device Research Conference (DRC)*, pp. 115–116.

22. Lo, C. F., Kang, T. S., Liu, L., Chang, C. Y., Pearton, S. J., Kravchenko, I. I., Laboutin, O., Johnson, J. W., and Ren, F. (2010). Isolation blocking voltage of nitrogen ion-implanted AlGaN/GaN high electron mobility transistor structure, *Appl. Phys. Lett.*, **97**, 262116-1–262116-3.

23. Lo, C. F., Kang, T. S., Liu, L., Ren, F., Pearton, S. J., Kim, J., Jang, S., Laboutin, O., Cao, Y., and Johnson, J. W. (2011). Effects of silicon nitride passivation on isolation-blocking voltage in AlGaN/GaN high electron mobility transistors, *J. Vac. Sci. Technol. B*, **29**, 031211-1–031211-5.

24. Sun, M., Lee, H.-S., Lu, B., Piedra, D., and Palacios, T. (2012). Comparative breakdown study of mesa- and ion-implantation-isolated AlGaN/GaN high-electron-mobility transistors on Si substrate, *Appl. Phys. Express*, **5**, 074202-1–074202-3.

25. Shiu, J.-Y., Lu, C.-Y., Su, T.-Y., Huang, R.-T., Zirath, H., Rorsman, N., and Chang, E. Y. (2010). Capacitance–voltage characteristics of Al_2O_3/AlGaN/GaN structures and state density distribution at Al_2O_3/AlGaN interface, *Jpn. J. Appl. Phys.*, **49**, 021001-1–021001-5.

26. Shiu, J.-Y., Huang, J.-C., Desmaris, V., Chang, C.-T., Lu, C.-Y., Kumakura, K., Makimoto, T., Zirath, H., Rorsman, N., and Chang, E. Y. (2007). Oxygen ion implantation isolation planar process for AlGaN/GaN HEMTs, *IEEE Electron Device Lett.*, **28**, 476–478.

27. Hanington, G., Hsin, Y. M., Liu, Q. Z., Asbeck, P. M., Lau, S. S., Khan, M. A., Yang, J. W., and Chen, Q. (1998). P/He ion implant isolation technology for AlGaN/GaN HFETs, *Electron. Lett.*, **34**, 193–195.

28. Miura, N., Nanjo, T., Suita, M., Oishi, T., Abe, Y., Ozeki, T., Ishikawa, H., Egawa, T., and Jimbo, T. (2004). Thermal annealing effects on Ni/Au based Schottky contacts on n-GaN and AlGaN/GaN with insertion of high work function metal, *Solid-State Electron.*, **48**, 689–695.

29. Lee, C., Saunier, P., Yang, J., and Khan, M. A. (2003). AlGaN–GaN HEMTs on SiC with CW power performance of >4 W/mm and 23% PAE at 35 GHz, *IEEE Electron Device Lett.*, **24**, 616–618.

30. Liu, Z. H., Ng, G. I., Arulkumaran, S., Maung, Y. K. T., Teo, K. L., Foo, S. C., Sahmuganathan, V., Xu, T., and Lee, C. H. (2010). High microwave-noise performance of AlGaN/GaN MISHEMTs on silicon with Al_2O_3 gate insulator grown by ALD, *IEEE Electron Device Lett.*, **31**, 96–98.

31. Mohammed, F. M., Wang, L., and Adesida, I. (2005). Noninterfacial-nitride formation ohmic contact mechanism in Si-containing Ti/Al/Mo/Au metallizations on AlGaN/GaN heterostructures, *Appl. Phys. Lett.*, **87**, 262111-1–262111-3.

32. Wang, W. K., Li, Y. J., Lin, C. K., Chan, Y. J., Chen, G. T., and Chyi, J. I. (2004). Low damage, Cl2-based gate recess etching for 0.3 μm gate-length AlGaN/GaN HEMT fabrication, *IEEE Electron Device Lett.*, **25**, 52–54.

33. Jacobs, B., Kramer, M. C. J. C. M., Geluk, E. J., and Karouta, F. (2002). Optimisation of the Ti/Al/Ni/Au ohmic contact on AlGaN/GaN FET structures, *J. Cryst. Growth*, **241**, 15–18.

34. Xin, H. P., Poust, S., Sutton, W., Li, D., Lam, D., Smorchkova, I., Sandhu, R., Heying, B., Uyeda, J., Barsky, M., Wojtowicz, M., and Lai, R. (2010). Optimization of AlGaN/GaN HEMT ohmic contacts for improved surface morphology with low contact resistance, in *Proceedings of CS MANTECH*, Portland, OR, pp. 149–150.

35. Ruvimov, S., Liliental-weber, Z., Washburn, J., Qiao, D., Lau, S. S., and Chu, P. K. (1998). Microstructure of Ti/Al ohmic contacts for n-AlGaN, *Appl. Phys. Lett.*, **73**, 2582–2584.

36. Wang, L., Mohammed, F. M., and Adesida, I. (2008). Formation mechanism of Ohmic contacts on AlGaN/GaN heterostructure: electrical and microstructural characterizations, *J. Appl. Phys.*, **103**, 093516-1–093516-3.

37. Fay, M. W., Moldovan, G., Brown, P. D., Harrison, I., Birbeck, J. C., Hughes, B. T., Uren, M. J., and Martin, T. (2002). Structural and electrical characterization of AuTiAlTi/AlGaN/GaN ohmic contacts, *J. Appl. Phys.* **92**, 94–100.

38. Liu, Z., Sun, M., Lee, H.-S., Heuken, M., and Palacios, T. (2013). AlGaN/AlN/GaN high-electron-mobility transistors fabricated with Au-Free technology, *Appl. Phys. Express*, **6**, 096502-1–096502-3.

39. Liu, X., Zhan, C., Chan, K. W., Owen, M. H. S., Liu, W., Chi, D. Z., Tan, L. S., Chen, K. J., and Yeo, Y.-C. (2013). AlGaN/GaN metal–oxide–semiconductor high-electron-mobility transistors with a high breakdown voltage of 1400 V and a complementary metal–oxide–semiconductor compatible gold-free process, *Jpn. J. Appl. Phys.*, **52**, 04CF06-1–04CF06-5.

40. Tripathy, S., Kyaw, L. M., Dolmanan, S. B., Ngoo, Y. J., Liu, Y., Bera, M. K., Singh, S. P., Tan, H. R., Bhat, T. N., and Chor, E. F. (2014). InxAl1-xN/AlN/GaN high electron mobility transistor structures on 200 mm diameter Si(111) substrates with Au-free device processing, *ECS J. Solid-State Sci. Technol.*, **3**, Q84–Q88.

41. Herfurth, P., Maier, D., Lugani, L., Carlin, J.-F., Rösch, R., Men, Y., Grandjean, N., and Kohn, E. (2013). Ultrathin body InAlN/GaN HEMTs for high-temperature (600°C) electronics, *IEEE Electron Device Lett.*, **34**, 496–498.

42. Van Hove, M., Boulay, S., Bahl, S. R., Stoffels, S., Kang, X., Wellekens, D., Geens, K., Delabie, A., and Decoutere, S. (2012). CMOS process-compatible high-power low-leakage AlGaN/GaN MISHEMT on silicon, *IEEE Electron Device Lett.*, **33**, 667–669.

43. Firrincieli, A., Jaeger, B. D., You, S., Wellekens, D., Hove, M. V., and Decoutere, S. (2014). Au-free low temperature ohmic contacts for AlGaN/GaN power devices on 200 mm Si Substrates, *Jpn. J. Appl. Phys.*, **53**, 04EF01-1–04EF01-4.

44. Alomari, M., Maier, D., Carlin, J.-F., Grandjean, N., Poisson, M.-A., Delage, S., and Kohn, E. (2009). Au free ohmic contacts for high temperature InAlN/GaN HEMT's, *ECS Trans.*, **25**, 33–36.

45. Malmros, A., Blanck, H., and Rorsman, N. (2011). Electrical properties, microstructure, and thermal stability of Ta-based ohmic contacts annealed at low temperature for GaN HEMTs, *Semicond. Sci. Technol.*, **26**, 075006-1–075006-7.

46. Arulkumaran, S., Ng, G. I., Vicknesh, S., Wang, H., Ang, K. S., Kumar, C. M., Teo, K. L., and Ranjan, K. (2013). Demonstration of submicron-gate AlGaN/GaN high-electron-mobility transistors on silicon with complementary metal–oxide–semiconductor-compatible non-gold metal stack, *Appl. Phys. Express*, **6**, 016501-1–016501-3.

47. Jatal, W., Baumann, U., Tonisch, K., Schwierz, F., and Pezoldt, J. (2015). High-frequency performance of GaN high-electron mobility transistors on 3C-SiC/Si substrates with Au-free ohmic contacts, *IEEE Electron. Device Lett.*, **36**, 123–125.

48. Liu, Y., Kyaw, L. M., Bera, M. K., Singh, S. P., Ngoo, Y. J., Lo, G. Q., and Chorl, E. F. (2014). Low thermal budget Au-free Hf-based ohmic contacts on InAlN/GaN heterostructures, *ECS Trans.*, **61**, 319–327.

49. Lee, H.-S., Lee, D. S., and Palacios, T. (2011). AlGaN/GaN high-electron-mobility transistors fabricated through a Au-free technology, *IEEE Electron Device Lett.*, **32**, 623–625.

50. Liu, Z., Heuken, M., Fahle, D., Ng, G. I., and Palacios, T. (2014). CMOS-compatible Ti/Al ohmic contacts (Rc <0.3 Ωmm) for u-AlGaN/AlN/GaN HEMTs by low temperature annealing (<450°C), in *Proceedings of the Device Research Conference (DRC)*, Santa Barbara, USA, pp. 75–76.

51. Nanjo, T., Imai, A., Suzuki, Y., Abe, Y., Oishi, T., Suita, M., Yagyu, E., and Tokuda, Y. (2013). AlGaN channel HEMT with extremely high breakdown voltage, *IEEE Trans. Electron Devices*, **60**, 1046–1053.

52. Liu, Z., Huang, C. C., Fahle, D., Xing, W., Heuken, M., Ng, G. I., and Palacios, T. (2015). Si ion implantation technology to reduce the access resistance in GaN HEMTs on sapphire and Si substrates, in *International Symposium of Compound Semiconductors (ISCS)*, Santa Barbara, USA.

53. Liu, Z., Huang, C. C., Xing, W., Fahle, C., Heuken, M., Ng, G. I., and Palacios, T. (2015). Si ion implantation and activation in NH_3 for CMOS-compatible fabrication of GaN HEMTs on Si substrates, in *International Conference on Nitride Semiconductors (ICNS)*, Beijing, China.

54. Hove, M. V., Boulay, S., Bahl, S. R., Stoffels, S., Kang, X., Wellelkens, D., Geens, K., Delabie, A., and Decoutere, S. (2012). CMOS process-compatible high-power low-leakage AlGaN/GaN MISHEMT on silicon, *IEEE Electron Device Lett.*, **33**, 667–669.

55. Liu, Z., Fahle, D., Heuken, M., Ng, G. I., and Palacios, T. (2013). Low ohmic contact resistance R_C<0.1 Ωmm for GaN HEMTs using Au-free metallization and low-temperature annealing, in *International Conference on Nitride Semiconductors (ICNS)*, Washington, DC.

56. Heikman, S., Keller, S., DenBaars, S. P., and Mishra, U. K. (2001). Mass transport regrowth of GaN for ohmic contacts to AlGaN/GaN, *Appl. Phys. Lett.*, **78**, 2876–2878.

57. Seo, H.-C., Chapman, P., Cho, H.-I., Lee, J.-H., and Kim, K. K. (2008). Defect-induced color-tunable monolithic GaN-based light-emitting diodes, *Appl. Phys. Lett.*, **93**, 102102-1–102102-4.

58. Yue, Y., Hu, Z., Guo, J., Berardi, S.-R., Li, G., Wang, R., Faria, F., Fang, T., Song, B., Gao, X., Guo, S., Kosel, T., Snider, G., Fay, P., Jena, D., and Xing, H. (2012). InAlN/AlN/GaN HEMTs with regrown ohmic contacts and f_T of 370 GHz, *IEEE Electron Device Lett.*, **33**, 988–990.

59. Liu, Y., Kauser, M. Z., Nathan, M. I., Ruden, P. P., Dogan, S., Morkoc, H., Park, S. S., and Lee, K. Y. (2004). Effects of hydrostatic and uniaxial stress on the Schottky barrier heights of Ga-polarity and N-polarity n-GaN, *Appl. Phys. Lett.*, **84**, 2112–2114.

60. Qiao, D., Yu, L. S., Lau, S. S., Redwing, J. M., Lin, J. Y., and Jiang, H. X. (2000). Dependence of Ni/AlGaN Schottky barrier height on Al mole fraction, *J. Appl. Phys.*, **87**, 801–804.

61. Arulkumaran, S., Egawa, T., Zhao, G. Y., Ishikawa, H., Umeno, M., and Jimbo, T. (2001). Effects of annealing on Ti, Pd, and Ni/n-$Al_{0.11}Ga_{0.89}N$ Schottky diodes, *IEEE Trans. Electron Devices*, **48**, 573–580.

62. Arulkumaran, S., Egawa, T., Zhao, G. Y., Ishikawa, H., Jimbo, T., and Umeno, M. (2000). Electrical characteristics of Schottky contacts on GaN and $Al_{0.11}Ga_{0.89}N$, *Jpn. J. Appl. Phys., Part 2*, **39**, L351–L353.

63. Lv, Y. J., Lin, Z. J., Corrigan, T. D., Zhao, J. Z., Cao, Z. F., Meng, L. G., Luan, C. B., Wang, Z. G., and Chen, H. (2011). Extraction of AlGaN/GaN heterostructure Schottky diode barrier heights from forward current-voltage characteristics, *J. Appl. Phys.*, **109**, 074512-1–074512-6.

64. Shin, J.-H., Park, J., Jang, S. Y., Jang, T., and Kim, K. S. (2013). Metal induced inhomogeneous Schottky barrier height in AlGaN/GaN Schottky diode, *Appl. Phys. Lett.*, **102**, 243505-1–243505-4.

65. Ha, W. J., Chhajed, S., Chavan, A., Lee, J.-H., Kim, K.-S., and Kim, J. K. (2012). Carrier transport mechanism of AlGaN/GaN Schottky barrier diodes with various Al mole fractions, *Phys. Status Solidi C*, **9**, 851–854.

66. Miura, N., Nanjo, T., Suita, M., Oishi, T., Abe, Y., Ozeki, T., Ishikawa, H., Egawa, T., and Jimbo, T. (2004). Thermal annealing effects on Ni/Au based Schottky contacts on n-GaN and AlGaN/GaN with insertion of high work function metal, *Solid-State Electron.*, **48**, 689–695.

67. Schmitzdag, A. C., Pingdag, A. T., Asif Khanddag, M., Chenddag, Q., Yangddag, J. W., and Adesida, I. (1996). Schottky barrier properties

of various metals on n-type GaN, *Semicond. Sci. Technol.*, **11**, 1464–1467.

68. Yu, L. S., Qiao, D. J., Xing, Q. J., Lau, S. S., Boutros, K. S., and Redwing, J. M. (1998). Ni and Ti Schottky barriers on n-AlGaN grown on SiC substrates, *Appl. Phys. Lett.*, **73**, 238–240.
69. Wang, L., Nathan, M. I., Lim, T. H., Kan, M. A., and Chen, Q. (1996). High barrier height GaN Schottky diodes: Pt/GaN and Pd/GaN, *Appl. Phys. Lett.*, **68**, 1267–1269.
70. Guo, J. D., Feng, M. S., Guo, R. J., Pan, F. M., and Chang, C. Y. (1995). Study of Schottky barriers on n-type GaN grown by low-pressure metalorganic chemical vapor deposition, *Appl. Phys. Lett.*, **67**, 2657–2659.
71. Jeon, C. M., Jang, H. W., and Lee, J.-L. (2003). Thermally stable Ir Schottky contact on AlGaN/GaN heterostructure, *Appl. Phys. Lett.*, **82**, 391–393.
72. Kumar, V., Selvanathan, D., Kuliev, A., Kim, S., Flynn, J., and Adesida, I. (2003). Characterisation of iridium Schottky contacts on n-$Al_xGa_{1-x}N$, *Electron. Lett.*, **39**, 747–749.
73. Lin, Z. J., Lu, W., Lee, J., Liu, D. M., Flynn, J. S., and Brandes, G. R. (2003). Barrier heights of Schottky contacts on strained AlGaN/GaN heterostructures: determination and effect of metal work functions, *Appl. Phys. Lett.*, **82**, 4364–4366.
74. Venugopalan, H. S., and Mohney, S. E. (1998). Thermally stable rhenium Schottky contacts to n-GaN, *Appl. Phys. Lett.*, **73**, 1242–1244.
75. Sheu, J. K., Su, Y. K., Chi, G. C., Jou, M. J., and Chang, C. M. (1998). Effects of thermal annealing on the indium tin oxide Schottky contacts of n-n-GaN, *Appl. Phys. Lett.*, **72**, 3317–3319.
76. Ahn, W., Seok, O., Ha, M.-W., Kim, Y.-S., and Han, M.-K. (2013). An industrial 650V GaN DHEMT cascode technology, *ECS Trans.*, **53**, 171–183.
77. Hayes, J. R., Kim, D.-W., Meidia, H., and Mahajan, S. (2003). Thermal stability of TaN Schottky contacts on n-GaN, *Acta Mater.*, **51**, 653–663.
78. Kalinina, E. V., Kuznetsov, N. I., Dmitriev, V. A., Irvine, K. G., and Carter, C. H. (1996). Schottky barriers on n-GaN grown on SiC, *J. Electron. Mater.*, **25**, 831–834.
79. Ha, W. J., Chhajed, S., Chavan, A., Lee, J.-H., Kim, K.-S., and Kim, J. K. (2012). Carrier transport mechanism of AlGaN/GaN Schottky barrier diodes with various Al mole fractions, *Phys. Status Solidi C*, **9**, 851–854.

80. Mehandru, R., Kang, S., Kim, S., Ren, F., Kravchenko, I., Lewis, W., and Pearton, S. J. (2004). Effect of deposition conditions and annealing on W Schottky contacts on n-GaN, *Mater. Sci. Semicond. Process.*, **7**, 95–98.

81. Dumont, J., Caudano, R., Sporken, R., Monroy, E., Munoz, E., Beaumont, B., and Gibart, P. (2000). *GaN and Related Alloys, 1999. Symposium (Materials Research Society Symposium Proceedings)*, **595**, W11.79.1–6.

82. Esposto, M., Di Lecce, V., Bonaiuti, M., and Chini, A. (2013). The influence of interface states at the Schottky junction on the large signal behavior of copper-gate GaN HEMTs, *J. Electron. Mater.*, **42**, 15–20.

83. Kim, S., Seo, T. H., Suh, E.-K., Song, K. M., Kim, M. J., and Kim, H. (2015). Graphene-GaN Schottky diodes, *Nano Res.*, **8**, 1327–1338.

84. Fisichella, G., Greco, G., Roccaforte, F., and Giannazzo, F. (2014). Current transport in graphene/AlGaN/GaN vertical heterostructures probed at nanoscale, *Nanoscale*, **6**, 8671–8680.

85. Lee, C.-S., Chang, E.-Y, Chang, L., Fang, C.-Y., Huang, Y.-L., and Huang, J.-S. (2003). Study of titanium tungsten nitride and tungsten nitride Schottky contacts on n-GaN, *Jpn. J. Appl. Phys.*, **42**, 4193–4196.

86. Lu, C.-Y., Chang, E. Y., Huang, J.-C., Chang, C.-T., Lin, M.-H., and Lee, C.-T. (2008). Enhancement of the Schottky barrier height using a nitrogen-rich tungsten nitride thin film for the Schottky contacts on AlGaN/GaN heterostructures, *J. Electron. Mater.*, **37**, 624–627.

87. Lee, S.-H., Chun, J.-K., Hur, J.-J., Lee, J.-S., Rue, G.-H., Bae, Y.-H., Hahm, S.-H., Lee, Y.-H., and Lee, J.-H. (2000). RuO_2/GaN Schottky contact formation with superior forward and reverse characteristics, *IEEE Electron Device Lett.*, **21**, 261–263.

88. Kyaw, L. M., Dolmanan, S. B., Bera, M. K., Liu, Y., Tan, H. R., Bhat, T. N., Dikme, Y., Chor, E. F., and Tripathy, S. (2014). Influence of RuOx gate thermal annealing on electrical characteristics of $Al_xGa_{1-x}N$/GaN HEMTs on 200-mm silicon, *ECS Solid-State Lett.*, **3**, Q5–Q8.

89. Ao, J.-P., Suzuki, A., Sawada, K., Shinkai, S., Naoi, Y., and Ohno, Y. (2010). Schottky contacts of refractory metal nitrides on gallium nitride using reactive sputtering, *Vacuum*, **84**, 1439–1443.

90. Mohammada, S. N. (2005). Contact mechanisms and design principles for Schottky contacts to group-III nitrides, *J. Appl. Phys.*, **97**, 063703-1–063703-19.

91. Lin, Y.-C., Chang, C.-H., Li, F.-M., Hsu, L.-H., and Chang, E. Y. (2013). Evaluation of TiN/Cu gate metal scheme for AlGaN/GaN high-electron-

mobility transistor application, *Appl. Phys. Express*, **6**, 091003-1–091003-3.

92. Hu, J., Stoffels, S., Lenci, S., Bakeroot, B., Venegas, R., Groeseneken, G., and Decoutere, S. (2015). Current transient spectroscopy for trapping analysis on Au-free AlGaN/GaN Schottky barrier diode, *Appl. Phys. Lett.*, **106**, 083502-1–083502-4.

93. Arulkumaran, S., Ng, G. I., Tan, C. L., Liu, Z. H., Bu, J., Radhakrishnan, K., Aggerstam, T., Sjödin, M., and Lourdudoss, S. (2006). Enhancement of both direct-current and microwave characteristics of AlGaN/GaNAlGaN/GaN high-electron-mobility transistors by furnace annealing, *Appl. Phys. Lett.*, **88**, 023502-1–023502-3.

94. Lee, J., Liu, D., Kim, H., and Lu, W. (2004). Post-annealing effects on device performance of AlGaN/GaN HFETs, *Solid-State Electron.*, **48**, 1855–1859.

95. Nanjo, T., Miura, N., Oishi, T., Suita, M., Abe, Y., Ozeki, T., Nakatsuka, S., Inoue, A., Ishikawa, T., Matsuda, Y., Ishikawa, H., and Egawa, T. (2004). Improvement of DC and RF Characteristics of AlGaN/GaN high electron mobility transistors by thermally annealed Ni/Pt/Au Schottky gate, *Jpn. J. Appl. Phys.*, **43**, 1925–1929.

96. Hashizume, T., Ootomo, S., and Hasegawa, H. (2003). Suppression of current collapse in insulated gate AlGaN/GaN heterostructure field-effect transistors using ultrathin Al_2O_3 dielectric, *Appl. Phys. Lett.*, **83**, 2952–2954.

97. Saito, W., Kuraguchi, M., Takada, Y., Tsuda, K., Omura, I., and Ogura, T. (2005). Influence of surface defect charge at AlGaN-GaN-HEMT upon Schottky gate leakage current and breakdown voltage, *IEEE Trans. Electron Devices*, **52**, 159–164.

98. Sanabria, C., Chakraborty, A., Xu, H., Rodwell, M. J., Mishra, U. K., and York, R. A. (2006). The effect of gate leakage on the noise figure of AlGaN/GaN HEMTs, *IEEE Electron Device Lett.*, **27**, 19–21.

99. Karmalkar, S., and Mishra, U. K. (2001). Enhancement of breakdown voltage in AlGaN/GaN high electron mobility transistors using a field plate, *IEEE Trans. Electron Devices*, **48**, 1515–1521.

100. Dora, Y., Chakraborty, A., McCarthy, L., Keller, S., DenBaars, S. P., and Mishra, U. K. (2006). High breakdown voltage achieved on AlGaN/GaN HEMTs with integrated slant field plates, *IEEE Election Device Lett.*, **27**, 713–715.

101. Xing, H., Dora, Y., Chini, A., Heikman, S., Keller, S., and Mishra, U. K. (2004). High breakdown voltage AlGaN–GaN HEMTs achieved by multiple field plates, *IEEE Election Device Lett.*, **25**, 161–163.

102. Ando, Y., Wakejima, A., Okamoto, Y., Nakayama, T., Ota, K., Yamanoguchi, K., Murase, Y., Kasahara, K., Matsunaga, K., Inoue, T., and Miyamoto, H. (2005). Novel AlGaN/GaN dual-field-plate FET with high gain, increased linearity and stability, in *IEEE International Electron Devices Meeting, IEDM Technical Digest*, Washington, DC, pp. 576–579.

103. Karmalkar, S., and Mishra, U. K. (2001). Very high voltage AlGaN/GaN high electron mobility transistors using a field plate deposited on a stepped insulator, *Solid-State Electron.*, **45**, 1645–1652.

104. Saito, W., Nitta, T., Kakiuchi, Y., Saito, Y., Tsuda, K., Omura, I., and Yamaguchi, M. (2007). Suppression of dynamic on-resistance increase and gate charge measurements in high-voltage GaN-HEMTs with optimized field-plate structure, *IEEE Trans. Electron Devices*, **54**, 1825–1830.

Chapter 4

Conventional AlGaN/GaN Heterojunction Field-Effect Transistors

Jiang Lingli
South University of Science and Technology of China, 1088 Xueyuan Ave, Nanshan, Shenzhen, Guangdong, China, 518055
jiangll@sustc.edu.cn

4.1 Introduction

The AlGaN/GaN high-electron-mobility transistor (HEMT) has become a very promising candidate for next-generation semiconductor devices. Compared to SiC power devices, AlGaN/GaN can easily form a heterojunction. With a high-concentration two-dimensional electron gas (2DEG), the AlGaN/GaN HEMT can provide high electrical conductivity to reduce parasitic resistance and optimize high-frequency characteristics (Fig. 4.1).

Gallium Nitride Power Devices
Edited by Hongyu Yu and Tianli Duan

ISBN 978-981-4774-09-3 (Hardcover), 978-1-315-19662-6 (eBook)
www.panstanford.com

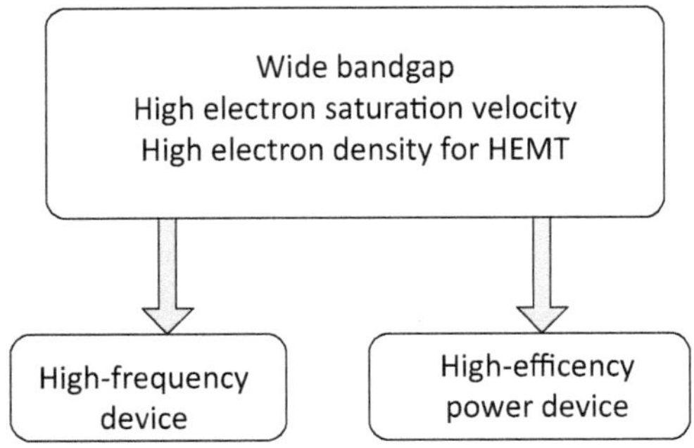

Figure 4.1 AlGaN/GaN HEMT devices.

4.2 Polarization and Generation of a 2DEG

4.2.1 Polarization

A GaN material has two different crystal structures, the wurtzite structure and the sphalerite structure. In the sphalerite structure crystal, the positive and negative ion centers are coincident. However, as shown in Fig. 4.2, the positive ion center of the wurtzite crystal structure is not coincident with the negative ion center, and there is a shift along the *c* axis. Thus GaN with wurtzite crystals has a spontaneous polarization characteristic. Along [0001] and [$000\overline{1}$], the Ga atom layer and the N atom layers alternately stack to constitute the crystal. Along [0001], the N atom layer is at the bottom and the Ga atom layer is at the top, and this type of material is named Ga-face; along [$000\overline{1}$], the Ga atom layer is at the bottom and the N atom layer is at the top, and this type of material is named N-face. Materials with different polarity faces have different physical and chemical characteristics, such as the Schottky barrier and the heterojunction interface state. Because of the shift along the *c* axis for the positive and negative ion centers, the N-face structure and Ga-face structure have exactly opposite spontaneous polarization directions. The spontaneous polarization vector ($\mathbf{P}_{SP}$) of the N-face is bottom up and that of the Ga-face is top down.

For the AlGaN/GaN heterojunction epitaxy, these two materials have different lattice constants. The lower layer tends to keep the origin lattice constant. Under the action of the lower layer lattice, the upper layer grows with the same lattice mode as the lower layer. However, the upper layer tends to keep the lattice constant of itself, and then the lattice of the upper layer is stretched or compressed. This type of stress on the lattice will cause piezoelectric polarization.

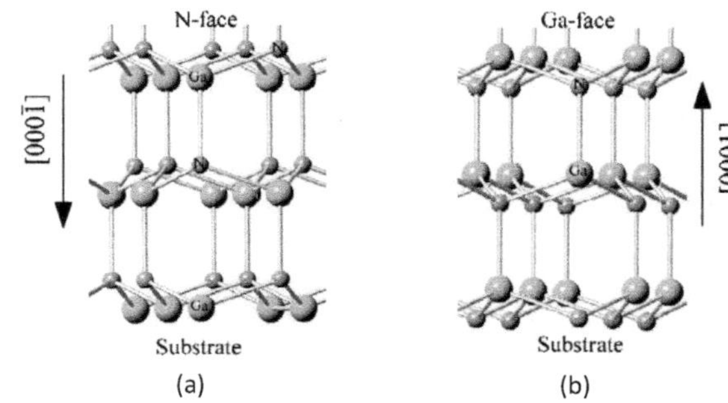

Figure 4.2 GaN lattice with a wurtzite crystal structure: (a) N-face and (b) Ga-face.

During the epitaxy layer growth, the stress between the lattices increases gradually. If the epitaxy layer is thicker than the critical thickness, this stress is relaxed, and more defects are induced in the materials. When the relaxation in AlGaN occurs, a 2DEG at the AlGaN/GaN heterojunction sharply decreases with the disappeared piezoelectric polarization.

For a conventional AlGaN/GaN HEMT device, the GaN layer is usually more than several micrometers; thus the GaN under AlGaN is always assumed to be relaxed. However, the AlGaN layer, which is usually several tens of nanometers, is assumed with tensile strain.

Figure 4.3 shows several typical polarization states for the AlGaN/GaN heterojunction. The Ga-face AlGaN/GaN heterojunction is shown in Fig. 4.3a, in which the spontaneous polarization vector is top down. As the lattice constant of AlGaN is smaller than that of GaN, the AlGaN layer is under tensile strain at the upper surface of GaN and the piezoelectric polarization vector ($\mathbf{P}_{\mathrm{PE}}$) is also top down. In this case, the fixed charges in AlGaN at the AlGaN/GaN interface are positive, and the charge density increases with the intensity of polarization. Because of the positive fixed charge, the free electrons move into the interface of the heterojunction. If the polarization charge density is high enough, the conduction band would bend down below the Fermi level, and a 2DEG is formed at the heterojunction. According to the production mechanism, the 2DEG in the AlGaN/GaN heterojunction is also named polarization-induced 2DEG.

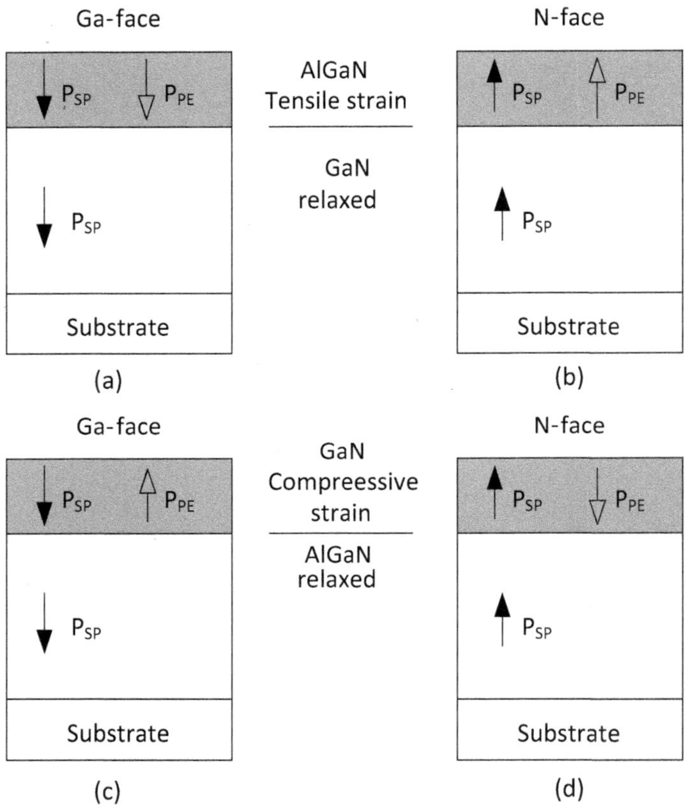

Figure 4.3 Polarization in AlGaN/GaN.

As shown in Fig. 4.3b, the spontaneous polarization vector of the N-face AlGaN/GaN heterojunction is bottom up, which is the same as the vector direction of the piezoelectric polarization. In this case, the fixed charges at the AlGaN/GaN interface are negative, which can induce the mobile positive charges. If the electric field caused by the fixed negative charges is strong enough, the valence band would bend up above the Fermi level and a two-dimensional hole gas (2DHG) is formed at the heterojunction.

4.2.2 Generation of a 2DEG

For conventional semiconductor materials, the free charges always come from the impurity ionization. However, in an AlGaN/GaN heterojunction, an areal density of 1×10^{13} cm^{-2} 2DEG can be obtained, even without intended doping. Thus, where the 2DEG comes from is a very interesting topic.

Smorchkova [1] proposed a surface state–tapped charge model, which declares that a 2DEG comes from a surface donor-like state. When the AlGaN barrier is too thin, the surface-state energy level is below the Fermi energy. In this case, the surface donor-like state is nonionized with the surface state electrically neutral, and the electrons from the surface state cannot form a high-density 2DEG. In the experiment [1], a 2DEG occurs at an AlGaN thickness of ~3 nm in the $Al_{0.27}Ga_{0.73}N$/GaN structure. With a further increase in the thickness of AlGaN, the density of 2DEG increases sharply and then gets into saturation. As shown in Fig. 4.4, before a 2DEG occurs, the surface barrier height $q\phi_S$ increases with the thickness of AlGaN, and the donor-like surface state E_{DS} rises up as well. When E_{DS} is at the same height as the Fermi energy, the surface donor-like state ionizes, and the free electrons generated constitute the 2DEG. With a further increase in the thickness of the AlGaN barrier, the E_{DS} keeps the same height as the Fermi energy, and the surface barrier height $q\phi_S$ almost keeps still. Thus, the density of the 2DEG turns into saturation.

According to this model, the surface donor-like state is located 1.42 eV below the conduction band edge of $Al_{0.27}Ga_{0.73}N$, and this value is higher than most reported Schottky barriers from experiments. That means the density of the 2DEG would increase if a Schottky barrier forms at the AlGaN surface. However, the experiment results indicate the Schottky barrier on top of AlGaN/GaN will deplete part of the 2DEG. In this case, the mechanism of typical thickness of the 2DEG occurring needs some other explanation [2].

With experiments on different GaN doping concentrations and AlGaN thicknesses, Koley [3] indicates that there are negative acceptor-like states at the GaN surface and in the AlGaN barrier as well. Therefore, in the AlGaN/GaN heterojunction, both the donor-like state and the acceptor-like state influence the generation of the 2DEG, and these states would alter with physical parameters such as the thickness of AlGaN and the alloy ingredient.

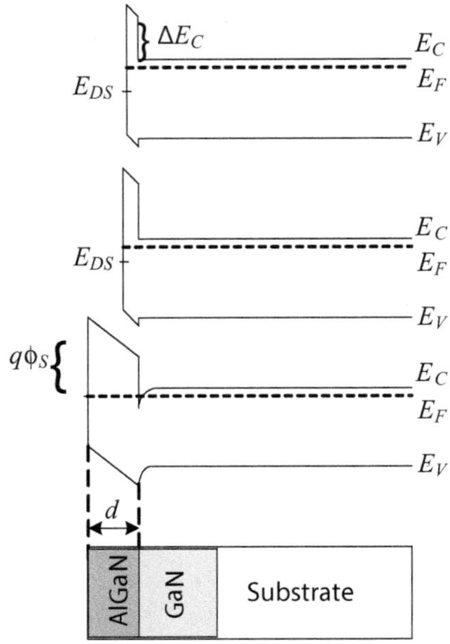

Figure 4.4 Schematic diagram of the band structure in AlGaN/GaN with increasing AlGaN barrier thickness.

4.3 GaN HEMT Operation Principle

A conventional Schottky gate GaN HEMT device structure is shown in Fig. 4.5. The source and drain metals constitute an ohmic contact, and the gate metal constitutes a Schottky contact. With a voltage stress V_{DS} on the drain, a lateral electrical field is built, and the 2DEG flows through the heterojunction interface to form the current I_{DS}.

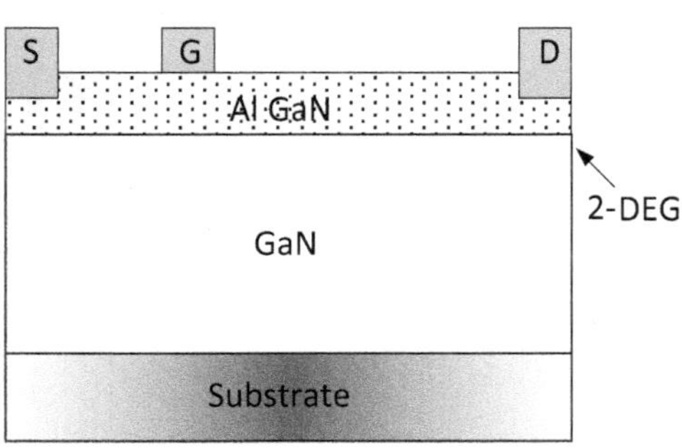

Figure 4.5 Schematic diagram of an AlGaN/GaN HEMT.

To discuss the gate voltage modulating function, a simplified model is proposed, as shown in Fig. 4.6 [2]. The AlGaN layer is supposed to have an N-type modulation doping, the concentration of which is N_D. ΔE_C is the valence band offset between GaN and AlGaN. $q\phi_b$ is the Schottky barrier height, and σ_{pol} is the polarization charge density caused by piezoelectric polarization $\mathbf{P}_{PE}$ and spontaneous polarization $\mathbf{P}_{SP}$.

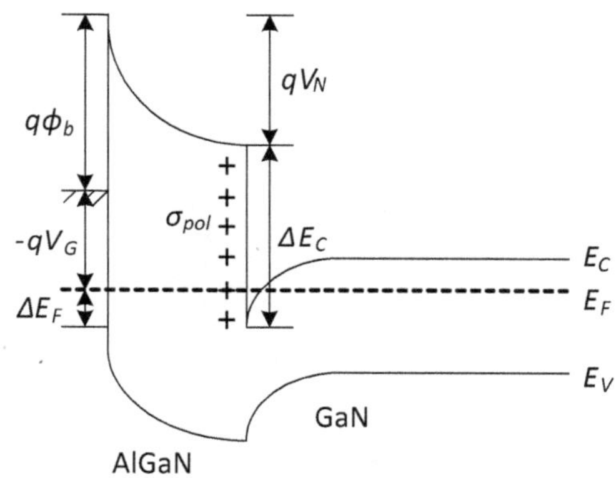

Figure 4.6 Heterojunction band with gate voltage.

The threshold voltage V_{th} is defined as the gate voltage under which the 2DEG in the channel is depleted and generated. At the threshold point, the 2DEG occupies the lowest energy level and coincides with the Fermi level. With voltage stress in the drain, if $V_G > V_{th}$, the 2DEG is in a continuous state and the device is turned on; if $V_G < V_{th}$, the 2DEG is depleted and the device is turned off. The threshold voltage of an AlGaN/GaN HEMT will be discussed in detail in Chapter 5, while in this chapter, the channel current characteristics of HEMTs will be discussed.

The channel current I_{DS} is defined as

$$I_{DS} = qWv(x)n_{2D}(x), \tag{4.1}$$

in which W is the channel width, $v(x)$ is the electron velocity, and n_{2D} is the density of the 2DEG. In this discussion, the device works when the channel electrical field E is below the critical field E_S. Thus $v(x)$ is directly proportional to E. With $E(x) = dV_C(x)/dx$,

$$I_{DS} = W\mu C_{ch}[V_{GT} - V_C(x)]dV_C(x)/d(x), \tag{4.2}$$

in which the channel capacitor per unit area C_{ch} is defined as

$$C_{\text{ch}} = \varepsilon_1/(d + \Delta d) \approx \varepsilon_1/d, \tag{4.3}$$

in which ε_1 is the AlGaN dielectric constant, Δd is the distance between the 2DEG and the AlGaN/GaN interface, and d is the AlGaN barrier depth. As the 2DEG is extremely thin, Δd can be ignored. At the channel interface, $V_C(0) = R_S I_{DS}$ and $V_C(L) = V_{DS} - R_D I_{DS}$. When the drain voltage V_{DS} is small enough, the variety along the channel can be ignored,

$$I_{DS} = \beta V_{GT}[V_{DS} - (R_S + R_D)I_{DS}], \tag{4.4}$$

as well as the transconductance coefficient

$$\beta = W\mu\varepsilon_1/dL. \tag{4.5}$$

The function, Eq. 4.4, is the linearization range current for the HEMT device. However, when the drain voltage V_{DS} is high enough, the electron velocity is saturated and the channel current is also saturated. The channel current is [2]

$$I_{\text{Dsat}} = \beta V_{\text{L}}[\sqrt{V_{\text{L}}^2 + (V_{\text{GT}} - R_S I_{\text{Dsat}})^2} - V_{\text{L}}]. \tag{4.6}$$

These above discussions for HEMT current are focused on the linearization and saturation area, and the simplifications in the function analysis are acceptable for principle analysis.

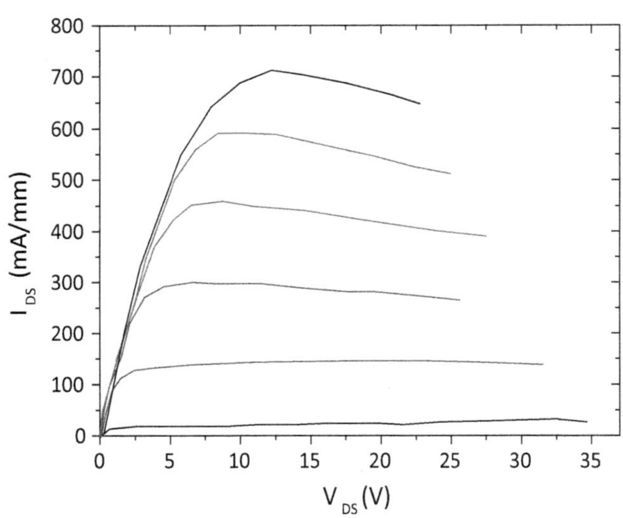

Figure 4.7 *I–V* characteristics with current collapse.

However, the saturation current of a GaN HEMT in Eq. 4.6 is just for an ideal case. During high-voltage stress, the GaN HEMT device has a current collapse problem. As shown in Fig. 4.7, the current and maximum transconductance in the saturation region decrease with drain voltage [4]. The detail principles of current collapse are described in a later chapter. Besides the current collapse, some other effects also influence the *I–V* characteristics of a GaN HEMT, such as the Schottky gate leakage [5], the self-heating effect, and the diffusion current outside the channel [6].

4.4 Breakdown for an AlGaN/GaN HEMT

If the gate voltage is below the threshold voltage and the drain is stressed with high voltage, the GaN HEMT device gets into the block region. In this case, the GaN region is depleted from the drain to the gate. The depletion region helps to sustain the high voltage. As the GaN material has a much higher critical field E_{crit} (3.3 MV/cm) than silicon (0.3 MV/cm), the GaN HEMT device has very good high-voltage ability and can be designed for high-efficiency power devices.

Under drain voltage stress, if the local electric field in the device is higher than the critical field E_{crit}, the device will break down. The contour lines are closer at the junction edge than in other region, such as the edge of the gate at the drain side and the edge of the drain contact (Fig. 4.8). These regions have a higher electric field and are much more susceptible to breakdown. If the breakdown occurs in the GaN or AlGaN regions, the electrons generated can destroy the 2DEG in the heterojunction. In this case, the on-resistance of the device would increase sharply, and the device is a failure [7].

Because of the Schottky gate contact, a GaN HEMT device has a 10^3–10^6 level higher leakage current than conventional silicon devices with reverse bias [8–15], and this leakage current would cause an early breakdown. To restrain the early breakdown phenomenon and to increase the breakdown voltage, several methods are proposed to depress the surface electric field.

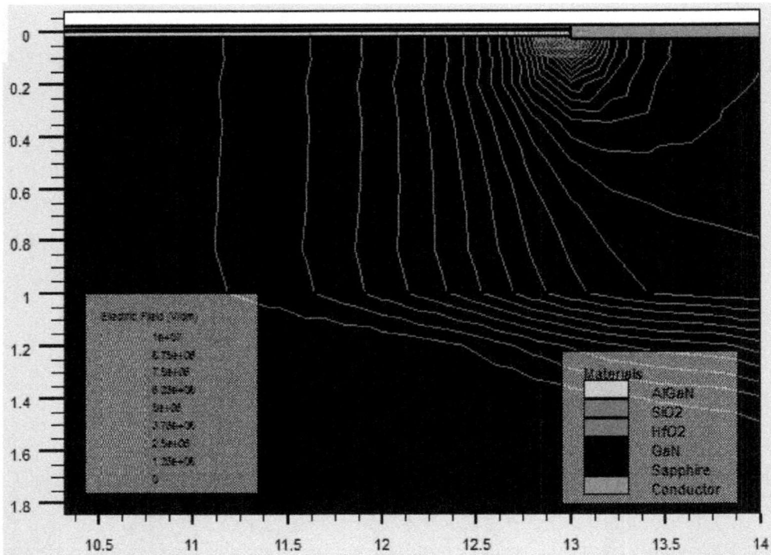

Figure 4.8 Electric field distribution with drain stress.

4.4.1 Gate Electric Field Plate

For GaN HEMTs, the gate electric field plate is a useful method to enhance the breakdown voltage [16–19]. When the device is under reverse stress, the gate field plate shares the same potential as the related electrode gate. This can modulate the local electric field distribution to reduce the peak field on the edge of the gate. A typical gate field plate structure is shown in Fig. 4.9. For a device without a gate field plate, the peak electric field always occurs at the edge of gate at the drain side. With the gate field plate, however, an additional electric field peak is induced at the terminal of the gate field plate, as shown in Fig. 4.9.

It's worth mentioning that the gate field plate is not only useful for breakdown voltage optimization but also helps to restrain the current collapse effect. When the device works in an on state, the electrons may inject from the gate into the surface state under a high electric field. When the electrons are trapped, the 2DEG density will be reduced and the on-resistance R_{ON} increases. However, the high electric field peak can be restrained with the gate field plate, and the current collapse effect can be optimized as well as the breakdown characteristics.

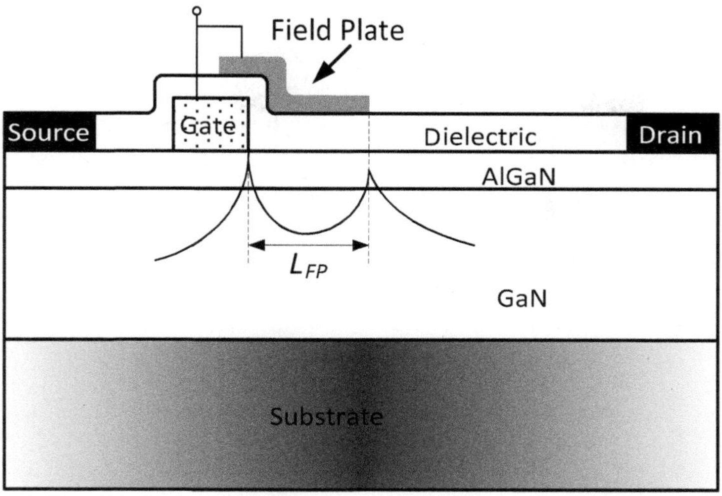

Figure 4.9 Gate field plate to restrain the electric field on the edge of the gate.

The key point in the field plate design is the dimension parameters' optimization for better electric field distribution. By altering the value of L_{FP}, the peak field at the gate edge is reduced and the peak field at the field plate edge is increased. With an appropriate L_{FP}, a higher breakdown voltage can be obtained. Figure 4.10 shows the typical optimization results for L_{FP} [2].

A multiple-gate-field-plate structure (Fig. 4.11) is also proposed for better breakdown characteristics [20, 21]. The first field plate can reduce the electric field peak of the edge of gate at the drain side, and the second field plate can restrain the surface electric field at the terminal of the first field plate. For a device with gate length = 1 μm and drift region length = 24 μm, the breakdown voltage of the conventional structure is about 250 V, while in a multiple-gate-field-plate structure, this can increase up to 900 V [20].

With the gate field plate structure, a higher breakdown voltage can be obtained. However, an additional parasitic capacitor is formed between the gate field plate and the AlGaN layer, which deteriorates the frequency characteristics. Structures with floating field plates are proposed to resolve the contradiction between breakdown voltage and frequency characteristics [2, 22]. As shown in Fig. 4.12, the device has a conventional field plate connected to the gate and has several floating field plates as well. With the floating field plates, additional electrical field peaks are built at the very terminal of the floating field plates. By optimizing the distance between the field plates, the main electric field peak at the edge of gate at the drain side can be restrained, and the device obtains a higher breakdown voltage without inducing a too-large C_{gd}. It is worth mentioning that there are no additional masks or process steps needed to constitute the floating field plates.

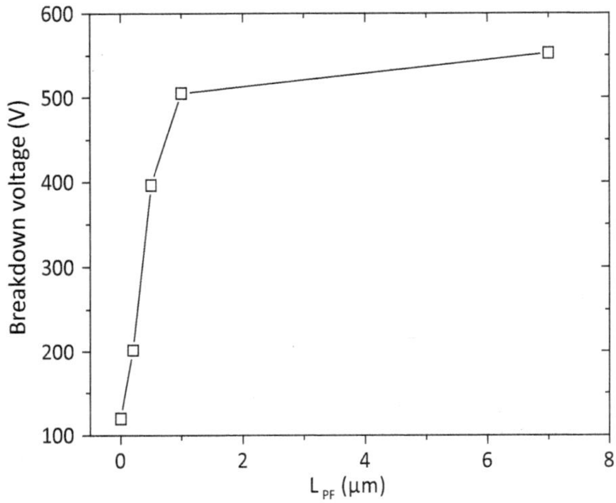

Figure 4.10 Breakdown voltage versus length of the gate field plate.

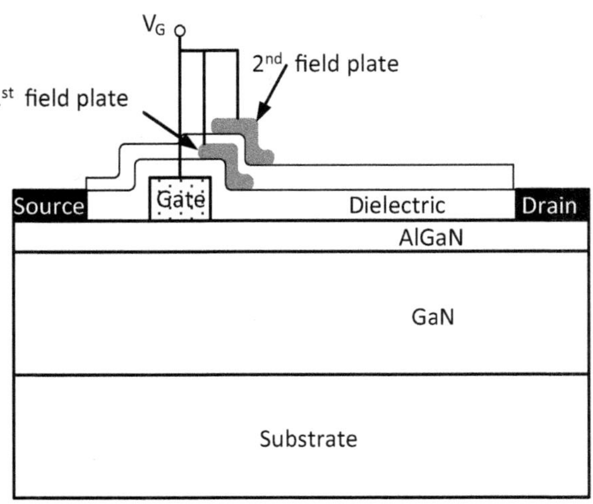

Figure 4.11 GaN HEMT with a multiple-gate-field plate.

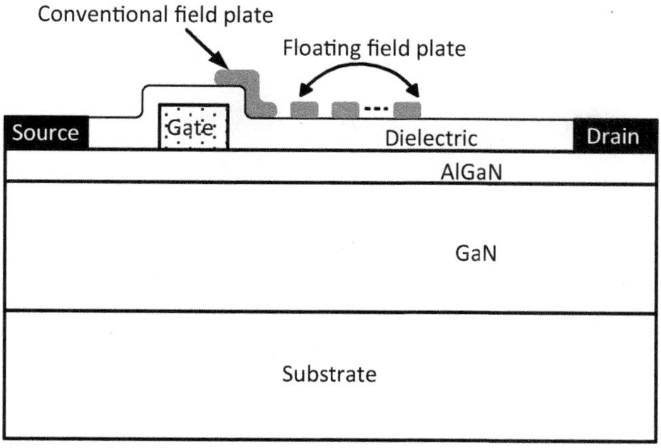

Figure 4.12 GaN HEMT with a floating gate field plate.

4.4.2 Source Electric Field Plate

A source field plate is another method to optimize the surface electric field peak (Fig. 4.13) [23]. The source field plate also induces an additional electric peak at the terminal of the plate, which can restrain the field peak at the edge of gate to the drain side. Compared with the gate field plate, the source field plate can reduce the miller capacitor C_{gd}, but it still induces additional C_{gs}.

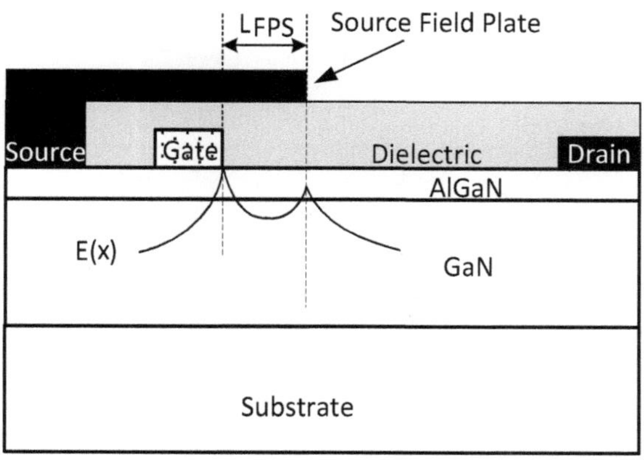

Figure 4.13 GaN HEMT with a source field plate.

With the same optimization principle, multiple source plates and a complex plate structure with both source and gate field plates are also proposed for further optimizations of the electrical distribution [24]. For all these structures, the electrical field distribution closely depends on the geometric parameters, including the length of the field plates, the depth of each dielectric layers, and the overlap width between the field plate layers.

4.4.3 Air Bridge Field Plate

The above-mentioned field plate technologies can effectively reduce the peak electric field at the gate edge of the drain side to enhance the breakdown voltage and restrain the current collapse. However, these field plate technologies all induce some parasitic capacitors, which will increase the switching power dissipation and restrict the application in the high-step-down-ratio convertor. Beside this, for the same physical dimension devices, the on-resistance of the structure with a field plate is higher than that without it [24, 25]. This phenomenon is caused by the extra vertical electric field, which influences the distribution of 2DEG in the on station.

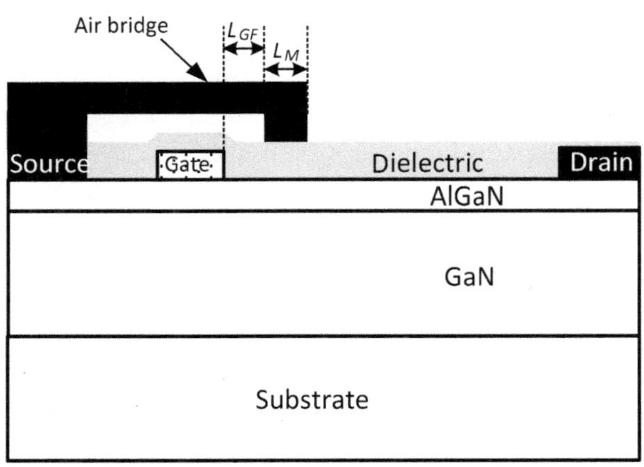

Figure 4.14 GaN HEMT with an air bridge field plate.

As shown in Fig. 4.14, the structure with an air bridge field plate [25] has a source field plate across the metal gate without contacting the dielectric at the gate region. In this structure, there is an air region between the air bridge and the metal gate. Compared with the conventional field plate structure, the dielectric constant of air $\varepsilon_r = 1$. Thus the input capacitance C_{gs} can be sharply reduced. Besides this, the air bridge structure weakens the field plate influence on the 2DEG, and a smaller on-resistance can be obtained.

Besides the field plate technique to optimize the surface electric field distribution, there are some other methods proposed to enhance the breakdown voltage of a GaN HEMT, including a metal-insulator-semiconductor HEMT (MIS-HEMT) structure to reduce the leakage current [26],fluoride plasma treatment to help expand the depletion region [27], and the reduced bulk field (REBULF) technique with Mg doping [28].

References

1. Smorchkova, I. P., Elsass, C. R., Ibbetson, J. P., et al. (1999). Polarization-induced charge and electron mobility in AlGaN/GaN heterostructures grown by plasma-assisted molecular-beam epitaxy, *J. Appl. Phys.*, **86**, 4520–4526.
2. Hao, Y., Zhang, J. F., and Zhang, J. C. (2013). *Nitride Wide Band Gap Semiconductor Materials and Electrical Devices* (Science Press, China), pp. 66–70, 171–184.
3. Koley, G., and Spencer, M. G. (2001). Surface potential measurements on GaN and AlGaN/GaN heterostructures by scanning Kelvin probe microscopy, *J. Appl. Phys.*, **90**(1), 337–344.
4. Khan, M. A., Shur, M. S., Chen, Q. C., and Kuznia, J. N. (1994). Current/voltage characteristic collapse in AlGaN/GaN heterostructure insulated gate field effect transistors at high drain bias, *Electron. Lett.*, **30**(25), 2175–2175.
5. Paul, R. P. (1990). Heterostructure FET model including gate leakage, *IEEE Trans. Electron Devices*, **37**(10), 2267–2270.
6. Albrecht, J. D., Paul, R. P., Binari, S. C., and Ancona, M. G. (2000). AlGaN/GaN heterostructure field-effect transistor model including thermal effects, *IEEE Trans. Electron Devices*, **47**(11), 2031–2036.
7. Joh, J., and del Alamo, J. A. (2008). Critical voltage for electrical degradation of GaN high-electron mobility transistors, *IEEE Electron Device Lett.*, **29**(4), 287–289.
8. Mizuno, S., et al. (2002). Large gate leakage current in AlGaN/GaN high electron mobility transistors, *Jpn. J. Appl. Phys.*, **41**(1), 5125–5126.
9. Hashizume, T., Kotani, J., and Hasegawa, H. (2004). Leakage mechanism in GaN and AlGaN Schottky interfaces, *Appl. Phys. Lett.*, **84**(24), 4884–4887.

10. Miller, E. J., Dang, X. Z., and Yu, E. T. (2000). Gate leakage current mechanisms in AlGaN/GaN heterostructure field-effect transistors, *J. Appl. Phys.*, **88**(10), 5951–5958.
11. Xu, C., Wang, J., Chen, H., et al. (2007). The leakage current of the Schottky contact on the mesa edge of AlGaN/GaN heterostructure, *IEEE Electron Device Lett.*, **28**(11), 942–944.
12. Miller, E. J., Yu, E. T., Waltereit, P., and Speck, J. S. (2004). Analysis of reverse-bias leakage current mechanisms in GaN grown by molecular-beam epitaxy, *Appl. Phys. Lett.*, **84**(4), 535–537.
13. Hasegawa, H., Inagaki, T., Ootomo, S., et al. (2003). Mechanisms of current collapse and gate leakage current in AlGaN/GaN heterostructure field effect transistors, *J. Vac. Sci. Technol., B*, **21**(4), 1844–1855.
14. Kozodoy, P., Ibbetson, J. P., Marchand, H., et al. (1998). Electrical characterization of GaN p-n junctions with and without threading dislocations, *Appl. Phys. Lett.*, **73**(7), 975–977.
15. Karmalkar, S., Sathaiya, D. M., and Shur, M. S. (2003). Mechanism of the reverse gate leakage in AlGaN/GaN high electron mobility transistors, *Appl. Phys. Lett.*, **82**(22), 3976–3978.
16. Karmalkar, S., and Mishra, U. K. (2001). Enhancement of breakdown voltage in AlGaN/GaN high electron mobility transistors using a field plate, *IEEE Trans. Electron Devices*, **48**(8), 1515–1521.
17. Chini, A., Buttari, D., Coffie, R., Heikman, S., Keller, S., and Mishra, U. K. (2004). 12 W/mm power density AlGaN/GaN HEMTs on sapphire substrate, *IEEE Electron Device Lett.*, **40**(1), 73–74.
18. Okamoto, Y., Ando, Y., Nakayama, T., Hataya, K., Miya-moto, H., Inoue, T., Senda, M., Hirata, K., Kosaki, M., Shi-bata, N., and Kuzuhara, M. (2004). High-power recessed-gate AlGaN GaN HFET with a field-modulating plate, *IEEE Trans. Electron Devices*, **51**(12), 2217–2222.
19. Dora, Y., Chakraborty, A., McCarthy, L., Keller, S., DenBaars, S. P., and Mishra, U. K. (2006). High breakdown voltage achieved on AlGaN/GaN HEMTs with integrated slant field plates, *IEEE Electron Device Lett.*, **27**(9), 713–715.
20. Xing, H., Dora, Y., Chini, A., et al. (2004). High breakdown voltage AlGaN-GaN HEMTs achieved by multiple field plates, *IEEE Electron Device Lett.*, **25**(4), 161–163.
21. Petru, A. (2010). Breakdown voltage enhancement in lateral AlGaN/GaN heterojunction FETs with multiple field plates, *IEEE International Conference ICSICT*, 1344–1346.

22. Bahat-Treidel, E., Hilt, O., Brunner, F., Sidorov, V., Wurfl, J., and Trankle, G. (2010). AlGaN-GaN HEMTs DH-HEMTs breakdown voltage enhancement using multiple grating field plates (MGFPs), *IEEE Trans. Electron Devices*, **57**(6), 1208–1216.

23. Saito, W., Takada, Y., Kuraguchi, M., Tsuda, K., Omura, I., Ogura, T., and Ohashi, H. (2003). High breakdown voltage AlGaN-GaN power-HEMT design and high current density switching behavior, *IEEE Trans. Electron Devices*, **50**(12), 2528–2531.

24. Chu, R., Corrion, A., Chen, M., Li, R., Wong, D., Zehnder, D., Hughes, B., and Boutros, K. (2011). 1200 V normally-off GaN-on-Si field-effect transistor with low dynamic on-resistance, *IEEE Electron Device Lett.*, **32**(5), 632–634.

25. Xie, G., Xu, E., Lee, J., Hashemi, N., Zhang, B., Fu, F. Y., and Ng, W. T. (2012). Breakdown voltage enhancement technique for RF based AlGaN/GaN HEMTs with a source-connected air-bridge field plate, *IEEE Electron Device Lett.*, **33**(5), 670–672.

26. Yagi, S., Shimizu, M., Inada, M., et al. (2005). High breakdown voltage AlGaN/GaN MIS-HEMT with SiN and TiO2 gate insulator, *Solid-State Electron.*, **50**(6), 1057–1061.

27. Cho, K.-H., Choi, Y.-H., Lim, J., et al. (2008). High breakdown voltage AlGaN/GaN HEMTs employing fluoride plasma treatment, *Phys. Scr.*, **78**(6), 065802.

28. Xie, G., Xu, E., Lee, J., Hashemi, N., Zhang, B., Fu, F. Y., and Ng, W. T. (2012). An AlGaN/GaN HEMT with reduced surface electric field and improved breakdown voltage, *Chin. Phys. B*, **21**(8), 086105.

Chapter 5

Original Demonstration of Depletion-Mode and Enhancement-Mode AlGaN/GaN Heterojunction Field-Effect Transistors

Zhigang Wang and Bing Wang
The School of Information Science and Technology, Southwest Jiaotong University, Chengdu, China
zhigangwang@swjtu.edu.cn

The developments of enhancement-mode and depletion-mode HFETs have been reviewed in this chapter for power electronics applications. A comprehensive insight into enhanced threshold voltage and the charge control model are discussed. The charge control model reveals that the enhanced channel for depleting a 2DEG is due to the variation of space potential, and threshold voltage shift is mainly due to a different charge-transferring effect. How to improve stability of the threshold voltage can be obtained in aforementioned sections, for example, reduction of defects can decrease the charge variation for a threshold voltage shift.

Gallium Nitride Power Devices
Edited by Hongyu Yu and Tianli Duan

ISBN 978-981-4774-09-3 (Hardcover), 978-1-315-19662-6 (eBook)
www.panstanford.com

5.1 Introduction

For a better transistor for power applications, five basic properties are required: (i) lower specific on-resistance, (ii) higher switching speeds, (iii) better thermal conductivity, (iv) higher breakdown voltage (BV), and (v) lower cost. Owing to the remarkable properties of III nitride materials, such as GaN with a wide bandgap (3.4 eV) and a large critical electric field (3 MV/cm), a wide-bandgap semiconductor as a third-generation semiconductor is an ideal substitute for silicon to meet these five requirements. Gallium nitride, as one kind of compound semiconductor, also has some features similar to GaAs. For example, GaN and its alloy (AlGaN) can form an AlGaN/GaN heterojunction structure to obtain a high two-dimensional electron gas (2DEG) density at this heterointerface. Therefore, on the basis of this heterostructure, AlGaN/GaN HFETs can be expected to realize a current scope of 10 A–100 A and a voltage range of 100 V–2 KV, as shown in Fig. 5.1. We can observe that GaN-based high-electron-mobility transistors (HFETs) are attractive candidates for high-power switching and high-frequency applications because of their high sheet density of the 2DEG. However, the conventional AlGaN/GaN HFET is typically a depletion-mode (D-mode) device with normally on characteristics. There is a desirable interest in high-performance enhancement-mode (E-mode) channel devices with a normally off channel for automotive electronics, power transformation, motor drives, industrial control, and wireless communication. The E-mode HFETs as the main issue must be demonstrated for the development of GaN-based HFETs.

The aim of this chapter is to demonstrate the original mechanism of E-mode and D-mode HFETs. In Section 5.2, the main development of E-mode GaN-based HFETs is concluded. From this section, various kinds of E-mode HFETs can be easily gained. In Section 5.3, we demonstrate the charge control model (CCM), which is used to demonstrate the main mechanism of 2DEG depletion for enhancing threshold voltage. Section 5.4 reveals the reliability of threshold voltage and its shift phenomena.

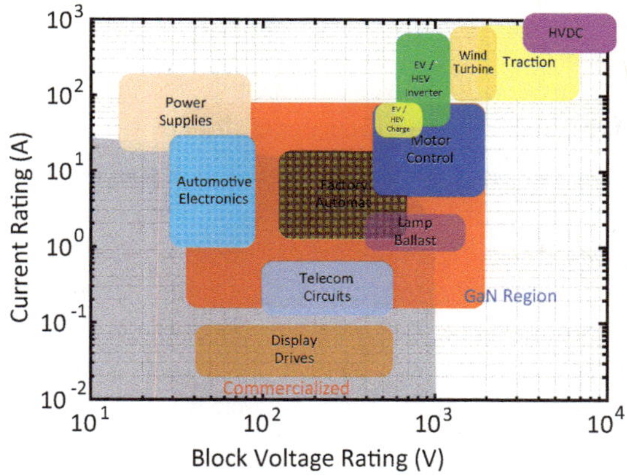

Figure 5.1 The application scope of GaN-based devices.

5.2 Development of E-Mode AlGaN/GaN HFETs

Compared to the normally on devices, normally off devices have several advantages. For example (i) without a gate drive during the off state, it can partly decrease the static power loss caused by the leakage current, and (ii) instead of the normally on devices' requirement for a normally off property as the secondary circuit, the normally off devices can also be applied in the inverter circuit directly, which decreases the number of devices and improves the level of integration.

In the past two decades, several kinds of E-mode AlGaN/GaN HFETs have been fabricated successfully, which make HFETs possible for commercial power applications. To be special, the E-mode devices are mainly summarized as follows:

- E-mode HFETs with a P-type cap layer
- E-mode HFETs with a recessed-barrier layer
- E-mode HFETs with a double-barrier layer
- Metal-insulator-semiconductor HFETs (MIS-HFETs)
- N-face GaN-based E-mode HFETs
- E-mode HFETs with fluorine ion implantation
- GaN-based metal-oxide-semiconductor field-effect transistors (MOSFETs) and AlGaN/GaN metal-oxide-insulated-semiconductor HFETs (MOS-HFETs)
- Other types of E-mode HFETs

5.2.1 E-Mode HFET with a P-Type Cap Layer

In 2000, an E-mode AlGaN/GaN HFET with a selectively grown p/n junction gate was fabricated successfully by Hu et al. [1]. In this device, the p/n junction, for example, a Mg-doped GaN cap layer as a P-type area and a silicon-doped $Al_{0.2}Ga_{0.8}N$ barrier layer as an N-type area, can deplete the high-concentration 2DEG in the channel; thus a threshold voltage of about 2.5 V is finally achieved. In 2006, a kind of device named gate injection transistor (GIT) [2], as shown in Fig. 5.2, was proposed by Panasonic in the International Electron Devices Meeting (IEDM). The hole in the P-type region of GIT can inject into the 2DEG region to generate conductivity modulation as a mechanism observed in an insulated gate bipolar transistor (IGBT). The fabricated GIT exhibits the threshold voltage of 1.0 V and the $R_{on,sp}$ of 2.6 mΩ·cm^2 with a BV of 640 V. Figure 5.3 shows the transconductance characteristics of GITs and MESFETs. The second g_m peak is obtained with an increase in the V_{gs}. This phenomenon is mainly due to the hole injection from the P-AlGaN barrier. Hence, the on-state resistance in this GIT can be achieved to be at a low level with a threshold voltage of 1.0 V and a high maximum drain–source current of 200 mA/mm.

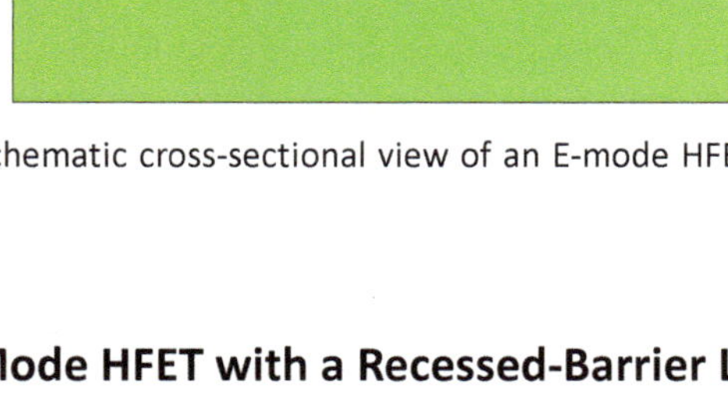

Figure 5.2 Schematic cross-sectional view of an E-mode HFET with a P-type cap layer.

5.2.2 E-Mode HFET with a Recessed-Barrier Layer

In 2004, a recessed-gate AlGaN/GaN HFET with a modulation field plate [3], as shown in Fig. 5.4, was proposed by Okamoto et al. The technique used in this recessed-gate structure leads to a threshold voltage shift from −4.2 V to −1.7 V. In 2006, another E-mode HFET with a recessed-gate structure was proposed by W. Saito et al. By etching the AlGaN barrier layer in the device, a threshold voltage of 1 V with an $R_{on,sp}$ of 4 mΩ·cm^2 and a BV of 435 V are obtained [4]. However, the method of etching the barrier layer can yield surface defects, resulting in a serious lack of reliability. Therefore, to overcome the unreliable limitations of ion etching, a selective area growth technique was put forward by Liu from Sun Yat-Sen University in 2010 [5]. With the selective area growth technique, the implantation damage is dramatically reduced. Thus an E-mode HFET with a maximum output current of 300 mA/mm, a maximum transconductance of 135 mS/mm, and a threshold voltage of 0.4 V is finally achieved.

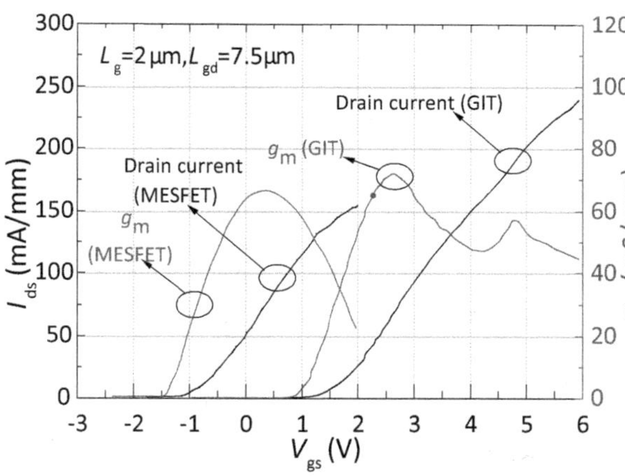

Figure 5.3 I_{ds}–V_{gs} and g_m–V_{gs} characteristics of GITs and MESFETs [2].

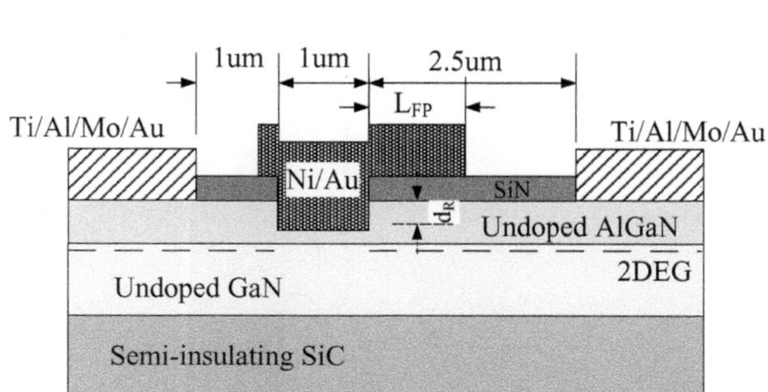

Figure 5.4 Schematic cross-sectional view of an E-mode HFET with a recessed-barrier layer.

5.2.3 E-Mode HFET with a Double-Barrier Layer

In 2006, a double-barrier-layer AlGaN/GaN HFET using the recessed-gate process [6], as shown in Fig. 5.5, was proposed by Ohmaki et al. from the Japanese company Nichia. The existence of a double-barrier-layer structure maintains the device with an $R_{on,sp}$ of 1.9 mΩ·cm^2, a BV of 435 V, and a threshold voltage of −0.1 V. In 2007, another high-performance E-mode AlN/GaN HFET was proposed by the Japan National Institute of Information and Communications Technology and the Fujitsu R&D Center [7]. It achieves several remarkable performances, such as the threshold voltage shifting from 0.14 V to 0.55 V, a maximum current density of 700–920 mA/mm, and a maximum transconductance of 362–400 mS/mm. In 2010, another E-mode AlN/GaN/AlGaN HFET with a double-heterostructure-barrier layer was presented by the Interuniversity Microelectronics Centre (IMEC) [8]. It features a double-heterostructure HFET with an ultrathin AlN barrier layer growing on the 100 nm silicon-type substrate and a high-concentration 2DEG at the surface of the heterostructure. And these features contribute to this HFET with an $R_{on,sp}$ of 1.25 mΩ·cm^2, a BV of 580 V, and a threshold voltage of about 0 V.

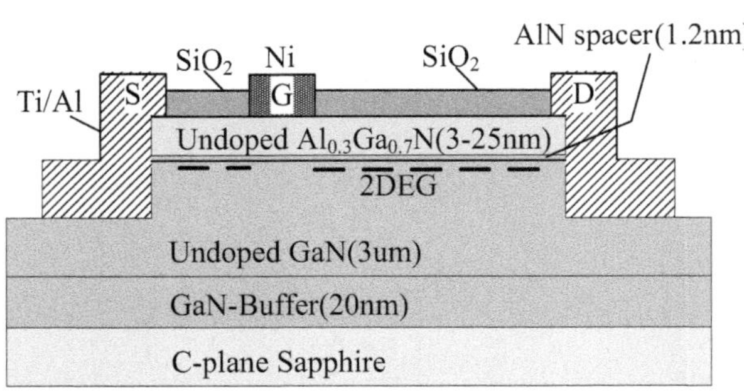

Figure 5.5 Schematic cross-sectional view of an E-mode HFET with a double-barrier layer.

In 2007, an E-mode AlGaN/GaN HFET with a thin InGaN cap layer [9], given by Fig. 5.6, was proposed by Mizutani et al. Its key idea is to employ the polarization-induced field in the InGaN cap layer so that it raises the conduction band of the AlGaN/GaN, which can shift the threshold voltage to the positive direction. Specifically, the fabricated HEMT with a 5 nm thick $In_{0.2}Ga_{0.8}N$ cap layer shows normally off operation with a threshold voltage of 0.4 V and a maximum transconductance of 85 mS/mm, as given in Fig. 5.7.

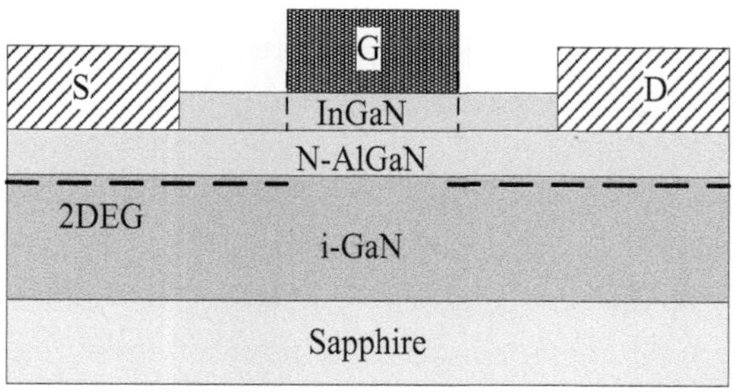

Figure 5.6 Schematic cross-sectional view of an E-mode HFET with a thin InGaN barrier layer.

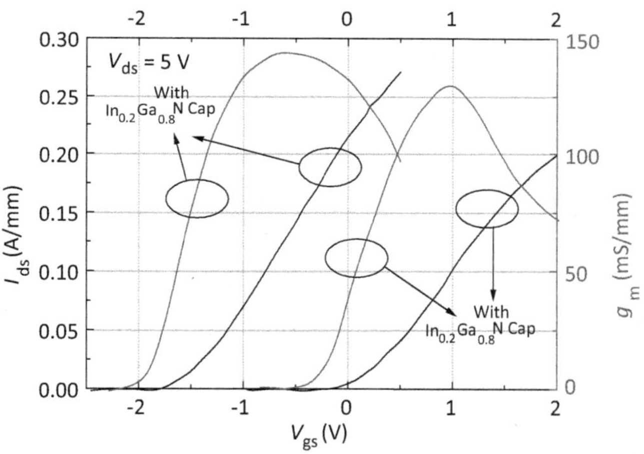

Figure 5.7 Transfer characteristics of AlGaN/GaN HEMT with and without an $In_{0.2}Ga_{0.8}N$ cap layer at V_{ds} = 5 V [9].

In 2009, an ultrathin InAlN/AlN barrier HEMT with high performance in a normally off operation [10] was developed by Ostermaier et al. Its main idea is to control the width of the ultrathin barrier layer by selectively etching the n++ GaN cap layer structure. As a result, a series of performances, such as a maximum output current of 800 mA/mm, a maximum transconductance of 400 mS/mm, a cutoff frequency of 33.7 GHz, and a threshold voltage of + 0.7 V, are obtained in this ultrathin InAlN/AlN barrier HEMT. Moreover, another E-mode HFET with a double-barrier layer by way of selective-area molecular beam epitaxy (MBE) regrowth was proposed by Hughes Research Laboratories on the basis of support of the Next Generation of Nitrides Electronics Project from the US Defense Advanced Research Projects Agency (DARPA) [11]. On the basis of the method of selective-area MBE regrowth, E-mode and D-mode AlN/GaN/AlGaN double-heterojunction field-effect transistors (DHFETs) can also be integrated on a single SiC substrate successfully, as shown in Fig. 5.8. This method also makes it possible to get an E-mode channel, avoiding additional process steps and electrical characteristics' degeneration. At the same time, the emergence of a monolithic integrated E-mode and D-mode device also lays a foundation for direct-coupled field-effect transistor (FET) logic circuits. Figure 5.9 shows I_{ds}–V_{gs} and g_m–V_{gs} characteristics for D-mode and E-mode DHFETs, respectively. The D-mode device has a peak drain current I_{max} of 900 mA/mm at a gate bias V_{gs} of 1.5 V. At a drain bias V_{ds} of 6 V, the peak transconductance g_m is 640 mS/mm and the extrapolated threshold voltage V_{th} is −0.10 V. The corresponding E-mode HFET has an I_{max} value of 720 mA/mm and a peak g_m value of 525 mS/mm. Its extrapolated V_{th} value of 0.24 V is obtained for a normally off channel.

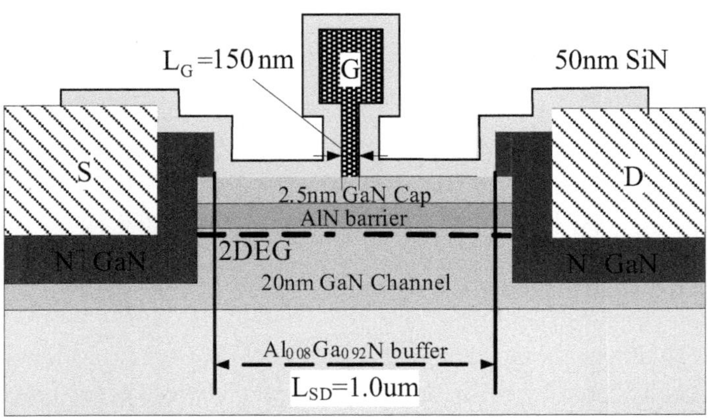

Figure 5.8 Schematic cross-sectional view of an E-mode HEMT with an ultrathin InAlN/AlN barrier.

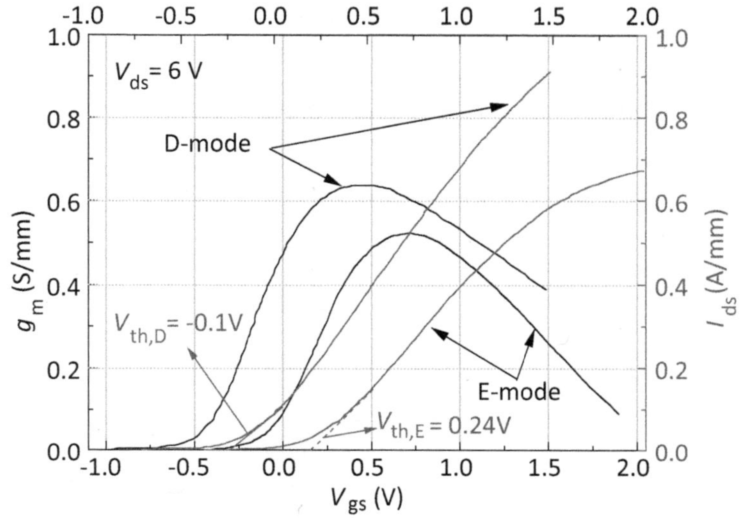

Figure 5.9 I_{ds}–V_{gs} and g_m–V_{gs} characteristics measured at V_{ds} = 6 V of a D-mode and an E-mode DHFET with an L_g value of 150 nm.

5.2.4 Metal-Insulator-Semiconductor HFET

In 2010, an E-mode AlGaN/AlN/GaN HFET with a high Al composition (>70%) was published in [12]. It has a thick AlGaN layer of 17 nm, an AlN layer of 0.6 nm, and an Al_2O_3 layer of 4 nm. The Schottky gate metal is Ni/Au substituting for Al/Au. The threshold voltage of the AlGaN/AlN/GaN HFET can be thereby enhanced from –1.0 V to –0.13 V.

In 2011, an E-mode HFET with a composite dielectric layer using the N_2O plasma oxidation technology was proposed by Chiu et al. in *Applied Physics Letters* [13]. In this device, an Al_2O_3/Ga_2O_3 composite dielectric layer after oxidizing in the N_2O beyond the AlGaN barrier layer can result in an improvement of the threshold voltage from –3.6 V to 0.17 V. In 2012, on the basis of the above structure, an E-mode HFET with a high-*k* composite dielectric layer, as shown in Fig. 5.10, was proposed in *IEEE Electron Device Letters* [14]. In this device, the gate AlGaN barrier layer is oxidized by nitric oxide (NO) gas before deposition of the Schottky metal gate, in which the oxidized AlGaN barrier layer mainly composed of Al_2O_3/Ga_2O_3 stack is formed. At the last, a gadolinium oxide (Gd_2O_3) dielectric film is deposited. This HFET with a composite dielectric layer enables the device threshold voltage change from −3.15 V to +0.6 V. Figure 5.11 shows I_{ds}– V_{gs} and g_m– V_{gs} characteristics for this metal-oxide-semiconductor high-electron-mobility transistor (MOS-HEMT) (E-mode) with a Gd_2O_3 stacked gate and the traditional (D-mode) HEMTs with a Schottky gate. The maximum I_{ds} of the E-mode MOS-HEMT is 364 mA/mm at V_{gs} = 8 V, and the maximum g_m is 105 mS/mm.

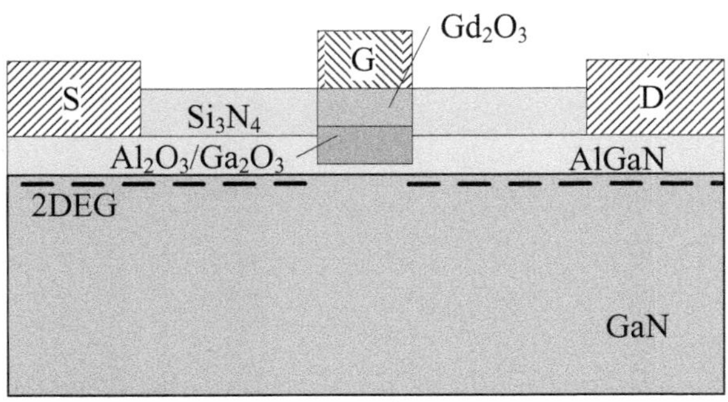

Figure 5.10 Schematic cross-sectional view of an E-mode HFET with a high-*k* dielectric gate.

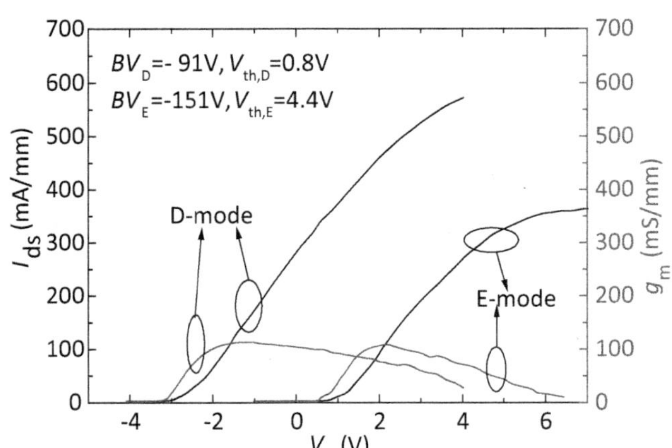

Figure 5.11 I_{ds}–V_{gs} and g_m–V_{gs} characteristics measured at V_{ds} = 8 V of a D-mode and an E-mode DHFET [14].

In 2012, another E-mode HFET with a high-*k* dielectric scandium oxide (Sc_2O_3) was presented by the Massachusetts Institute of Technology and Cambridge collaboration [15]. A high-*k* gate dielectric layer is introduced in this HFET to reduce the invert leakage current so that the switch current ratio (I_{ON}/I_{OFF}) can reach 10^8. Furthermore, the threshold voltage of the MIS-HFET can be enhanced by changing the fabricated condition of the gate dielectric by altering the ultraviolet light and surface NH_4OH.

5.2.5 N-Polar GaN-Based E-Mode HFETs

In 2011, an E-mode N-polar GaN HFET, given by Fig. 5.12, was proposed by the University of California, Santa Barbara [16]. Compared to a conventional HFET, the method of depositing gate metal and annealing process is taken in this HFET, and then a source and drain n+ layer is grown through self-alignment. Specifically, this N-polar GaN-based E-mode HFET takes advantage of the high gate barrier to deplete the 2DEG, finally resulting in several remarkable performances, such as a threshold voltage of 1 V, a maximum current density of 740 mA/mm, and a maximum transconductance of 225 mS/mm. Figure 5.13 shows the output characteristics of a 0.62 μm gate length and 0.18 μm gate length HFET. For the 0.62 μm gate length HFET, a high gate leakage current of 6 mA/mm can be measured at V_{gs} = 4.0 V and V_{ds} = 4.0 V.

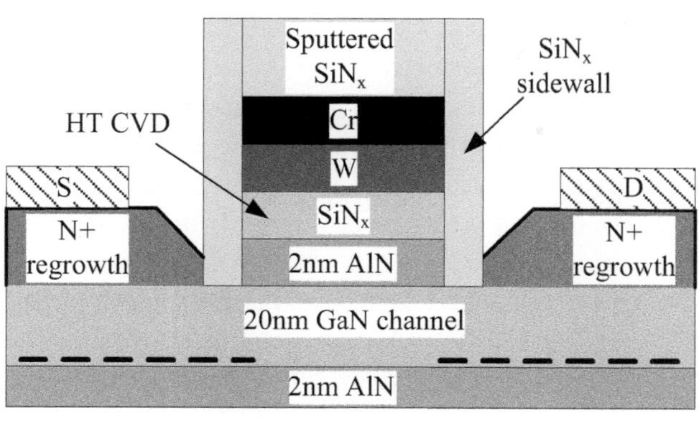

Figure 5.12 Schematic cross-sectional view of an N-polar GaN-based E-mode HFET.

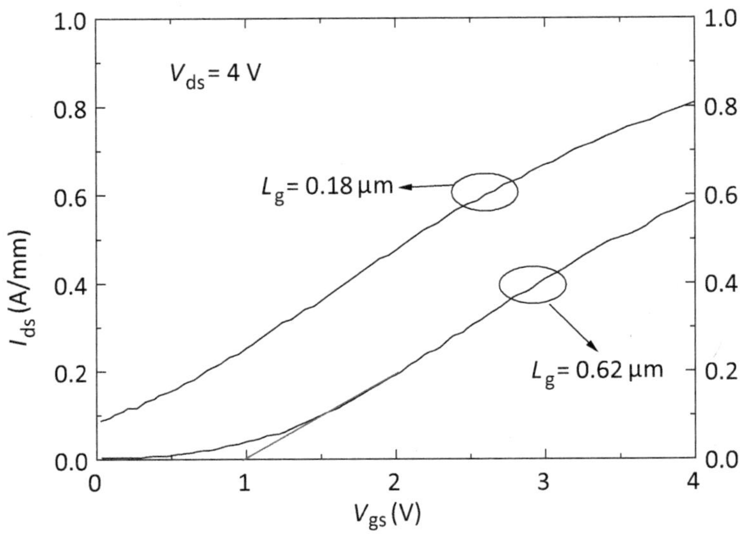

Figure 5.13 I_{ds}–V_{gs} characteristics of a self-aligned 0.62 μm gate length HFET and a self-aligned 0.18 μm gate length HFET [16].

5.2.6 E-Mode HEMTs by Fluoride-Based Plasma Treatment

In 2005, an E-mode AlGaN/GaN HFET using the fluoride-based plasma-treatment technique, as shown in Fig. 5.14, was proposed, for the first time by Chen's team from the Hong Kong University of Science [17, 18]. Fluorine ions are incorporated into the AlGaN barrier layer and then activated by a post-gate-annealing process at about 500°C. Usually, the activated fluoride ion can deplete the 2DEG. Therefore, with improvement of the threshold voltage drifting from −4.0 V to 0.9 V, a maximum transconductance of 148 mS/mm is obtained, as given in Fig. 5.15. Shortly afterward, the fluoride-based treatment technique was successfully applied by this team for fabricating MIS-HFETs [19], as shown in Fig. 5.16, resulting in an E-mode channel to get a threshold voltage of 2 V. In 2012, for pursuing a higher threshold voltage, a composite-gate-insulated layer has been successfully introduced in a MIS-HFET for shifting a threshold voltage to be 5.1 V [20].

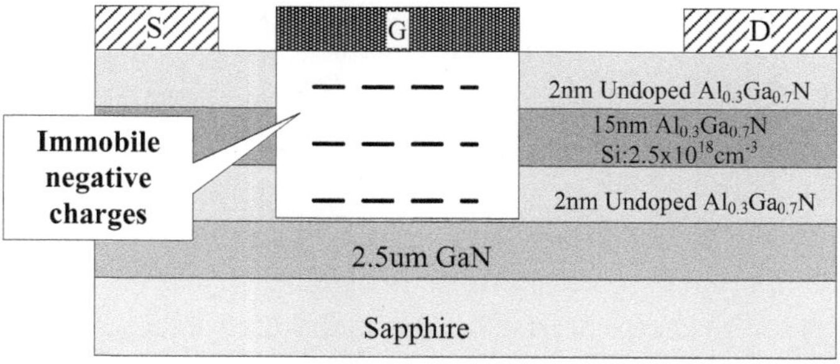

Figure 5.14 Cross section and dimensions of an E-mode HFET with fluoride-based plasma treatment.

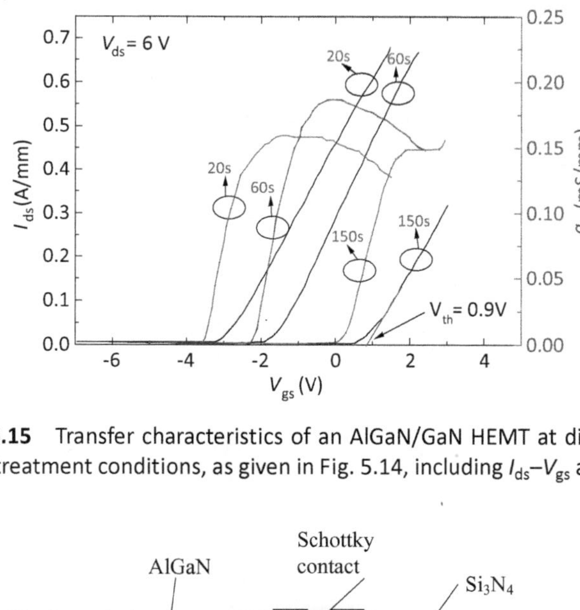

Figure 5.15 Transfer characteristics of an AlGaN/GaN HEMT at different CF_4 plasma treatment conditions, as given in Fig. 5.14, including I_{ds}–V_{gs} and g_m–V_{gs}.

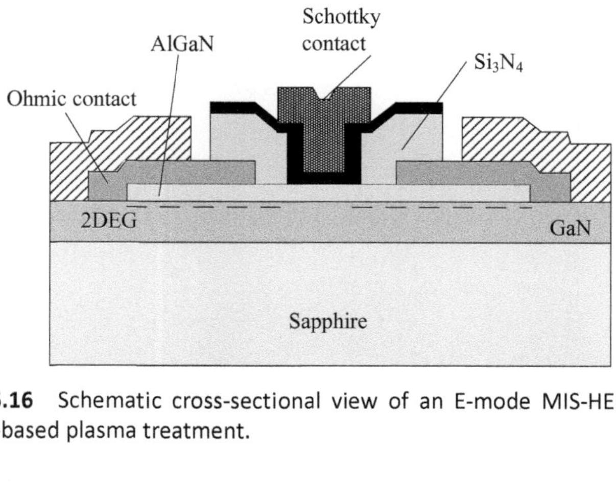

Figure 5.16 Schematic cross-sectional view of an E-mode MIS-HEMT with fluoride-based plasma treatment.

5.2.7 GaN-Based MOSFETs and AlGaN/GaN MOS-HFETs

In 2006, a kind of GaN-based high-voltage MOSFET on a sapphire substrate was proposed in the International Symposium on Power Semiconductor Devices and ICs (ISPSD) international conference [21]. Its design idea takes into account the AlGaN/GaN HFET and the conventional power MOSFET. A high density of the 2DEG with high mobility at the interface of the AlGaN/GaN enables a superior conduction characteristic. A high BV and a positive threshold voltage can be obtained in a conventional MOSFET. For example, the P-type MOSFET threshold voltage is 3.3 V and the N-type MOSFET threshold voltage is +1.5 V, but with a low maximum turn-on current of 12 mA/mm at a gate bias of 20 V. Therefore, on the basis of the insufficiency of the GaN-based MOSFET structure, an ideal combination of the AlGaN/GaN HFET and MOSFET for realizing a normally off channel device and low on-state resistance was proposed at the ISPSD international conference in 2009, as shown in Fig. 5.17 [22]. Figure 5.18 shows the I_{ds}–V_{gs} characteristics and BV–L_{gd} of this fabricated MOS-HFET. Its BV of 500 V is obtained with the length of L_{gd} increasing to over 15 μm. Metal-organic chemical vapor deposition (MOCVD) has been used to grow an epitaxial AlGaN/GaN heterostructure on the surface of the Si(111) substrate. The plasma CVD method has also been taken to deposit a SiO_2 gate dielectric layer. The measured results show that a threshold voltage of +2.8 V at a channel width of 340 mm, a current of 70 A, an on-resistance of 16.5 mΩ·cm^2, and a BV of 500 V are finally obtained.

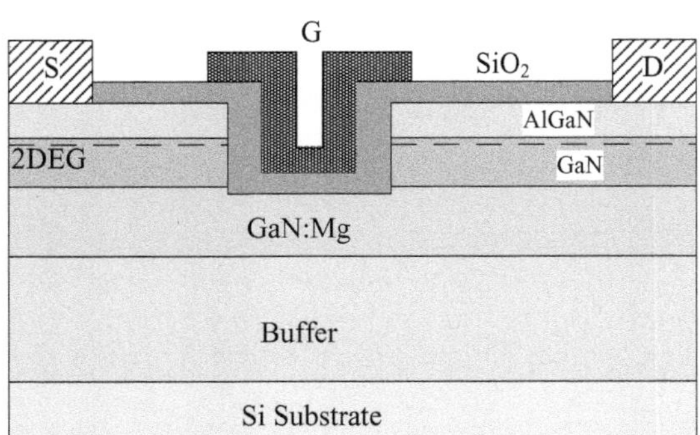

Figure 5.17 Schematic cross-sectional view of a MOS-HFET.

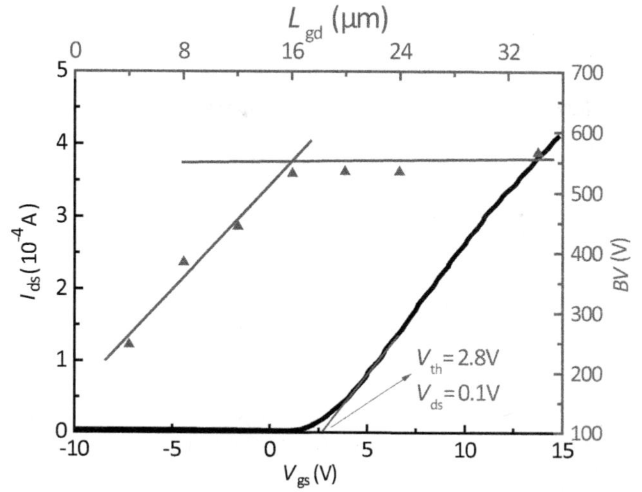

Figure 5.18 I_{ds}–V_{gs} characteristics and BV–L_{gd} of a GaN hybrid MOS-HFET with a channel width of 200 μm [21].

In 2010, another insulated-gate AlGaN/GaN heterojunction MOSFET was proposed by Im et al. [23]. By controlling GaN polarization through the epitaxial growth of the III–N heterojunction, a high-concentration 2DEG (>10^4 cm^{-2}) is obtained in this device. Especially, after removing the AlGaN barrier layer by the reactive ion at a Cl_3/Cl_2 atmosphere, a 30 nm Al_2O_3 layer is deposited by the high-quality atomic layer deposition (ALD) as its gate oxide layer, which can reduce the off-state leakage current (~10^{-7} A/mm) successfully.

5.2.8 Other Types of E-Mode HFETs

In addition to the seven kinds of E-mode HFETs mentioned above, there are also other series of E-mode HFETs with novel mechanisms, for example, AlGaN/GaN HFETs with a vertical current path (current aperture vertical electron transistor [CAVET]) [24], AlGaN/GaN HFET with a charged floating gate [25], and E-mode AlGaN/GaN with a nanorod gate structure [26].

To sum up, in this section, various kinds of E-mode HFETs have been fabricated for switching performance. The main mechanism of these E-mode channels should be illustrated. To reveal the E-mode mechanism, we must model the process of the 2DEG from enhancement to depletion. In the following section, a charge control model (CCM) is given for revealing the E-mode mechanism.

5.3 Charge Control Models

As mentioned in previous chapters, the existence of spontaneous polarization and piezoelectric polarization introduces a high-density 2DEG, which leads to a remarkable output characteristic in the AlGaN/GaN heterostructure [27]. However, these polarizations feature conventional D-mode AlGaN/GaN HEMTs with an inherent normally on behavior. This normally on characteristic limits their operation in power application. In Section 5.2, we have listed the main proposed approaches achieving E-mode HFETs, for example, techniques of the recessed-barrier layer, P-type cap layer, and double-barrier layer. The enhancement of threshold voltage in these devices is mainly attributed to space charge alteration under the gate.

On the basis of the aforementioned mechanism of the 2DEG, the primary subject of the CCM is to address the fundamental relationship between space charges and threshold voltage (V_{th}) in AlGaN/GaN HFETs.

5.3.1 CCM in a Heterojunction with a Single Barrier

In a GaN-based HFET with a single barrier, as shown in Fig. 5.19, if there is no Schottky contact at the surface of the AlGaN barrier, the 2DEG is mainly from the ionized donor surface states. Then, the positively charged donor level on the surface is above the Fermi level, as shown in Fig. 5.19. The positive charges offered by this level are against the surface negative polarization charges and then offset the polarization-induced electric field. At the same time, the ionized surface state leads to the barrier the mobile charges trapped in the triangular quantum well.

If the Fermi level in the surface of the Schottky metal is pinned, the surface barrier height is fixed, and then a 2DEG is derived from the acceptor-like states at the metal side surface. As a degenerate semiconductor, metal can provide enough mobile charges to the barrier. With the width of the barrier decreasing, the metal surface states are occupied by the 2DEG in a triangular quantum well and the 2DEG is depleted. Thus, an E-mode channel is achieved.

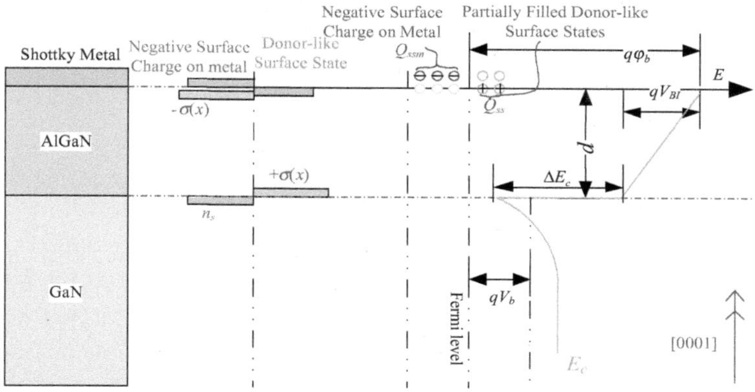

Figure 5.19 Schematic diagram of a single heterojunction: structure, conduction band structure, and space charge distribution.

The expression of the threshold voltage [28] is given by

$$V_{\text{th}} = \varphi_{\text{b}} - \frac{\Delta E_{\text{c}}(x)}{q} - \frac{\sigma(x)}{\varepsilon_0 \varepsilon(x)} d = \frac{\sigma(x)}{\varepsilon_0 \varepsilon(x)} (d_{\text{c}} - d), \tag{5.1}$$

where $\sigma(x)$ is the polarization charge density calculated in the former chapter, d is the thickness of barrier layer, $\varepsilon(x)$ is the dielectric constant of the barrier layer and ε_0 is the vacuum dielectric constant, $\Delta E_{\text{C}}(x)$ is the offset value of the GaN and AlGaN conduction band, x is Al fraction in the AlGaN barrier layer, $q\varphi_{\text{b}}(x)$ is the surface barrier height, and d_{c} is the critical thickness. And d_{c} can be shown as

$$d_{\text{c}} = \frac{\varepsilon_0 \varepsilon(x)}{\sigma(x)} [\varphi_{\text{b}}(x) - \Delta E_{\text{c}}(x)/q]. \tag{5.2}$$

From Eq. 5.1, several approaches, as given in Section 5.2, can partly be analyzed by this CCM. For example, the enhancement of the threshold voltage by the gate recess etching technique is mainly attributed to the reduction of the barrier thickness *di,* as given in Eq. 5.1.

5.3.2 CCM in a Heterojunction with Double Barriers

AlGaN/GaN HFETs with different barrier layers also have different energy band structures in these barriers. So there is a band discontinuity between two adjacent barriers. For example, the Al fraction x in AlGaN is different. The $Al_{x2}Ga_{1-x2}N$ layer and the $Al_{x1}Ga_{1-x1}N$ layer can also be seen as one heterojunction. Figure 5.20 shows the heterostructures of the double barriers. The $Al_{x2}Ga_{1-x2}N/Al_{x1}Ga_{1-x1}N/GaN$ structure can generate two heterojunctions, namely the $Al_{x2}Ga_{1-x2}N/Al_{x1}Ga_{1-x1}N$ heterojunction and the $Al_{x1}Ga_{1-x1}N/GaN$ heterojunction.

If $x_1 < x_2$, the lattice constant of the first barrier layer is larger than the second barrier, so the second heterojunction, namely the $Al_{x2}Ga_{1-x2}N/Al_{x1}Ga_{1-x1}N$ heterojunction, will introduce a piezoelectric polarization effect with the same direction as the first heterojunction (the $Al_{x1}Ga_{1-x1}N/GaN$ junction), which strengthens the polarization effect of the first heterojunction. Thus, the amount of polarization charge in the first heterojunction is raised and the concentration of the 2DEG is thereby lifted up. The threshold voltage is declined.

If $x_1 > x_2$, in the second heterojunction an opposite polarization effect is generated to neutralize the polarization effect of the first heterojunction. Then, the 2DEG is sacrificed partly and the threshold voltage is correspondingly raised. Moreover, with reduction of the Al fraction in the second barrier, the modulation of the $Al_{x1}Ga_{1-x}1N/$ GaN layer can bring more negative charges in the heterointerface to counteract the negative charges in the first barrier for reducing the barrier height, and then, the threshold voltage is modulated accordingly.

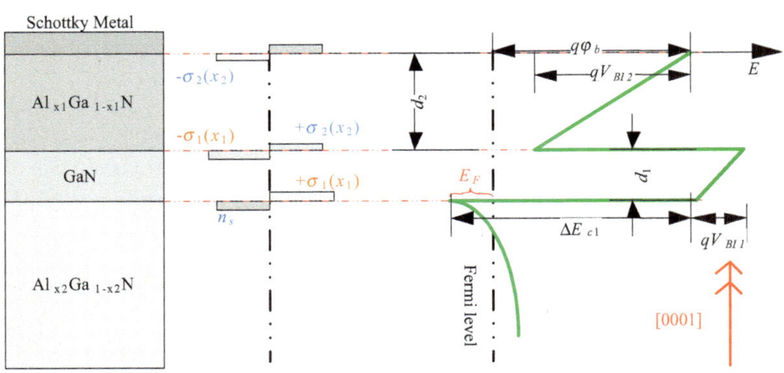

Figure 5.20 Schematic diagram of a double heterojunction: structure, conduction band structure, and space charge distribution.

In this structure, the expression of the threshold voltage can be given as

$$V_{th} = \varphi_b - V_{B12} + \Delta E_{C2}(x_2)/q - V_{B11}(n_s \to 0) - \Delta E_{C1}(x_1)/q, \quad (5.3)$$

where V_{BI1} is the built-in potential of the first barrier and V_{BI2} is the built-in potential of the second barrier, n_s is the concentration of 2DEG, $\Delta E_{C1}(x_1)$ is the offset value of the first barrier's conduction band, and $\Delta E_{C2}(x_2)$ belongs to the second barrier. Especially, there are other equations as follows:

$$\begin{cases} V_{BI1} = \int_0^{d_1} \varepsilon_0 \varepsilon_1 E_1 dt = \dfrac{\sigma_{1/0}}{C_1} \\ V_{BI2} = \int_0^{d_2} \varepsilon_0 \varepsilon_2 E_2 dt = \dfrac{\sigma_{2/1} + \sigma_{1/0}}{C_2} \end{cases} \quad (5.4)$$

and

$$\begin{cases} \varepsilon_0 \varepsilon_1 E_1 = \sigma_{1/0} \\ \varepsilon_0 \varepsilon_2 E_2 = \varepsilon_0 \varepsilon_1 E_1 + \sigma_{2/1} \end{cases}, \quad (5.5)$$

where $C_1 = \dfrac{\varepsilon_0 \varepsilon_1 (x_1)}{d_1}$ is the first barrier layer capacitance, $C_2 = \dfrac{\varepsilon_0 \varepsilon_2 (x_2)}{d_2}$ is the second barrier layer capacitance, and E_1 and E_2 are the electric fields of the first barrier layer and the second barrier layer, respectively. $\sigma_{1/0}$ is the interface charge between the first barrier layer and the GaN-based substrate. $\sigma_{2/1}$ is the interface charge between the first barrier layer and the second barrier layer.

Decreasing d_1 or increasing d_2 can result in improvement of threshold voltage. Decreasing $\sigma_{1/0}$ or increasing $\sigma_{2/1}$ can also lift the V_{th}. All of these can check the aforementioned analysis.

5.3.3 CCM in a Heterojunction with Multibarriers

On the basis of the former analysis, a CCM in a heterojunction with multibarriers can be modeled in the same method. Thus, the main idea of a multibarrier CCM is also formulized to control the polarization charge by improving the Al component and the thickness of the barrier, resulting in a V_{th} shift.

In the structure shown in Fig. 5.21, on the basis of the pinned mechanism and boundary conditions and charge balance equation, the potential can be given by

$$q\varphi_{\text{b}} = E_{\text{F}} - \sum_{i}^{n} qV_{\text{B}Ii} - \sum_{i}^{n} \Delta E_{\text{Ci}}, \tag{5.6}$$

where V_{BI} is the built-in potential of the barrier layer and ΔE_{Ci} is the offset of a different barrier with a different material. And the boundary conditions of the multibarrier layer can be shown as

$$\begin{cases} \varepsilon_0\varepsilon_1 \dfrac{dV_{\text{BI1}}}{dt} = \varepsilon_0\varepsilon_1 E_1 = \sigma_{1/0} + q(N_{\text{A1}} - N_{\text{D1}})t + q(N_{\text{TA1}} - N_{\text{TD1}})t \\ \varepsilon_0\varepsilon_2 \dfrac{dV_{\text{BI2}}}{dt} = \varepsilon_0\varepsilon_2 E_2 = \varepsilon_0\varepsilon_1 E_1 + \sigma_{2/1} + q(N_{\text{A2}} - N_{\text{D2}})t \\ +q(N_{\text{TA2}} - N_{\text{TD2}})t \\ \varepsilon_0\varepsilon_3 \dfrac{dV_{\text{BI3}}}{dt} = \varepsilon_0\varepsilon_3 E_3 = \varepsilon_0\varepsilon_2 E_2 + \sigma_{3/2} + q(N_{\text{A3}} - N_{\text{D3}})t \\ +q(N_{\text{TA2}} - N_{\text{TD2}})t \\ \cdots \\ \varepsilon_0\varepsilon_i \dfrac{dV_{\text{BI}i}}{dt} = \varepsilon_0\varepsilon_i E_i = \varepsilon_0\varepsilon_{i-1} E_{i-1} + \sigma_{i/i-1} + q(N_{\text{A}i} - N_{\text{D}i})t \\ +q(N_{\text{TA}i-1} - N_{\text{TD}i-1})t \\ \cdots \\ \varepsilon_0\varepsilon_n \dfrac{dV_{\text{BI}n}}{dt} = \varepsilon_0\varepsilon_n E_n = \varepsilon_0\varepsilon_{n-1} E_{n-1} + \sigma_{n/n-1} + q(N_{\text{A}n} - N_{\text{D}n})t \\ +q(N_{\text{TA}n-1} - N_{\text{TD}n-1})t \end{cases}, \tag{5.7}$$

where $\sigma_{i/i-1}$ is the polarization charge density between the ith and the $(i-1)$th barrier and E_i, $N_{\text{A}i}$, $N_{\text{D}i}$, $N_{\text{TA}i}$, and $N_{\text{TD}i}$ are the ith electric field, acceptor doping concentration, donor doping concentration, and concentrations of acceptor-type trap and donor-type trap, respectively. Then, we get the following V_{BI} expression:

$$\begin{cases} V_{\mathrm{BI1}} = \int_0^{\mathrm{d}_1} E_1 dt \\ = \dfrac{\sigma_{1/0} + q(N_{\mathrm{A1}} - N_{\mathrm{D1}})d_1/2 + q(N_{\mathrm{TA1}} - N_{\mathrm{TD1}})d_1/2}{C_1} \\ V_{\mathrm{BI2}} = \int_0^{\mathrm{d}_1} E_2 dt = \int_0^{\mathrm{d}_2} E_1 dt \\ + \dfrac{\sigma_{2/1} + q(N_{\mathrm{A2}} - N_{\mathrm{D2}})d_2/2 + q(N_{\mathrm{TA2}} - N_{\mathrm{TD2}})d_2/2}{C_2} \\ V_{\mathrm{BI3}} = \int_0^{\mathrm{d}_1} E_3 dt = \int_0^{\mathrm{d}_2} E_2 dt \\ + \dfrac{\sigma_{3/2} + q(N_{\mathrm{A3}} - N_{\mathrm{D3}})d_3/2 + q(N_{\mathrm{TA2}} - N_{\mathrm{TD2}})d_3/2}{C_3} \\ \cdots \\ V_{\mathrm{BI}i} = \int_0^{\mathrm{d}_i} E_i dt = \int_0^{\mathrm{d}_i} E_{i-1} dt \\ + \dfrac{\sigma_{i/i-1} + q(N_{\mathrm{A}i} - N_{\mathrm{D}i})d_i/2 + q(N_{\mathrm{TA}i-1} - N_{\mathrm{TD}i-1})d_i/2}{C_i} \\ \cdots \\ V_{\mathrm{BI}n} = \int_0^{\mathrm{d}_n} E_n dt = \int_0^{\mathrm{d}_n} E_{n-1} dt \\ + \dfrac{\sigma_{n/n-1} + q(N_{\mathrm{A}n} - N_{\mathrm{D}n})d_n/2 + q(N_{\mathrm{TA}n-1} - N_{\mathrm{TD}n-1})d_n/2}{C_n} \end{cases} \tag{5.8}$$

And then the built-in potential can be summarized as

$$V_{\mathrm{BI}n} = \frac{1}{C_n}\begin{bmatrix} \sum_i^n \sigma_{i/i-1} + q\sum_i^{n-1}(N_{\mathrm{A}i} - N_{\mathrm{D}i} + N_{\mathrm{TA}i} - N_{\mathrm{TD}i})d_i \\ + \frac{1}{2}q(N_{\mathrm{A}n} - N_{\mathrm{D}ni} + N_{\mathrm{TA}n} - N_{\mathrm{TD}n})d_n \end{bmatrix}. \tag{5.9}$$

On the basis of the above expressions, the total threshold voltage is finally given by

$$V_{\mathrm{th}} = \varphi_{\mathrm{b}} + \sum_i^n V_{\mathrm{BI}i} + \sum_i^n \Delta E_{\mathrm{C}i}/q. \tag{5.10}$$

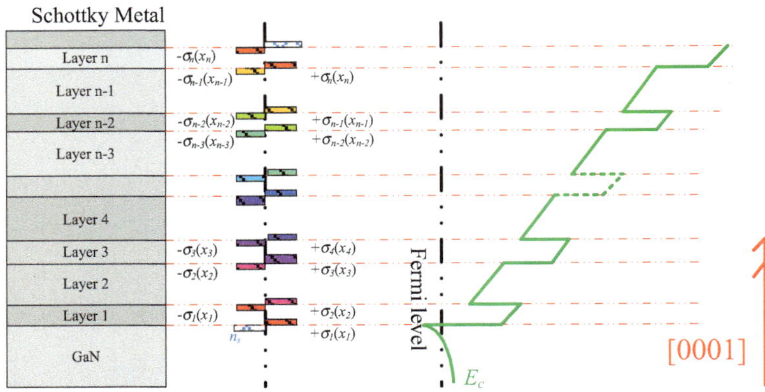

Figure 5.21 Schematic diagram of a multiheterojunction: structure, conduction band, and space charge distribution.

From Eq. 5.9 and Eq. 5.10, we can improve the threshold voltage by controlling the amount of polarization charge through the method of changing the thickness d_i and the material of the barriers $\sigma_{i/i-1}$.

5.4 Reliability of the Threshold Voltage

AlGaN/GaN HEMTs have been fabricated with the E-mode channel for a depleting 2DEG, but instabilities of the threshold voltage become a major limitation for switching performance. The instability and unreliability of the threshold voltage are not attributed to a single relevant reason. As per the classification summary of normally off devices given in Section 5.2, the cause of the enhanced threshold voltage is not generally clarified; thus a theoretical analysis is given in Section 5.3 to explain the physical mechanism of threshold variation by a CCM. In these normally off devices, the enhanced threshold voltage is sometimes reduced with temperature, time, stress condition, and frequency. For instance, in an unpassivated HFET, the 2DEG can be retrapped or detrapped by the donor-like surface states and the acceptor-like interface states at the surface or interface, and then, the output current is collapsed as a virtual gate effect occurs. Obviously, the threshold voltage is unstable for drifts and hysteresis at the frequency-bias driver, possibly due to the 2DEG trapped or detrapped. Meanwhile, the threshold voltage shift is always accompanied with a current collapse.

To date, realization of normally off HFET can be achieved, but how to get a stable threshold voltage is the main issue, and its mechanism is under discussion. In preceding sections, the general form of the threshold voltage is formulized in the CCM. In this model, we illustrated the threshold voltage with respect to space charges and structure characteristics. However, during measurement, the threshold voltage is not identical after some measurement. Especially, in the E-mode channel, the 2DEG charge density and displacement can be changed due to charge shifting into the substrate, barrier, heterointerface, surface, etc. Hence, the threshold voltage drift can be simplified as charge extraction of capacitance–voltage (*CV*) measurement as

$$\Delta V_{\text{th}} = \frac{q\Delta N_{\text{it}}}{C_{\text{eq}}}, \tag{5.11}$$

Where ΔV_{th} is the shift of the threshold voltage, ΔN_{it} is the charge variation at the gate area, and C_{eq} is equivalent capacitance. In this formula, the charge variation can be generalized with charge variation and capacitance. Generally, the threshold voltage variation is mainly attributed to several types of charges extracted by the *CV* measurement for space charge variation. The overall threshold voltage drift depends on the defects in III–N barrier layers, the dielectric/III–N heterointerface and dielectric, etc. The phenomenon of threshold voltage drift is comprehensively illustrated into four aspects, as given in the following sections.

5.4.1 Traps Exist in III–N Barrier Layers

Conventionally, defects as traps exist not only in dielectric but also in III-N barriers. These charge-trapped states as fixed charges in a III-N barrier layer can offer potential enhancement to realize an unstable threshold voltage. Acceptor-like traps can absorb electrons for depleting 2DEG to enhance the threshold voltage, and donor-like traps intend to release electrons for reducing the threshold voltage. For instance, a 4 nm AlN passivation by plasma-enhanced atomic layer deposition (PEALD) is grown for AlGaN/GaN HFETs. In a PEALD-grown AlN layer, a positive polarization charge of $\sim 3.2 \times 10^{12}$ cm^{-2} can compensate the high density of slow-response acceptor-like interface traps [29]. The high-density positive polarization charges can suppress deep-level trapping electrons without sacrificing the 2DEG. If the traps at the interface under the gate are not compensated, electrons in the 2DEG channel can be trapped in deep-level states to generate a threshold voltage shift.

Figure 5.22 shows a schematic diagram of a conduction band and defects' distribution in a GaN/AlGaN/GaN HFET. The fixed defects in the GaN layer and the AlGaN layer, as shown in Fig. 5.19, are neutral without electrons charged beyond the Fermi level. These defects as traps can offer an occupied path for transferring electrons between barrier and substrate. The charge movement in the barrier can easily generate a potential shift and, then, the threshold voltage drift at this stress condition. For instance, as in a P-GaN/AlGaN/GaN HFET, the donor-like trap generation close to the P-GaN/AlGaN interface can cause conduction energy band bowing and provide a threshold voltage shift. Figure 5.23 shows that the threshold voltage is varied during stress and after stress. With the gate bias increases from 0 V to 20 V, the threshold voltage is varied from −1.36 V to −1.17 V in a span of 40 minutes. After several days, the fluctuation of threshold voltage tends to be relatively small. This phenomenon of a threshold voltage shift may be mainly attributed to the charges transferring in defects, as illustrated in Fig. 5.22.

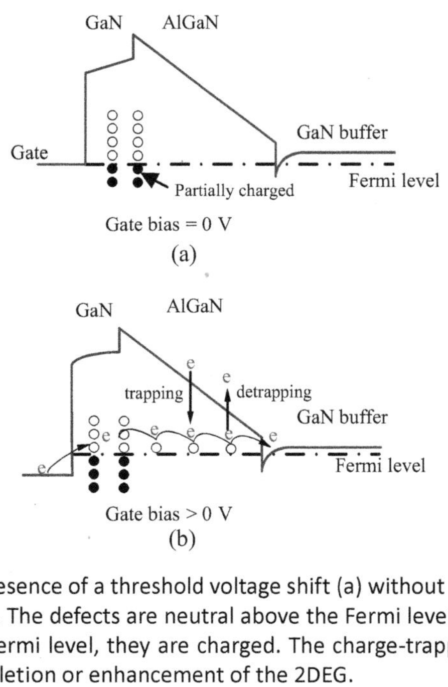

Figure 5.22 Presence of a threshold voltage shift (a) without gate bias and (b) during gate bias. The defects are neutral above the Fermi level, and when they are below the Fermi level, they are charged. The charge-trapping process can result in the depletion or enhancement of the 2DEG.

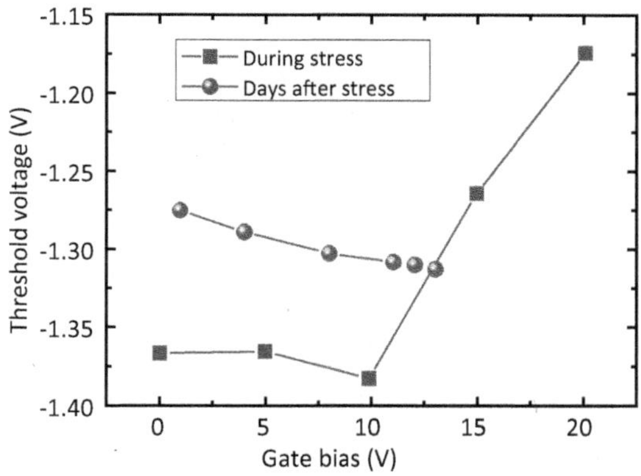

Figure 5.23 Threshold voltage during stress and after stress of a GaN/AlGaN/GaN HFET [30].

5.4.2 Fixed Charges Exist at the Dielectric/III–N Heterointerface or in the Dielectric

Conventionally, the dielectric material under the gate of a MIS-HEMT, as shown in Fig. 5.16, is one kind of MIS heterointerface. If this dielectric is a charge-free layer, a positive threshold voltage can be obtained as the thickness of the dielectric increases to a certain value. The main mechanism can be predicted in Eq. 5.10. However, high-density fixed charges can't be avoided by various methods of manufacture, such as atomic layer deposition (ALD), reactive-ion-sputtered (RIS) deposition, and MOCVD. Former experimental results show that a fixed charge density can achieve more than 10^{12} cm^{-3} at the interface of the dielectric and semiconductor. These fixed charges can enhance or reduce the barrier built-in potential, resulting in a shift of the threshold voltage. Figure 5.24 shows a schematic diagram of a conduction band and space charge distribution. The increase of positive charge the Q_f at the Al_2O_3/AlInN/GaN interface from $Q_f < Q_{sp2}$ to $Q_f > Q_{sp2}$ can enhance the conduction energy band. These increased charges may be from the 2DEG channel or free carriers of III–N semiconductors. Figure 5.25 shows the shift of the threshold voltage at a gate bias −8 V and 2 V. The charge density ΔN_{it}, as given in Eq. 5.11, is extracted to be about 3.5×10^{12} cm^{-2} at the Al_2O_3/GaN heterointerface by *CV* measurement. Positive charges can elevate the energy band for explaining the positive shift.

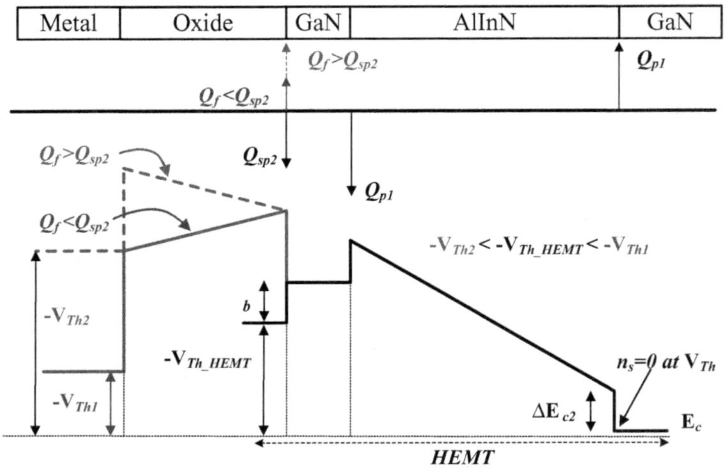

Figure 5.24 Conduction band diagram and charge distribution at the gate area. When charges at the oxide/GaN interface increase from $Q_f < Q_{sp2}$ to $Q_f > Q_{sp2}$, the threshold voltage is enhanced [31].

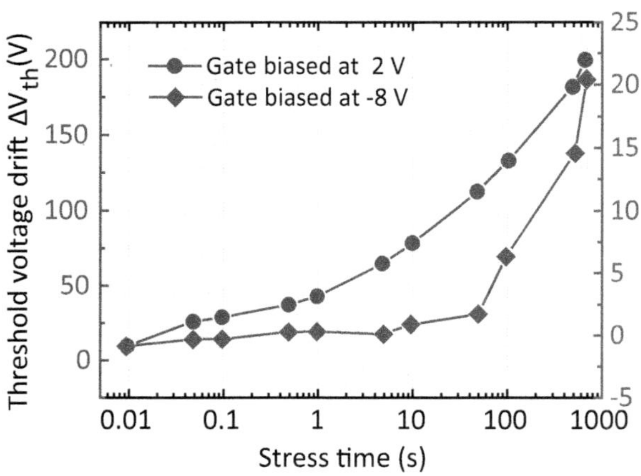

Figure 5.25 Conduction band diagram and charge distribution at the gate area. When charges at the oxide/GaN interface increase from $Q_f < Q_{sp2}$ to $Q_f > Q_{sp2}$, the threshold voltage is enhanced [31].

A forward gate bias is applied in a MIS-HFET, which can induce the threshold voltage drift, as given in Table 5.1, with different dielectric materials. During this process, the electrons can flux into defects from the 2DEG channel for a positive shift and a net increase of conduction band. The electron transport is dominated by thermic emission or trap-assisted tunneling from band to band. These free electrons can be captured by dielectric/III–N heterointerface states or defect states in the dielectric. Table 5.1 gives that the threshold voltage drift of various dielectrics is attributed to the electron-trapped density. The high electron-trapped density extracted for estimating the number of defects in the order of $\sim 10^{12}$ cm^{-2} is relatively smaller than the defect states density ($\sim 10^{13}$ cm^{-2}) for availably capturing electrons. In fact, the deep-level defect states increase with a biased condition and their location is also movable. The tendency of the breakdown of the dielectric material can generate defect states in the dielectric for a threshold voltage shift.

Table 5.1 Overview of samples with different dielectric materials in an AlGaN/GaN MIS-HFET [32]

Material	t_D[a] (nm)	ε_D[b]	t_B[c] (nm)	V_{th}[d] (V)	$\Delta N_{it}/\Delta V_{th}$[e] (cm^{-2}V^{-1})
Si_3N_4	25	7.0	21	-6.6	1.55×10^{12}
Si_3N_4	25	7.0	10	-5.7	1.55×10^{12}
Si_3N_4	40	7.0	10	-8.8	9.67×10^{11}
Si_3N_4	75	7.0	10	-16.4	5.16×10^{11}
HfO_2	30	16.7	21	-6.8	3.04×10^{12}
$HfSiO_2$	30	19.1	21	-6.7	3.60×10^{12}
$HfSiO_2$	30	13.6	21	-8.0	2.32×10^{12}
SiO_2	30	3.9	29	-10.1	7.18×10^{11}
Al_2O_3	25	8.0	21	-8.2	1.77×10^{12}

[a]Thickness of the dielectric layer
[b]Dielectric constant
[c]Thickness of the barrier layer
[d]Threshold voltage
[e]Density of trapped electrons per threshold voltage drift according to Eq. 5.11

5.4.3 Dynamic Recovery of the Threshold Voltage Shift by Trapping Speed

The dynamic drift of the threshold voltage in a MIS-HFET is always induced by stress. When forward gate bias is applied with an oscilloscope signal, the electron under the gate can be retrapped or detrapped with the respective AC signal, as the test circuit given in Fig. 5.26. This storage effect as the capacitance-charging process doesn't respond with the relative high-speed transient signal, as given in Fig. 5.26. The standard measurement conventionally generates a long delay response, and the threshold voltage drift is underestimated. An explanation of the stress-induced threshold voltage drift is that the defects in the distribution of the device offer capture-cross sections and transport path for charges. The defects can capture movable carriers at gate or drain biased, but their capturing frequency is not synchronized with the driven signal.

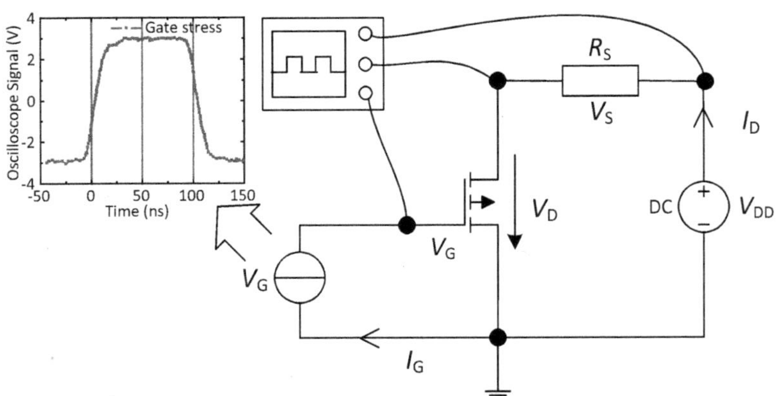

Figure 5.26 Measurement circuit of an HFET at an oscilloscope signal [33].

Figure 5.27 shows the recovery trace of the threshold voltage with different intervals, from 100 ns to 10 µs. The logarithm of the recovery time is linear to the threshold voltage drift. This phenomenon reveals that the electron capture and emission process is logarithmic dynamic with stress spikes at 100 ns/10 µs/1 µs gate bias pulses. The repetitive-recovery stress process sets an insight into the degradation of the threshold voltage instability.

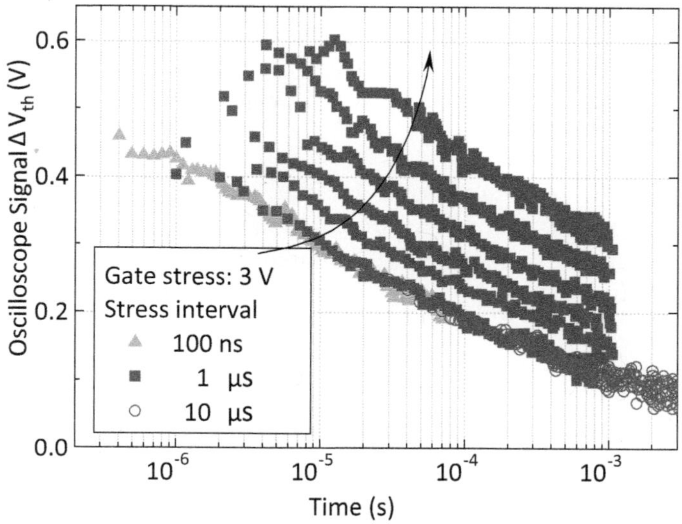

Figure 5.27 Recovery trace of threshold voltage drift at gate bias of 3 V after stress pulses. The value of the threshold voltage drift is extracted from the same traps filled or emptied [33].

5.4.4 Lattice-Mismatch-Induced Reduction of Strain or Stress

A comprehensive study of trap-states-induced threshold voltage shift has been introduced in Section 5.4.1 and Section 5.4.2. In III–V heterostructures, lattice-mismatch-induced strain or stress relaxation is another reason for the shifting threshold voltage. The mismatch process is generated in switching performance at a high reversed-bias condition. For example, the gate drain bias is highly reversed approaching or beyond the critical voltage for conversing polarization, which can generate cracks and pits as mismatch in III–N semiconductors. The generation of defects and percolations is strongly dependent on gate bias and drain bias. Electrons accelerated by the electric field become hot carriers to inject into an AlGaN barrier or a back barrier. And then, these cracks and pits offer defects as new trap centers for electrons charging. Hence, the existence of a threshold voltage drift is inevitable, as shown in Fig. 5.23. Hence, in AlGaN/GaN HFET with high polarization, the mismatch between two materials with different lattice constants can also shift the threshold voltage.

Acknowledgments

The project was supported by the National Natural Science Foundation of China (61404110) and the National Higher-Education Institution General Research and Development Project (2682014CX097).

References

1. Hu, X., Simin, G., Yang, J., Asif Khan, M., Gaska, R., and Shur, M. S. (2000). Enhancement mode AlGaN/GaN HFET with selectively grown pn junction gate, *Electron. Lett.*, **36**(8), 753–754.
2. Uemoto, Y., Hikita, M., Ueno, H., Matsuo, H., Ishida, H., Yanagihara, M., Ueda, T., Tanaka, T., and Ueda, D. (2006). *A Normally-off AlGaN/GaN Transistor with $R_{on}A$=2.6 mΩ•cm^2 and BV_{ds}=640V Using Conductivity Modulation.* Paper presented at the International Electron Devices Meeting (in San Francisco).
3. Okamoto, Y., Ando, Y., Nakayama, T., Hataya, K., Miyamoto, H., Inoue, T., Senda, M., Hirata, K., Kosaki, M., Shibata, N., and Kuzuhara, M. (2004). High-power recessed-gate AlGaN–GaN HFET with a field-modulating plate, *IEEE Trans. Electron Devices*, **51**, 2217–2224.
4. Saito, W., Takada, Y., Kuraguchi, M., Tsuda, K., and Omura, I. (2006). Recessed-gate structure approach toward normally off high-Voltage AlGaN/GaN HEMT for power electronics applications, *IEEE Trans. on Electron Devices*, **53**(2), 356–362.
5. Wen, Y., He, Z., Li, J., Luo, R., Xiang, P., Deng, Q., Xu, G., Shen, Z., Wu, Z., Zhang, B., Jiang, H., Wang, G., and Liu, Y. (2011). Enhancement mode AlGaN/GaN heterostructure field effect transistors fabricated by selective area growth technique, *Appl. Phys. Lett.*, **98**, 072108(1–3).
6. Ohmaki, Y., Tanimoto, M., Akamatsu, S., and Mukai, T. (2006). Enhancement-mode AlGaN/AlN/GaN high electron mobility transistor with low on-state resistance and high breakdown voltage, *Jpn. J. Appl. Phys. 2: Lett. Express Lett.*, **45**(42–45), L1168–L1170.
7. Higashiwaki, M., Mimura, T., and Matsui, T. (2007). Enhancement-mode AlN/GaN HFETs using Cat-CVD SiN, *IEEE Trans. Electron Devices*, **54**(6), 1566–1570.

8. Medjdoub, F., Derluyn, J., Cheng, K., Leys, M., Degroote, S., Marcon, D., Visalli, D., Van Hove, M., Germain, M., and Borghs, G. (2010). Low on-resistance high-breakdown normally off AlN/GaN/AlGaN DHFET on Si substrate, *IEEE Electron Device Lett.*, **31**(2), 111–113.

9. Mizutani, T., Ito, M., Kishimoto, S., and Nakamura, F. (2007). AlGaN/GaN HEMTs with thin InGaN cap layer for normally off operation, *IEEE Electron Device Lett.*, **28**(7), 549–551.

10. Ostermaier, C., Pozzovivo, G., Carlin, J.-F., Basnar, B., Schrenk, W., Douvry, Y., Gaquiere, C., DeJaeger, J.-C., Cico, K., Froehlich, K., Gonschorek, M., Grandjean, N., Strasser, G., Pogany, D., and Kuzmik, J. (2009). Ultrathin InAlN/AlN barrier HEMT with high performance in normally off operation, *IEEE Electron Device Lett.*, **30**(10), 1030–1032.

11. Brown, D. F., Shinohara, K., A, W., Milosavljevic, I., Grabar, R., Hashimoto, P., Willadsen, P. J., Schmitz, A., Corrion, A. L., Kim, S., Regan, D., Butler, C. M., Burnham, S. D., and Micovic, M. (2011). Monolithic integration of enhancement- and depletion-mode AlN/GaN/AlGaN DHFETs by selective MBE regrowth *IEEE Trans. Electron Devices* **58**, 1063–1067.

12. Li, G., Zimmermann, T., Yu, C., Chuanxin, L., Xiu, X., Ronghua, W., Fay, P., Xing, H. G., and Jena, D. (2010). Threshold voltage control in $Al_{0.72}Ga_{0.28}N$/AlN/GaN HEMTs by work-function engineering, *IEEE Electron Device Lett.*, **31**(9), 954–956.

13. Chiu, H.-C., Yang, C.-W., Chen, C.-H., Fu, J. S., and Chien, F.-T. (2011). Characterization of enhancement-mode AlGaN/GaN high electron mobility transistor using N_2O plasma oxidation technology, *Appl. Phys. Lett.*, **99**(15), 153508.

14. Chiu, H.-C., Wu, J.-H., Yang, C.-W., Huang, F.-H., and Kao, H.-L. (2012). Low-frequency noise in enhancement-mode GaN MOS-HEMTs by using stacked Al_2O_3/ Ga_2O_3/Gd_2O_3 gate dielectric, *IEEE Electron Device Lett.*, **33**(7), 958–960.

15. Wang, X., Saadat, O. I., Xi, B., Lou, X., Molnar, R. J., Palacios, T., and Gordon, R. G. (2012). Atomic layer deposition of Sc_2O_3 for passivating AlGaN/GaN high electron mobility transistor devices, *Appl. Phys. Lett.*, **101**(23), 232109.

16. Singisetti, U., Wong, M. H., Dasgupta, S., Nidhi, Swenson, B., Thibeault, B. J., Speck, J. S., and Mishra, U. K. (2011). Enhancement-mode N-polar GaN MISFETs with self-aligned source/drain regrowth, *IEEE Electron Device Lett.*, **32**(2), 137–139.

17. Cai, Y., Zhou, Y., Lau, K. M., and Chen, K. J. (2006). Control of threshold voltage of AlGaN/GaN HEMTs by fluoride-based plasma treatment:

from depletion mode to enhancement mode, *IEEE Trans. Electron Devices*, **53**(9), 2207–2215.

18. Yong, C., Yugang, Z., Chen, K. J., and Lau, K. M. (2005). High-performance enhancement-mode AlGaN/GaN HEMTs using fluoride-based plasma treatment, *IEEE Electron Device Lett.*, **26**(7), 435–437.
19. Wang, R., Cai, Y., Tang, C.-W., Lau, K. M., and Chen, K. J. (2006). Enhancement-mode Si_3N_4/AlGaN/GaN MISHFETs, *IEEE Electron Device Lett.*, **27**(10), 793–795.
20. Chang, C. T., Hsu, T. H., Chang, E. Y., Chen, Y. C., Trinh, H. D., and Chen, K. J. (2010). Normally-off operation AlGaN/GaN MOS-HEMT with high threshold voltage, *Electron. Lett.*, **46**(18), 1280–1281.
21. Huang, W., Khan, T., and Chow, T. P. (2006). *Enhancement-Mode n-Channel GaN MOSFETs on p and n-GaN/Sapphire Substrates.* Paper presented at the 18th IEEE International Symposium on Power Semiconductor Devices and IC's (in Italy).
22. Kambayashi, H., Satoh, Y., Niiyama, Y., Kokawa, T., Iwami, M., Nomura, T., Kato, S., and Chow, T. P. (2009). *Enhancement-Mode GaN Hybrid MOS-HFETs on Si Substrates with over 70 A Operation.* Paper presented at the 21st International Symposium on Power Semiconductor Devices & ICs (in Spain).
23. Im, K.-S., Ha, J.-B., Kim, K.-W., Lee, J.-S., Kim, D.-S., Hahm, S.-H., and Lee, J.-H. (2010). Normally off GaN MOSFET based on AlGaN/GaN heterostructure with extremely high 2DEG density grown on silicon substrate, *IEEE Electron Device Lett.*, **31**(3), 192–194.
24. Chowdhury, S., Swenson, B. L., and Mishra, U. K. (2008). Enhancement and depletion mode AlGaN/GaN CAVET with Mg-ion-implanted GaN as current blocking layer, *IEEE Electron Device Lett.*, **29**(6), 543–545.
25. Hasegawa, F., Kambayashi, H., Li, J., Ikeda, N., Nomura, T., Kato, S., and Yoshida, S. (2009). Proposal and simulated results of a normally off AlGaN/GaN HFET structure with a charged floating gate, *Physica Status Solidi (C)*, **6**(S2), S940–S943.
26. Xuan, R., Kuo, W.-H., Hu, C.-W., Lin, S.-F., and Chen, J.-F. (2012). Enhancing threshold voltage of AlGaN/GaN high electron mobility transistors by nano rod structure: from depletion mode to enhancement mode, *Appl. Phys. Lett.*, **101**(11), 112105.
27. Jang, H. W., Lee, J. H., and Lee, J. L. (2002). Characterization of band bendings on Ga-face and N-face GaN films grown by metalorganic chemical-vapor deposition, *Appl. Phys. Lett.*, **80**(21), 3955–3957.

28. Wang, Z., Zhang, B., Chen, W., and Li, Z. (2013). A closed-form charge control model for the threshold voltage of depletion- and enhancement-mode AlGaN/GaN devices, *IEEE Trans. Electron Devices*, **60**(5), 1607–1612.

29. Huang, S., Jiang, Q., Yang, S., Tang, Z., and Chen, K. J. (2013). Mechanism of PEALD-grown AlN passivation for AlGaN/GaN HEMTs: compensation of interface traps by polarization charges, *IEEE Electron Device Lett.*, **34**(2), 193–195.

30. Rao, H. and Bosman, G. (2010). Device reliability study of high gate electric field effects in AlGaN/GaN high electron mobility transistors using low frequency noise spectroscopy, *J. Appl. Phys.*, **108**(5), 053707.

31. Dutta, G., Turuvekere, S., Karumuri, N., DasGupta, N., and DasGupta, A. (2014). Positive shift in threshold voltage for reactive-ion-sputtered Al_2O_3//AlInN/GaN MIS-HEMT, *IEEE Electron Device Lett.*, **35**(11), 1085–1087.

32. Lagger, P., Steinschifter, P., Reiner, M., Stadtmüller, M., Denifl, G., Naumann, A., Müller, J., Wilde, L., Sundqvist, J., Pogany, D., and Ostermaier, C. (2014). Role of the dielectric for the charging dynamics of the dielectric/barrier interface in AlGaN/GaN based metal-insulator-semiconductor structures under forward gate bias stress, *Appl. Phys. Lett.*, **105**(3), 033512.

33. Lagger, P., Schiffmann, A., Pobegen, G., Pogany, D., and Ostermaier, C. (2013). Very fast dynamics of threshold voltage drifts in GaN-based MIS-HEMTs, *IEEE Electron Device Lett.*, **34**(9), 1112–1114.

Chapter 6

Surface Passivation and GaN MIS-HEMTs

Tianli Duan[a] and Liu Zhihong[b]
[a]*Materials Characterization and Preparation Center, South University of Science and Technology of China, Shenzhen, 518055, China*
[b]*Singapore-MIT Alliance for Research and Technology, Singapore, 117543, Singapore*
duantl@sustc.edu.cn; liuzhihong@ieee.org

Surface states, existing at the AlGaN/GaN surface, will be charged and scatter the electrons at the channel, decreasing the electron mobility. Moreover, the surface states will be charged as the gate, which will be a virtual gate and lead to the depletion of the channel electrons, thereby decreasing the channel 2DEG density. A dielectric layer could passivate the surface states and increase the device performance. However, the parasitic capacitance is induced after surface passivation. Therefore, dielectric-free passivation technology should be developed. On the basis of the surface passivation, a MIS-HEMT is developed to increase the device performance.

6.1 Introduction

The most attractive advantage of GaN is to form a two-dimensional electron gas (2DEG) with GaN-based materials (AlGaN, InGaN,

Gallium Nitride Power Devices
Edited by Hongyu Yu and Tianli Duan

ISBN 978-981-4774-09-3 (Hardcover), 978-1-315-19662-6 (eBook)
www.panstanford.com

etc.). The merit of a 2DEG is its high electron mobility, which arises because channel electrons are separated from the dopants, thereby reducing ionized impurity scattering. The formation of a 2DEG is due to the polarization-induced electric field and surface states [1, 2]. Surface donor states could donate electrons to the quantum well at the AlGaN/GaN interface to form a 2DEG in this electric field [1], as shown in Fig. 6.1.

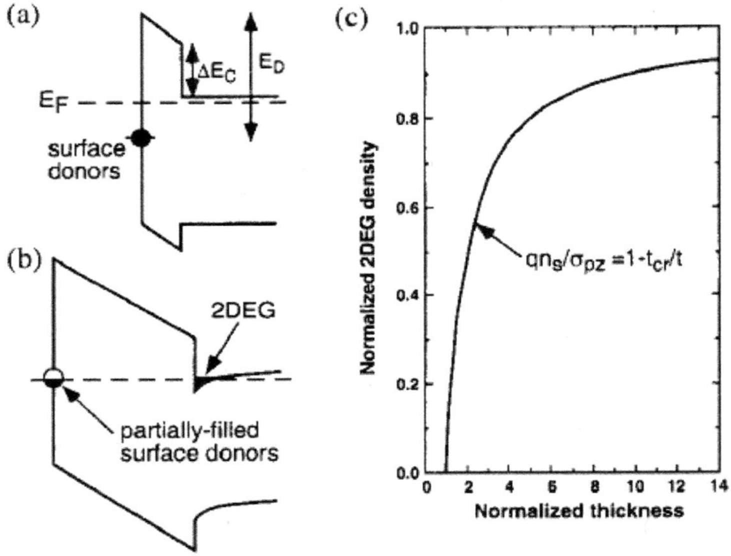

Figure 6.1 Schematic band diagram illustrating the surface donor model with the undoped AlGaN barrier thickness (a) less than and (b) greater than the critical thickness for the formation of the 2DEG. Note the position of the Fermi level relative to the surface state in each case. (c) Calculated 2DEG density as a function of barrier thickness according to the surface donor model. Reproduced from Ref. [1], with the permission of AIP Publishing.

GaN-based high-electron-mobility transistors (HEMTs), taking advantage of 2DEGs, have been widely used in high-power devices and satellite communication for military and commercial purposes [3]. However, surface-defect-related gate leakage currents and drain current collapse occur at high frequency, and the intrinsically high leakage current of the Schottky gate degrades the performance of the traditional HEMT [4–8].

Recently, it has been demonstrated that the adoption of a high κ as the gate dielectric and surface passivation layer in metal-insulator-semiconductor HEMTs (MIS-HEMTs) could help to mitigate the above issues [9–12]. MIS-HEMTs, which combine the merits of HEMTs and MIS devices, offer the advantages of a positive threshold voltage, a low leakage current, and excellent reliability [13, 14]. They are deemed the most promising next-generation high-power transistors.

6.2 Surface Passivation

Polarization charges generate an internal electric field in the AlGaN layer [2]. With this electric field, as shown in Fig. 6.1, when the thickness of AlGaN is beyond the critical thickness, electrons transfer from the surface donor states into the quantum well at the AlGaN/GaN interface, and thus a 2DEG is formed, leaving positively charged states on the surface. Moreover, strong polarization charges need external compensation, indicating a large number of surface states. In addition, some fabrication processes also damage the surface and bring more states [15]. Therefore, surface states are thought to be unavoidable in this material system [1, 2, 7, 15]. These surface states may come from the dangling bonds of the surface atoms, the dislocations/defects at the surface as grown and generated by the damage during the processes, or foreign contaminations. These surface states will work as carrier traps and seriously deteriorate the device performance. The charged surface states will work as a virtual gate and lead to depletion of the channel electrons, thereby decreasing the channel 2DEG density. The trap charges at the surface also contribute to the remote coulomb scattering to the channel electrons and decrease the 2DEG mobility and velocity. In actual applications, large surface states cause serious effects such as current collapse, frequency dispersion, and large dynamic on-resistance.

Surface-state passivation is found to be an effective way to mitigate the state-related issues. Recently, a few materials have been used as a surface passivation layer in GaN HEMTs, such as Si_3N_4 [10, 16], SiO_2 [17], ZrO_2 [6], HfO_2 [18], and Al_2O_3 [9, 19]. Among these materials, SiN is most commonly used for surface passivation in GaN HEMTs. This section will give an example of the typical passivation effects by SiN in an AlGaN/GaN HEMT. A layer of 120 nm SiN was deposited by a plasma-enhanced chemical vapor deposition (PECVD) system at 300°C using SiH_4, NH_3, and N_2 plasma under a chamber pressure of 1000 mTorr and at a radio frequency (RF) power of 20 W. The deposition rate is around 0.9 nm/s. After a 120 nm SiN film was deposited on the whole wafer, the probing pad area was opened after patterning by normal photolithography and CF_4 plasma etch.

The SiN surface passivation increases the DC performance. For example, the DC output (I_d–V_d) and transfer (I_d–V_g) characteristics of an AlGaN/GaN HEMT before and after surface passivation are plotted in Figs. 6.2a and 6.2b. Table 6.1 summarizes the DC characteristics affected by the SiN surface passivation. After passivation, the maximum drain current I_{dmax} significantly increased by ~70%, from 300 mA/mm to 490 mA/mm (at V_g = +1 V). The channel resistance in the linear region (R_{on}) decreased by 29%, from 9.3 Ω·mm to 6.6 Ω·mm, and the maximum extrinsic transconductance g_{mmax} increased by 30%, from 120 mS/mm to 160 mS/mm. A slightly negative shift in the threshold voltage (ΔV_{th} = ~−0.15 V) was observed after passivation. The gate voltage swing (GVS) of the transconductance g_m, defined as the gate voltage range in that the values of g_m are ≥90% of g_{mmax}, increased from 1.05 V to 1.60 V. The higher GVS for g_m enables the device to have a higher output power and better linearity [20]. The improved DC characteristics by SiN surface passivation are consistent with other published reports [20–22], which is believed to be due to the increased sheet carrier density (n_s) and electron mobility (μ) of the 2DEG by the SiN thin film deposited at the source-to-gate and drain-to-gate region. We have confirmed the increase of n_s after SiN passivation through Hall measurement [23].

Table 6.1 Summary of the DC characteristics affected by the SiN surface passivation

Parameters	**Before passivation**	**After passivation**
n_s (cm^{-2})	0.83×10^{12}	1.37×10^{13}
I_{dmax} (mA/mm)	298	489
R_{on} (Ω·mm)	9.3	6.6
g_{mmax} (mS/mm)	120	156
V_{th} (V)	−2.39	−2.55

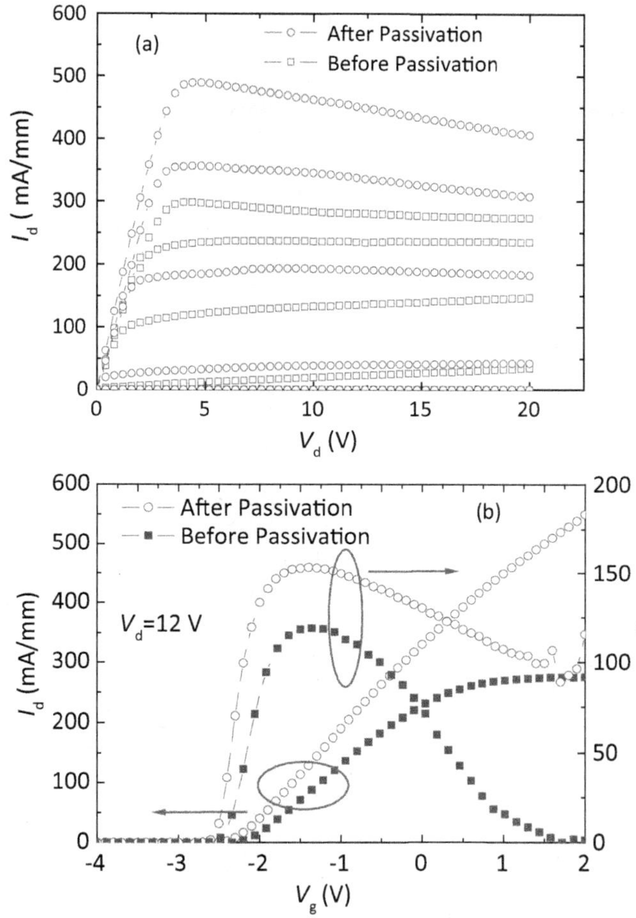

Figure 6.2 DC characteristics of (a) output I_d–V_d and (b) transfer I_d–V_g before and after passivation for an AlGaN/GaN HEMT on a high-resistivity Si substrate.

The current collapse caused by the trapping effect can be characterized by a pulse measurement. When electrons are captured in the trap centers located at the device surface and/or at the buffer, a virtual gate with negative charges will be formed and depletes the electrons in the 2DEG channel. Thus the electron density in the 2DEG will decrease and the mobility also degrades due to the enhanced scattering. The output current will correspondingly be lower. The electrons captured in the trap center will be released after a certain time related to the time constant of the traps. Usually there exists a large number of traps at the surface due to the dangling bonds and various defects and also in the buffer due to the compensation dopants such as Fe or C to keep the buffer semi-insulating. When the gate is biased at a negative voltage, especially at a voltage smaller than the threshold voltage (or the quiescence gate voltage $V_{qg} < V_{th}$), electrons injected from the gate electrode will fill the surface traps through hopping conduction and then deplete the 2DEG, so the drain current will decrease during the following measurement compared to the normal case. Likewise, when the drain is biased at a high voltage (or at a high quiescence drain voltage V_{dq}), the electrons will obtain a high energy and become hot electrons. They will spill into the buffer and be captured by the traps inside the buffer. Then a virtual gate will be formed below the channel and deplete the 2DEG. The drain current will be lower during the following measurement compared to the normal.

The current–voltage (I–V) characteristics under gate stress bias (e.g., quiescent gate bias $V_{gq} < V_{th}$, quiescent drain bias $V_{dq} = 0$) can be used to check the effect of the surface passivation on the reduction of the surface states [24]. Figure 6.3 shows the pulsed I–V measurements for four samples with different surface passivation statuses: (a) unpassivated, (b) passivated with 120 nm SiN, (c) passivated with a 10 nm Al_2O_3/120 nm SiN stack, and (d) passivated with 10 nm Al_2O_3. The SiN was deposited by PECVD, and Al_2O_3 was deposited by atomic layer deposition (ALD). The pulsed I–V measurement was performed with a pulse width of 200 ns and a duty cycle of 0.02. The values of the maximum drain current (I_{dmax}) and on-resistance (R_{on}) at $V_{gq} = -6$ V and $V_{dq} = 0$ V have been compared with those values at $V_{gq} = 0$ and $V_{dq} = 0$ V for all the samples, respectively. The results at $V_g = 0$ V are listed in Table 6.2. It can be seen from Fig. 6.3a and Table 6.2 that I_{max} at gate stress has dropped by 59% and R_{on} has increased by 72% for the unpassivated device. This indicates the existence of the large-density surface states. The Al_2O_3 passivated device shows no drop of I_{dmax} but a 14% increase of R_{on}, whereas the SiN passivated device and the Al_2O_3/SiN bilayer passivated device show no change of I_{dmax} and R_{on}, indicating that only a thin Al_2O_3 film is not sufficient to serve as a passivation layer.

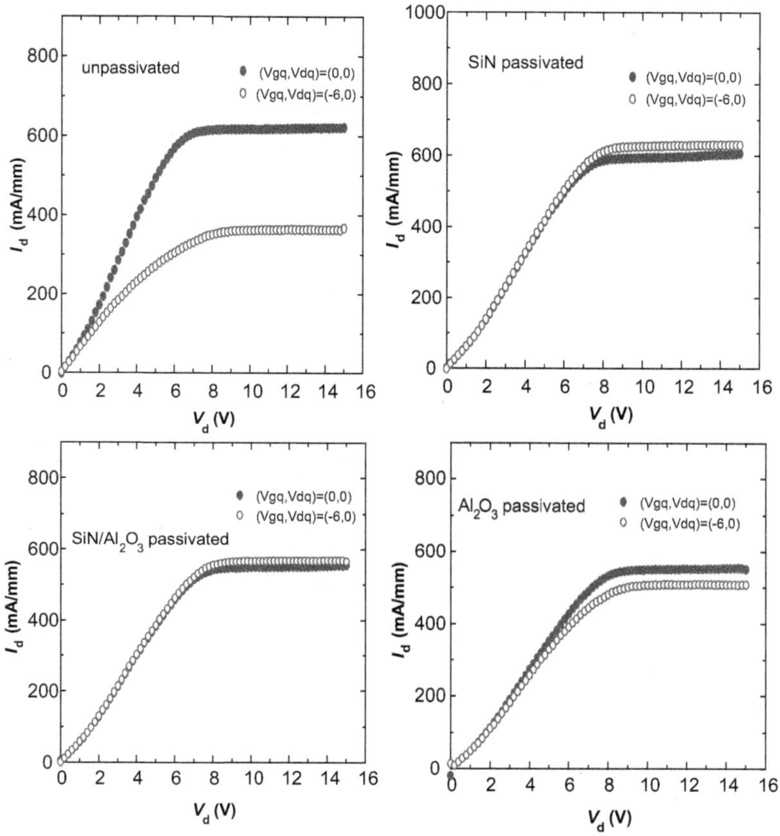

Figure 6.3 Pulsed I_d–V_d characteristics of a GaN HEMT on a Si substrate with different surface passivation statuses at the biases of (V_{gq}, V_{gq}) = (0, 0) and (V_{gq}, V_{gq}) = (−6, 0). Clockwise from top left: unpassivated, passivated with 120 nm SiN by PECVD, passivated with 10 nm ALD Al_2O_3/120 nm PECVD SiN stack layers, and passivated with 10 nm ALD Al_2O_3.

Table 6.2 Various characteristics for the GaN HEMTs with different passivation schemes

Sample no.	Passivation scheme	R_{sh} (Ω/ϒ)	I_{dmax} drop at pulsed DC	R_{on} increase at pulsed DC
A	Without passivation	367	59%	72%
B	120 nm PECVD SiN	324	0	0
C	SiN/Al_2O_3 (120/10 nm)	328	0	0
D	10 nm ALD Al_2O_3	325	0	14%

As for the high-frequency characteristics, some reports show that the dielectric surface passivation can increase the device speed, such as the cutoff frequency (f_T) and the maximum oscillation frequency (f_{max}), and reduce the device's microwave noise [25], due to the enhanced 2DEG mobility by suppressing the surface charge scattering.

However, the dielectric will also increase the parasitic capacitance between the gate and source (C_{pgs})/drain (C_{pgd}) metals and thus decrease the device's speed [26]. Therefore, it is necessary to develop a technology of dielectric-free passivation, such as oxygen plasma oxidization [27, 28], ozone oxidization [29], chemical oxidization [30], or SiH_4 treatment [31], for high-frequency GaN HEMTs. Dong et al. [32] have proposed another approach to reducing the surface states by changing the GaO_x mophology via an elevated-temperature treatment. The disordered native GaO_x layer can thus be transformed into an ordered oxide with stoichiometry and a structure similar to those of bulk Ga_2O_3, thereby reducing the surface states. Bae et al. [33] have used a low-temparature treatment to enhance the quality of the GaO_x formed at the SiO_2/GaN interface and thus reduce the interface defect density, whereas Therrien et al. [34] have demonstrated a microscopic mechanism that can lower the defect density at the high-k/GaN interface through the formation of a high-quality Ga_2O_3 layer during remote plasma-assisted oxidation. But since power devices are working at much lower frequencies, typically <10 MHz, the issue of the increase of parasitic capacitance C_{pgs} and C_{pgd} is not so serious for the passivation layer with low and moderate dielectric constants (such as SiN, SiO_2, Al_2O_3, or AlN).

For power devices, besides the current collapse, another important point is the breakdown voltage. There are also different reports for the SiN passivation effect on the breakdown voltage in GaN HEMTs. We have observed that the breakdown voltage decreased after the surface passivation by PECVD SiN [19], and so did a few other groups [35]. But some groups reported the opposite trend, that is, the breakdown voltage increased after the PECVD SiN passivation [36]. In addition, there are a few groups that found that the breakdown voltage did not have any obvious change after surface passivation [36]. In terms of other passivation materials, such as SiO_2 [37], Al_2O_3 [38], AlN [39], Sc_2O_3 [40], and Ga_2O_3 [41], in most cases the device breakdown voltage was found to increase after surface passivation.

There is still no strong theory to explain the mechanisms behind the breakdown increase or decrease after surface passivation in GaN HEMTs. Most of the breakdown enhancement was attributed to reducing the surface-state density and thereby suppressing the surface conduction and ionization in the states on the surface [42]. The second explanation is that the vacancies introduced in the passivation layer will capture electrons injected from the gate at high reverse gate drain bias and form negative charges, which flatten the electrical field distribution in the channel and suppress the electrical peak at the gate edge close to the drain [41]. Some researchers found that the SiN deposited at low temperature with low quality can enhance the breakdown more effectively than that deposited at high temperature with high quality [36]. The third mechanism to explain why the breakdown voltage could increase with a passivation layer is that the electrical field distribution at the GaN channel between the gate and drain will become more uniformly distributed due to the high dielectric constant of the passivation layer [38, 39, 42]. The higher or thicker the dielectric constant of the passivation layer is, the higher the breakdown will be [43].

On the other side, there are also a couple of different interpretations for the breakdown decrease after the surface passivation. The first mechanism is similar to the phenomenon of "walk out" in GaAs HEMTs [44], that is, before passivation there exist a lot of trap states on the nitride surface and these traps will be negatively charged at a high reverse gate drain voltage. These negative charges deplete the channel electrons and render the electrical field to distribute along the channel more uniformly. After passivation, the surface trap density is greatly reduced and thus the negative charges at a high reverse gate drain voltage are subsequently reduced. Thereby the electrical field peak at the gate edge of the drain side becomes sharper and higher and then the breakdown voltage of the device will be lower [45]. The second mechanism is that during SiN deposition, the nitride surface is damaged, which increases the surface leakage current. This surface leakage lowers down the breakdown voltage [46].

6.3 Metal-Insulator-Semiconductor High-Electron-Mobility Transistors

Compared to the conversional HEMT, a MIS-HEMT has the advantages of low gate leakage, low power consumption, and improved drain current collapse performance. These merits enable GaN-based MIS-HEMTs to be popular for high-power and high-frequency device applications.

The initial-version MIS-HEMT is present below. A 13 nm SiO_2 layer deposited by PECVD is inserted between the GaN substrate and the gate dielectric, as shown in Fig. 6.4. It is found that the gate leakage current of a MIS-HEMT is 6 orders of magnitude smaller than for the HEMT [47], resulting in lower power consumption and larger breakdown voltage. Moreover, the MIS-HEMT exhibits a stable operation speed at an elevated temperature of 300°C [17], which enables MIS-HEMTs to be applied in high-temperature working conditions.

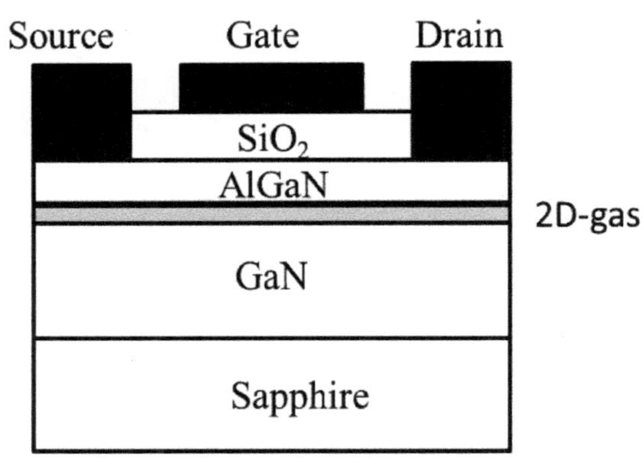

Figure 6.4 Schematic of an AlGaN/GaN MIS-HFET structure.

6.3.1 Characteristics of Various Gate Dielectrics

The criteria to choose the insulator material include its bandgap, band offset to III nitrides, dielectric constant, and breakdown field. A wide bandgap is necessary for the gate insulator to keep a low leakage current in the MIS structure. A high dielectric constant can reduce the equivalent oxide thickness (EOT) of the gate insulator and enable the device to have a low gate leakage current, while keeping a high gate capacitance, which is needed for high gate modulation efficiency on the channel carrier concentration. A high breakdown field is required for the gate insulator to enhance the device off-state breakdown voltage and reliability. In addition, low trap density and low fixed/mobile charge density are also required.

Table 6.3 lists the band offsets of the commonly used dielectrics on GaN, AlGaN, or AlN by theoretical calculation or experiments. GaN MIS-HEMTs with a few types of dielectric materials, such as SiO_2 [17, 48–50], SiN [51–54], Al_2O_3 [19, 55, 56], HfO_2 [57, 58], Ga_2O_3 [59, 60], Sc_2O_3 [61, 62], TiO_2 [63, 64], ZnO_2 [65], NiO [66, 67], ZrO_2 [68, 69], Ta_2O_5 [70], and AlN [71], have been demonstrated in literature. Table 6.4 lists the bandgaps, dielectric constants, and breakdown fields of a few commonly used dielectrics in GaN MIS-HEMTs.

6.3.2 Atomic Layer Deposition of Al_2O_3

Among the various materials for the gate dielectric in GaN MIS-HEMTs, Al_2O_3 have attracted great interest due to its excellent physics properties, including a wide bandgap (~9.0 eV), a high dielectric constant (~9.0), and a high breakdown field (5–30 mV/cm). In the following sections we will give an introduction on the process and characterization of a GaN MIS structure using Al_2O_3 as the gate dielectric. There are various techniques to deposit an Al_2O_3 thin film on a GaN wafer, including electron beam evaporation, sputtering, MOCVD, Al metal oxidization, and ALD. Among these techniques, ALD has the advantages of good film quality, high uniformity, nanoscale controllability, etc.

ALD is a method of depositing thin films on the substrate with atomic-scale precision. The difference between ALD and chemical vapor deposition (CVD) exists in that an ALD reaction breaks the CVD reaction into two parts and keeps the precursor materials separate during the reaction. The first precursor gas is introduced alone into the process chamber and produces a monolayer of gas on the wafer surface. Then the chamber and the pipe line are pumped. Then the second precursor of gas is released and introduced into the chamber and reacts with the first precursor on the wafer surface. A monolayer of film is thus formed. The steps of ALD growth procedure of Al_2O_3 are described below:

Table 6.3 Band offsets of various dielectrics on GaN, AlGaN, and AlN by calculation and experiments

Dielectric	Nitride	Deposition method	Characterization method	CBO[a] (eV)	VBO[b] (eV)	Ref.[c]
Al_2O_3	GaN	–	CNL[o] calculation	2.1	3.4	[72]
	GaN	PEALD[e]	In situ XPS[p]/UPS[r]	1.3	1.8	[73]
	GaN	ALD[d]	XPS/UPS	3.0	0.1	[74]
	GaN	ALD	XPS	2.0	1.2	[75]
	GaN	ALD	C–V[t]	2.1	–	[76]
	$Al_{0.3}Ga_{0.7}NN$	MBE[f]	In situ XPS/EELS[q]	2.1	0.8	[77]
	$Al_{0.2}5Ga_{0.75}N$	ALD	XPS	1.8	0.9	[78]
	$Al_{0.25}Ga_{0.75}N$	ALD	XPS	1.8	1.2	[79]
	AlN	–	CNL calculation	–0.5	3.1	[72]
AlN	GaN	–	CNL calculation	2.4	0.4	[72]
	GaN	ECR MBE	In situ XPS	2.0	0.8	[80]
	GaN	MBE	In situ XPS	1.4	1.4	[81]
CaO	GaN	MOCVD[h]	XPS	2.5	1.0	[82]
Ga_2O_3	GaN	–	CNL	0.5	1.1	[72]
	GaN	Therm. oxid.[k]	XPS	0.1	1.4	[83]
	$Al_{0.3}Ga_{0.7}N$	–	CNL calculation	1.4	0.6	[72]
	AlN	–	CNL calculation	–2.1	0.7	[72]
Gd_2O_3	GaN	–	CNL calculation	1.9	0.7	[72]
	AlN	–	CNL calculation	–0.7	0.3	[72]

Dielectric	Nitride	Deposition method	Characterization method	CBO[a] (eV)	VBO[b] (eV)	Ref.[c]
$GdScO_3$	$Al_{0.3}Ga_{0.7}N$	–	CNL calculation	2.3	–0.2	[72]
$(Gd,Ga)_2O_3$	GaN	Eb. eva.[m]	XPS	0.9	1.1	[84]
$GdScO_3/SrTiO_3$	GaN	MBE	XPS	2.0	0.4	[85]
Gd_2O_3	GaN	–	CNL calculation	1.9	0.7	[72]
$GdScO_3$	GaN	–	CNL calculation	2.6	0.1	[72]
HfO_2	GaN	–	CNL calculation	1.1	1.6	[72]
	GaN	PEALD	In situ XPS/UPS	1.0	1.4	[73]
	GaN	ALD	XPS/UPS	4.3	–1.9	[74]
	GaN	ALD	XPS	1.5	0.5	[75]
	GaN	Oxid. Hf	In situ XPS/UPS	2.1	0.3	[18]
	GaN	Sputtering	XPS/UV[u]	1.7	0.6	[11]
	$Al_{0.3}Ga_{0.7}N$	–	CNL calculation	0.6	1.6	[72]
	$Al_{0.25}Ga_{0.75}N$	ALD	XPS	1.1	0.9	[80]
	AlN	–	CNL calculation	–1.5	1.3	[72]
HfAlO	GaN	ALD	XPS	1.6	0.8	[75]
$HfSiO_4$	GaN	–	CNL calculation	1.6	1.7	[72]
	AlN	–	CNL calculation	–1.1	1.4	[72]
Ta_2O_5	GaN	–	CNL calculation	0.1	1.1	[72]
	AlN	–	CNL calculation	–2.5	0.7	[72]

(*Continued*)

Table 6.3 (*Continued*)

Dielectric	Nitride	Deposition method	Characterization method	CBO[a] (eV)	VBO[b] (eV)	Ref.[c]
La_2O_3	GaN	–	CNL calculation	1.9	0.8	[72]
	GaN	MBE	XPS	0.9	1.9	[86]
	AlN	–	CNL calculation	–0.6	0.4	[72]
$(La_2O_3)_{0.5}(SiO_2)_{0.5}$	GaN	PLD[j]	XPS/EELS	1.4	0.9	[87]
$LaAlO_3$	GaN	–	CNL calculation	1.1	1.3	[72]
	AlN	–	CNL calculation	–1.5	0.9	[72]
$LaAlO_3/SrTiO_3$	GaN	MBE	XPS	2.5	–0.3	[82]
MgO	GaN	–	CNL calculation	2.6	2.0	[72]
	GaN	MBE	XPS	3.2	1.2	[88]
	GaN	PEMBE[g]	XPS	3.3	1.1	[89]
	AlN	MOCVD	XPS	1.5	0.2	[90]
$Mg_{0.5}Ga_{0.5}O$	GaN	PEMBE	XPS/PL[s]	3.4	0.7	[91]
Sc_2O_3	GaN	–	CNL calculation	1.9	0.8	[72]
	GaN	MBE	XPS	2.1	0.8	[72]
	GaN	PEMBE	XPS	2.1	0.4	[92]
	GaN	PLD	XPS	2.0	0.8	[11]
	$Al_{0.3}Ga_{0.7}N$	–	CNL calculation	1.4	0.8	[72]
	AlN	–	CNL calculation	–0.6	0.3	[72]

Dielectric	Nitride	Deposition method	Characterization method	CBO[a] (eV)	VBO[b] (eV)	Ref.[c]
$SrTiO_3$	GaN	–	CNL calculation	−0.1	0.2	[72]
	GaN	MBE	XPS	−0.2	0.1	[11]
	AlN	–	CNL calculation	−2.7	−0.2	[72]
SiO_2	GaN	–	Calculation	2.5	3.2	[72]
	GaN	ALD	XPS	3.0	2.4	[75]
	GaN	MBE	In situ XPS/UPS	3.6	2.0	[93]
	$Al_{0.3}Ga_{0.7}N$	–	CNL calculation	1.9	3.4	[72]
	AlN	–	CNL calculation	−0.1	2.9	[72]
Si_3N_4	GaN	–	CNL calculation	1.3	0.8	[72]
	GaN	MBE	In situ XPS	2.5	−0.6	[94]
	GaN	PEMBE	XPS	–	1.0–1.2	[95]
	$Al_{0.3}Ga_{0.7}N$	–	CNL calculation	0.9	0.6	[96]
	AlGaN	ECR-CVD[i]	In situ XPS/EELS	0.7	0.1	[77]
	AlN	–	CNL calculation	−1.3	0.4	[72]
TiO_2	GaN	Oxid.[l] Ti	XPS	−0.34	0.09	[97]
	$Al_{0.33}Ga_{0.67}N$	Oxid. Ti	XPS	−1.1	0.01	[97]
$PbTiO_3$	GaN	–	CNL calculation	0.4	−0.2	[72]
Y_2O_3	GaN	–	CNL calculation	1.9	0.8	[72]

(Continued)

Table 6.3 *(Continued)*

Dielectric	Nitride	Deposition method	Characterization method	CBO[a] (eV)	VBO[b] (eV)	Ref.[c]
ZnO						[72]
	GaN	–	CNL calculation	–0.7	0.9	[98]
	GaN	PEMBE	XPS/UPS	–0.8	0.8	[99]
	GaN	Sputtering	XPS/PL/UV-Vis[v]	–0.6	0.5	[100]
	GaN	Sputtering	T. dep. *I–V*[w]	–0.6	0.5	[101]
	GaN	Sputtering	XPS	–0.7	0.7	[102]
	$Al_{0.22}Ga_{0.78}N$	Vap. Cool. Cond.[n]	XPS	0.8	–1.5	[103]
	AlN	PEMBE	XPS	–3.3	–0.4	[104]
ZrO_2	GaN	–	CNL calculation	1.1	1.6	[72]
	GaN	ALD	XPS	1.2	1	[105]
	AlN	–	CNL calculation	–1.5	1.1	[72]

[a]CBO: conduction band offset; [b]VBO: valence band offset; [c]Ref.: reference; [d]ALD: atomic layer deposition; [e]PEALD: plasma-enhanced atomic layer deposition; [f]MBE: molecular beam epitaxy; [g]PEMBE: plasma-enhanced molecular beam epitaxy; [h]MOCVD: metal-organic chemical vapor deposition; [i]ECR-CVD: electron cyclotron resonance chemical vapor deposition; [j]PLD: pulsed laser deposition; [k]Therm. oxid.: thermal oxidization; [l]Oxid.: oxidization; [m]Eb. eva.: electron beam evaporation; [n]Vap. cool. cond.: vapor cooling condensation; [o]CNL: charge neutrality level; [p]XPS: X-ray photoelectron spectroscopy; [q]EELS: electron energy loss spectroscopy; [r]UPS: ultraviolet photoelectron spectroscopy; [s]PL: photoluminescence measurement; [t]*C–V*: capacitance–voltage; [u]UV: ultraviolet absorption; [v]UV-Vis: UV-visible absorption; [w]T. dep. *I–V*: temperature-dependent current–voltage measurement

Table 6.4 Bandgaps, dielectric constants, and breakdown fields of various dielectrics for GaN MIS-HEMTs

Dielectric	Bandgap E_g (eV)	Dielectric constant ε	Breakdown field E_{BD} (MV/cm)
Al_2O_3	6.8–9.4 [106–108], 8.7 [15]	8–10 [106, 109], 9 [15]	10 [108], 10 [110], 5 [111], 10 [112]
HfO_2	5.6–5.8 [113], 5.7 [15]	25 [107], 16.5–24 [114–116], 25 [15]	4 [117], 4 [118]
$HfSiO_4$	6.5 [106]	11 [106]	>10 [119]
$LaAlO_3$	5.6–6.2 [106]	30 [106]	4.3 [120], 4 [121]
La_2O_3	6 [122]	30 [122]	13.5 [123], 3.2 [124]
Ga_2O_3	4.8 [125]	10.2–14.2 [125]	2.8 [126], 3.85 [127]
Gd_2O_3	5.8 [128, 129]	17 [128, 129]	0.3 [12]
$Ga_2O_3(Gd_2O_3)$	-	-	>12 [130]
$PbTiO_3$	3.4 [131]	78–104 [106]	-
$SrTiO_3$	3.3 [131]	2000 [106]	-
Ta_2O_5	4.4 [106]	22 [106]	5.4 [132]
TiO_2	3.5 [106]	80 [106]	3.2 [133], 1 [134]
SiO_2	8.9 [15], 8.8 [135]	3.9 [15], 3.9 [135]	9.6 [136], 23 [118]
Si_3N_4	5.3 [106], 4.5 [15]	7 [106], 7.5 [15]	5.7 [137]
Sc_2O_3	6.3 [89]	14.5 [89]	5 [138]
ZrO_2	5.8 [106], 7.8 [15]	23 [106], 18–25 [15], 23 [139] 23 [6], 20.7 [140]	4 [141] 4 [119]
MgO	7.2 [15]	9.6 [15]	[142], 2.3 [143]

Step 1: After placing the sample in the reactor and pumping the chamber to a low enough pressure, trimethylaluminum (TMA) is pulsed into the reaction chamber.

Step 2: The TMA reacts with the absorbed hydroxyl groups, producing methane as the reaction product:

$$Al(CH_3)_3 + :Ga\text{-}O\text{-}H \longrightarrow :Ga\text{-}O\text{-}Al(CH_3)_2 + CH_4 \uparrow \quad (6.1)$$

Step 3: The TMA reacts with the absorbed hydroxyl groups until the surface is passivated. TMA does not react with itself, terminating the reaction to one layer. This causes the perfect uniformity of ALD. The excess TMA is pumped away, together with the methane reaction product.

Step 4: After the TMA and methane reaction product is pumped away, water vapor (H_2O) is pulsed into the reaction chamber.

Step 5: H_2O reacts with the dangling methyl groups on the new surface, forming aluminum oxygen (Al-O) bridges and hydroxyl surface groups, waiting for a new TMA pulse. Again methane is the reaction product:

$$2H_2O + :Ga\text{-}O\text{-}Al(CH_3)_2 \longrightarrow :Ga\text{-}O\text{-}Al(OH)_2 + 2CH_4 \uparrow \quad (6.2)$$

Step 6: The reaction product methane is pumped away. Excess H_2O vapor does not react with the hydroxyl surface groups, again causing perfect passivation to one atomic layer.

Step 7: One TMA and one H_2O vapor pulse form one cycle. Here three cycles are shown, with approximately 1 Å per cycle. Each cycle, including pulsing and pumping, takes approximately 3 sec.

There are two fundamental mechanisms during the growth: the chemisorption saturation process and the sequential surface chemical reaction process. Since each pair of gas pulses, that is, one cycle, produces exactly one monolayer of film, the thickness of the resulting film can be precisely controlled by the number of deposition cycles. During the growth, the precursors are kept separate all the time so that the thin-film layer thickness can be obtained as fine as atomic/molecular scale per monolayer.

ALD has the following advantages over other deposition techniques for a high-*k* dielectric thin film:

- Stoichiometric films.
- Perfect step coverage and 3D conformability. Typically the thin film by ALD has an aspect ratio as high as 60:1. The bottoms and sidewalls of the mesa/trenches are covered uniformly.

- High large-area uniformity.
- High thin-film quality, low leakage, and high reliability.
- Precise thickness control. The thickness of the film is determined simply by the number of deposition cycles.
- Low-temperature deposition possibility (from room temperature to 400°C).

The gate dielectric quality and the GaN MIS-HEMT performance can be further improved by postdeposition annealing. We have found that annealing at 600°C for 5 min in a N_2 atmosphere after Al_2O_3 deposition can reduce the fixed charge and positively shift the threshold voltage toward the expected value from the theoretical calculation in GaN MIS-HEMTs with ALD-Al_2O_3 as the gate dielectric [144]. Other researchers have reported various annealing conditions, from 400°C–600°C, at N_2 or forming gas (H_2/N_2) to reduce the interface trap density, interface fixed charge, or gate leakage current [105, 145–149].

6.3.3 Characterization of the Interface Traps by Traditional *C–V* Measurement

The quality of the gate insulator and the interface between it and the nitride semiconductor is crucial for the MIS-HEMT to achieve excellent and reliable performance. Until now many characterization techniques have been developed to evaluate the gate insulator quality, including optical metrology, such as atomic force microscopy (AFM) and transmission electron microscopy (TEM), and electrical metrology, such as I–V and C–V and temperature-accelerated tests. Among these techniques, C–V measurement is most commonly used as it is simple and reliable to evaluate the thin-film quality both qualitatively and quantitatively.

The schematic and the equivalent circuit for a MIS diode with an AlGaN/GaN heterostructure are shown in Fig. 6.5. The traps located at those places other than the interface are ignored. The resistance R_s accounts for the series resistance from the ohmic contact to the channel below the gate through the access region between the source and gate electrodes. The resistance R_t accounts for the effect of gate leakage current. C_{IN} accounts for the capacitance of the gate insulator. C_b accounts for the capacitance of the barrier layer (here the term "barrier" includes the possible spacer layer and cap layer). C_d accounts for the capacitance of the channel depletion region. The C_{it} and R_{it} in series describe the capacitance effect of the traps located at the interface between the gate insulator and the III nitride barrier. When the gate bias changes, the number of trapped electrons in the interface states also varies so that an effective capacitor is formed. The R_{it}, together with the C_{it}, describes the time delay (τ) required for the electrons trapped at the interface to form an equilibrium with those in the channel, and C_{it} is related to the trap density (D_{it}), as the following expression shows [150]:

$$\tau = R_{it}C_{it}$$
$$D_{it} = \frac{C_{it}}{qA}, \tag{6.3}$$

where A is the area of the gate.

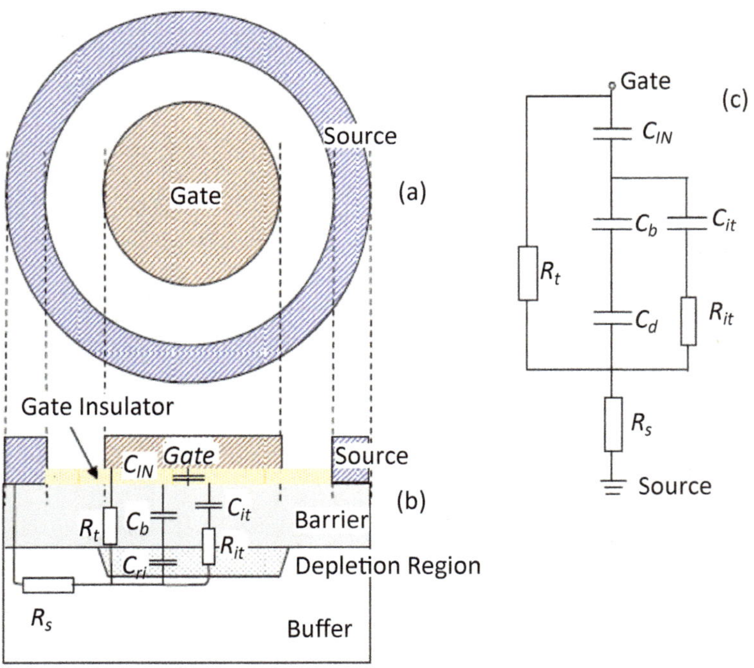

Figure 6.5 The schematic and equivalent circuit of a MIS diode: (a) the top view of the layout for a typical circular MIS diode; (b) the cross section and equivalent circuit element distribution; (c) the equivalent circuit topology.

Considering the circuit in Fig. 6.5c, the circuit can be simplified into Fig. 6.6a if (i) the leakage current of the diode is negligible, that is, R_t is very large; (ii) the gate bias is oscillating at such a high frequency that the emission and capture of the trapped electrons cannot catch up with the change of the gate bias and, therefore, the effect of C_{it} and R_{it} is negligible; and (iii) the device is working in the accumulation region. The equivalent circuit for the corresponding conventional Schottky diode (without the gate insulator) is simplified into Fig. 6.6b. The circuit of the *C–V* measurement is assumed to Fig. 6.6c by the equipment.

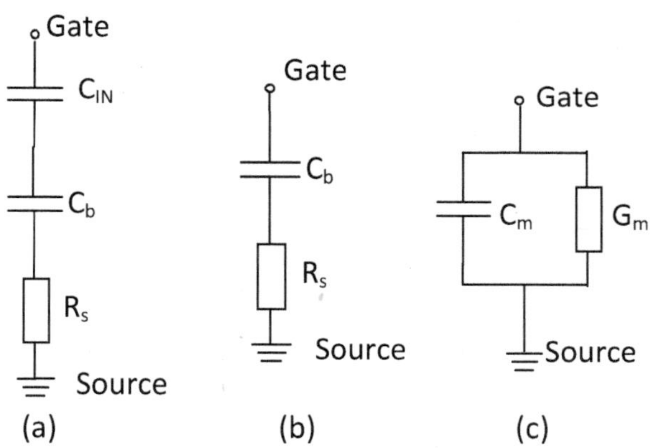

Figure 6.6 Simplified equivalent circuit of (a) a MIS diode, (b) a Schottky diode, and (c) measurement.

The C_m and G_m in Fig. 6.6c are the values of capacitance and conductance, respectively, directly measured. The R_s effect on the measured capacitance can be subtracted by

$$C_{m'} = \frac{C_m}{(1 - R_s G_m)^2 + (\omega R_s C_m)^2}. \tag{6.4}$$

Then for a conventional HEMT ($C_m' = C_{HEMT}$)

$$C_{HEMT} = C_b. \tag{6.5}$$

For a metal-insulator-semiconductor (MIS) diode ($C_m' = C_{MIS\text{-}HEMT}$)

$$\frac{1}{C_{MIS\text{-}HEMT}} = \frac{1}{C_{IN}} + \frac{1}{C_b}. \tag{6.6}$$

Afterward C_{IN} can be calculated from the measured overall capacitance for the corresponding MIS diode ($C_{MIS\text{-}HEMT}$) and conventional Schottky diode (C_{HEMT}):

$$C_{IN} = 1 \Big/ \left(\frac{1}{C_{MIS\text{-}HEMT}} - \frac{1}{C_{HEMT}} \right) \tag{6.7}$$

C_{IN} is determined by the dielectric constant ε and the thickness of the gate insulator d by the expression

$$C_{IN} = \frac{A\varepsilon}{d}, \tag{6.8}$$

where A is the area of the diode gate. Thus if anyone between ε or d is known by other means, the value of the other one can also be determined.

Like a Si metal-oxide-semiconductor (MOS) diode, the large amount of interface states in an Al_2O_3/AlGaN/GaN MIS diode will broaden the transition region from the depletion to the accumulation. For a trap-free depletion-mode MIS diode, when the gate bias is zero, the channel is open and the diode is working in the accumulation region with $C_{MIS\text{-}HEMT} = 1 \Big/ \left(\frac{1}{C_{IN}} + \frac{1}{C_b} \right)$. When the gate voltage is ramped down from zero until the threshold voltage V_{th}, the device is still working in the accumulation region and $C_{MIS\text{-}HEMT}$ remains nearly constant. When V_g reaches and goes below V_{th}, the channel is turned off and $C_{MIS\text{-}HEMT}$ is sharply lowered to a small value close to zero.

When a large amount of interface states exists at the interface between the gate insulator and the III nitride semiconductor, the traps will capture or release electrons when gate bias changes. The net charge change will change the threshold voltage V_{th}. Displayed on the C–V curve, this trap effect will lead to a wider transition from the depletion to the accumulation region. Thus the degree of the sharpness for this transition in the C–V curve can be used to roughly judge the quality of the gate insulator.

Figure 6.7 shows two C–V curves for two ALD Al_2O_3/AlGaN/GaN MIS structures with different Al_2O_3 qualities measured at high frequency (1 MHz). The curve in Fig. 6.7a was measured for the Al_2O_3 deposited earlier of poor quality and Fig. 6.7b for the Al_2O_3

whose quality is much improved. It can be clearly seen that the transition region from depletion to accumulation in the *C–V* curve is much sharper for the Al_2O_3/AlGaN/GaN MIS diode with high-quality Al_2O_3 (with small numbers of traps inside the oxide and/or at the interface) than that one with poor-quality Al_2O_3 (with large numbers of traps).

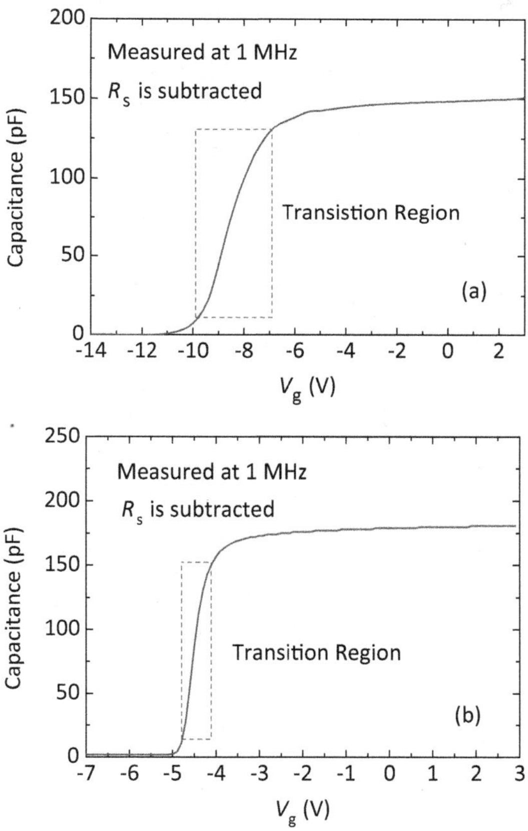

Figure 6.7 High-frequency (at 1 MHz) *C–V* characteristics of the ALD Al_2O_3/AlGaN/GaN MIS structure on Si with different Al_2O_3 qualities: (a) for the early deposited Al_2O_3 of poor quality and (b) for the Al_2O_3 of improved quality.

Another simple technique to directly evaluate the quality of the thin film is the low-frequency *C–V* measurement. At a low enough frequency, the numbers of the electrons at the interface states have enough time to reach equilibrium through electron exchange with the electrons at the conduction band in the channel, and hence C_{it} will add to the measured overall capacitance C_m, as the circuit in Fig. 6.6c shows. As a result, the measured overall capacitance at low frequency will be larger than that measured at high frequency, and their difference varies with the electrical potential at the interface, which depends on the gate bias. Figure 6.6 shows two groups of *C–V* curves for an ALD Al_2O_3/AlGaN/GaN MIS structure with different Al_2O_3 qualities measured at both high frequency (1 MHz) and low frequency (10 KHz). The curve shown in Fig. 6.8a is for the Al_2O_3 deposited early of poor quality, and Fig. 6.8b is for the Al_2O_3 whose deposition condition has been much improved. It can be seen that in the case of the gate insulator of poor quality, its *C–V* curve at low frequency shows much difference from that at high frequency. The big hump comes from the capacitance of the interface states with a peak density at the energy level corresponding to that gate bias. For the Al_2O_3 of good quality, the gap of the two curves at high and low frequencies is very small.

The techniques described above can be employed for the purpose of qualitatively evaluating the quality of the deposited Al_2O_3 thin film. To quantitatively determine the interface trap density, more calculations and other techniques are needed. A simple method to measure the interface trap density is the so-called *hi-lo* frequency method [151]. As described in the last paragraph, the difference between the measured overall capacitances at high and low frequencies includes the information of trap effects. But to work out the trap density D_{it}, the values of R_t and R_s are required to be determined first by other measurements. Then Fig. 6.5c will be simplified into Fig. 6.9a. The interface-state density can be calculated by [151]

$$C_{it} = \frac{C_{LF} - C_{HF}}{\left(1 - \dfrac{C_{LF}}{C_{IN}}\right)\left(1 - \dfrac{C_{HF}}{C_{IN}}\right)} \tag{6.9}$$

and

$$D_{it} = \frac{C_{it}}{q}, \tag{6.10}$$

where C_{LF} and C_{HF} are the measured capacitance values (C_m) at low and high frequencies, respectively, as shown in Fig. 6.9b.

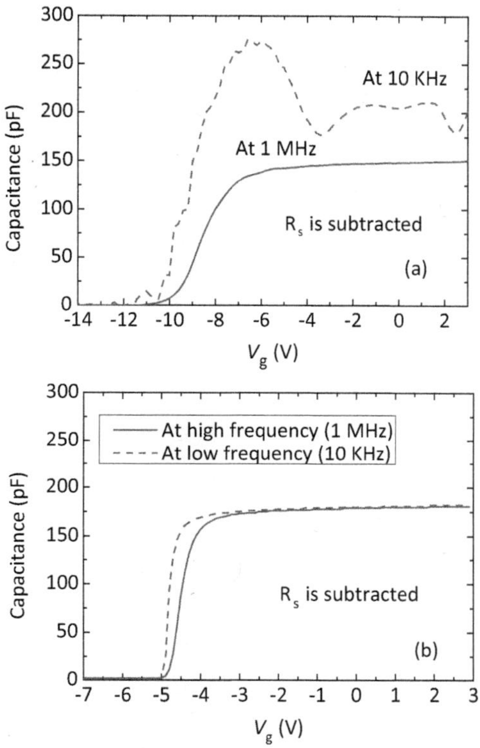

Figure 6.8 *C–V* characteristics for the ALD Al_2O_3/AlGaN/GaN MIS structure on Si with different Al_2O_3 qualities at high frequency (1 MHz) and low frequency (10 KHz): (a) for the Al_2O_3 deposited early of poor quality and (b) for the Al_2O_3 of improved quality.

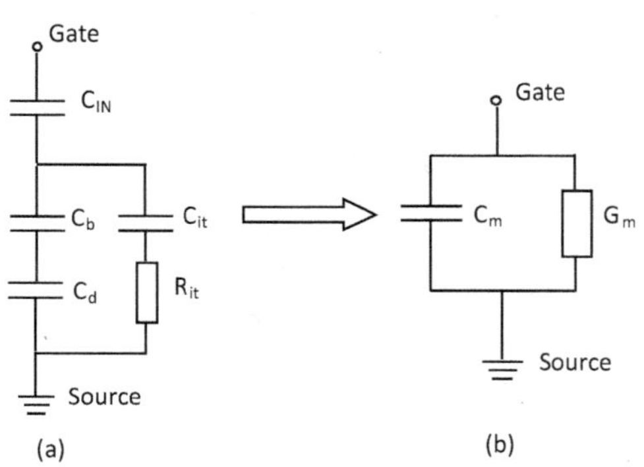

Figure 6.9 Simplified equivalent circuit of a MIS diode for interface-state density calculation by the hi-lo frequency method.

Another widely used and reliable method to measure the interface trap densities is the AC conductance technique [152]. To further simplify the analysis, the C_{IN} in the circuit of Fig. 6.10a can be subtracted and the part inside the dashed frame can be de-embedded. The C_b and C_d in series can be combined as one element C_d'. After Thevenin transformation, the circuit of Fig. 6.10b can be described as Fig. 6.10c, that is, a conductance G_p in parallel with a capacitance C_p. The frequency-dependent values of G_p can be used to calculate the interface trap densities.

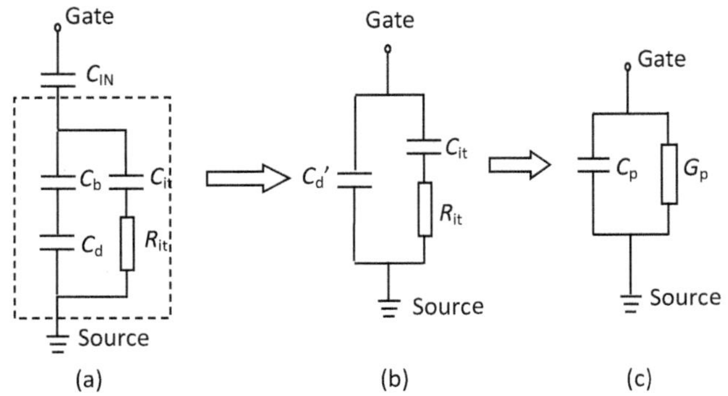

Figure 6.10 Simplified equivalent circuit for a MIS diode.

If the traps have a single energy level, the value of G_p/ω can be expressed as

$$\frac{G_p}{\omega} = \frac{q^2 N_{it}}{kT}\left(\frac{\omega\tau}{1+\omega^2\tau^2}\right), \tag{6.11}$$

where N_{it} is the trap density, k is the Boltzmann constant, and ω is the angular frequency. $\frac{G_p}{\omega}$ reaches its peak at $\omega\tau = 1$:

$$\left(\frac{G_p}{\omega}\right)_{\max} = \frac{q^2 N_{it}}{kT} \tag{6.12}$$

If the traps have an energy level continuum, G_p/ω can be expressed as

$$\frac{G_p}{\omega} = qD_{it}\left(\frac{\ln(1+\omega^2\tau^2)}{2\omega\tau}\right). \tag{6.13}$$

It will reach its peak at $\omega\tau = 1.95$:

$$\left(\frac{G_p}{\omega}\right)_{max} = 0.403qD_{it} \tag{6.14}$$

If the MIS diode has a large gate area and the fluctuation of the surface potential (u_s) is needed to be considered, the statistical model is needed to be adopted under the assumption of random fluctuation for u_s:

$$\frac{G_p}{\omega} = \frac{qD_{it}}{2}(2\pi\sigma_s^2)^{-\frac{1}{2}}\int_{-\infty}^{\infty}\exp\left(-\frac{(u_s-\overline{u_s})^2}{2\sigma_s^2}\right)\cdot\frac{1}{\omega\tau}\cdot\ln(1+\omega^2\tau^2)du_s, \tag{6.15}$$

where $\overline{u_s}$ and σ are, respectively, the mean value and standard deviation of u_s in units of kT/q, that is, $u_s = \frac{\Psi_s}{kT/q}$ and $\overline{u_s} = \frac{\overline{\Psi_s}}{kT/q}$. The values of D_{it} and τ can be obtained by fitting the curve of $G_p/\omega \sim \omega$ using the above equation.

Figure 6.11 shows the measured interface trap density using different techniques for an ALD Al_2O_3/AlGaN/GaN MIS structure, where Fig. 6.11a shows the results measured by the hi-lo frequency method and Fig. 6.11b shows those measured by the AC conductance method. E_T is the trap energy level from the conduction band bottom and is calculated by

$$E_T = -k_BT\ln(\tau\sigma_T N_C v_t), \tag{6.16}$$

where k_B is the Boltzmann constant, T is the temperature and here T = 300 K, σ_T is the capture cross section of the traps, N_C is the effective density of states at the bottom of the conduction band, and v_t is the average thermal velocity of the electrons. It is assumed that $\sigma_T = 3.4 \times 10^{-15}$ cm^2, $N_C = 4.3 \times 10^{14} \times T^{3/2}$ cm^{-3}, and $v_t = 2.6 \times 10^7$ cm/s [153].

As shown in Fig. 6.11b, regarding the AC conductance method with different models, the results of the multilevel model and the single-level model are consistent with those of the statistical model, indicating that the surface potential fluctuation can be ignored in the fabricated MIS diode even with a large gate area up to 300 μm diameter. This reveals the good uniformity of an ALD-deposited thin dielectric film. On the other hand, the trap density measured by the hi-lo method is lower than that from the AC conductance, which is believed to be from the inaccuracy of the hi-lo frequency method. This is partially due to the slight difference in the values of the threshold voltage at the high-frequency and low-frequency measurements [154]. However, it turned out that the magnitudes of the interface-state density measured by two methods are similar, indicating that the hi-lo frequency method is still valuable for a quick and rough estimation of the dielectric interface quality.

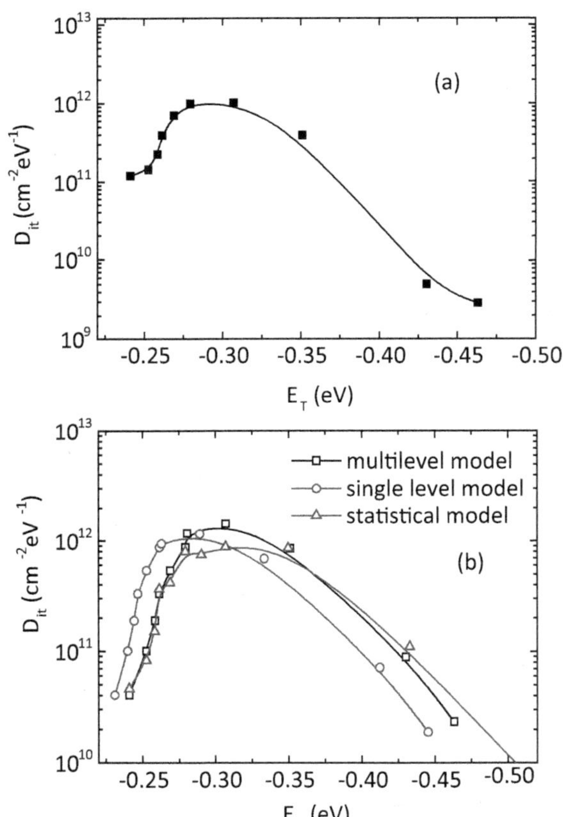

Figure 6.11 Measured interface trap density using different techniques for an ALD Al_2O_3/AlGaN/GaN MIS structure: (a) by the hi-lo frequency method and (b) by the AC conductance method.

6.3.4 Other Approaches to Measure the the Interface Trap Density

It is worth noting that the quantity of the trap density at the interface between the gate insulator and the nitride is inevitably underestimated using the above-mentioned *C–V* approaches. This is because due to the wide bandgaps of the GaN-based materials (AlGaN or InAlN), lots of traps have deep energy levels (>0.5 eV) and thus long time constants (>mS), which are beyond the traditional *C–V* measurement scope. Other methods are available to extract the interface trap density, including hysteresis, subthreshold swing (SS), AC transconductance, and photoassisted *C–V*.

6.3.4.1 Hysteresis method

Hysteresis characteristics are a widely used method to quickly check the interface quality of a MIS structure. Due to the time needed for the traps to release the captured carriers, there is some discrepancy between the threshold voltages (ΔV_{th}) extracted for the gate bias swept from a negative value lower than the threshold voltage to a positive value and the one for the gate bias swept back to the negative value. The interface trap density N_{it} can be estimated as [155]

$$N_{it} \approx \Delta V_{th} \times C_{ox}/q, \tag{6.17}$$

where C_{ox} is the capacitance of the gate insulator. However, this method will also underestimate the interface trap density value, as only the border traps within a certain energy range similar to the *C–V* measurement will respond during the hysteresis measurement.

6.3.4.2 Subthreshold swing method

If a GaN metal-insulator-semiconductor field-effect transistor (MISFET) structure is considered, its SS can be expressed as

$$\mathrm{SS} = \frac{kT}{q} \ln 10 \left(1 + \frac{C_d + C_{it}}{C_{ox}} \right), \tag{6.18}$$

where C_{ox} is the insulator capacitance in the MIS structure and C_d is the depletion capacitance of GaN. Combining Eq. 6.18 and Eq. 6.10, D_{it} can be estimated by [156]

$$D_{\text{it}} < qC_{\text{ox}}\left(\frac{q \cdot SS}{kT\ln 10} - 1\right). \tag{6.19}$$

6.3.4.3 AC transconductance method

A group from Yale University has proposed a method to measure the gate oxide traps in a GaN MIS-HEMT using an AC transconductance method [157]. This method is for a three-terminal HEMT device. The density of the oxide traps $N_t(E_T, x)$ as a function of the trap level E_T and the location in the oxide x from top can be extracted from the measured AC transconductance G_m:

$$N_t(E_T, x) = \frac{dG_m}{d\ln\omega}\left[\left(\frac{x}{t_{\text{ox}}}\right)^2 q\lambda V_{\text{ds}}\mu_{\text{dc}}\frac{W}{L}\right]^{-1}, \tag{6.20}$$

where ω is the angular frequency during the measurement, t_{ox} is the gate oxide thickness, q is the electron charge, μ_{dc} is the DC mobility, and W and L are the gate width and gate length, respectively, of the device. λ is the decay constant of the tunneling and can be expressed as

$$\lambda = \frac{\hbar}{\sqrt{8m^*(E_{\text{C, ox}} - E_{\text{F}})}}, \tag{6.21}$$

where $\hbar$ is the reduced Planck's constant, m^* is the effective mass in the gate oxide, $E_{\text{C,ox}}$ is the conduction band edge energy of the gate oxide, and E_F is the gate metal Fermi level. The location x can be related to the measurement frequency ω on the basis of the following equation:

$$x = \lambda\ln\left(\frac{n_0 v_T \sigma}{\omega}\right), \tag{6.22}$$

where n_0 is the carrier density in reservoirs, v_T is the thermal velocity, and σ is the capture cross section.

The corresponding trap level E_T can be calculated by

$$E_T = E_F + (V_g - V_{\text{th}})\frac{x}{t_{\text{ox}}}, \tag{6.23}$$

where V_g is the gate bias and V_{th} is the threshold voltage of the device.

6.3.4.4 Photoassisted *C–V*

Limited by the emission time of the captured electrons in the traps, the traditional *C–V* measurement, either the hi-lo method or the AC conductance method, can only characterize the traps in GaN MIS-HEMTs with a very narrow energy range (typically 0.2–0.5 eV below the conduction band of GaN at room temperature). To increase the trap level range that can be characterized by *C–V* measurement, one of two approaches can be utilized: to increase the measurement temperature [158] or to use photoillustration [159]. The time constant is dependent on the measurement temperature following the rewritten form of Eq. 6.16, as below:

$$\tau = \frac{1}{N_{\mathrm{C}} v_{\mathrm{T}} \sigma} \exp\left(\frac{E_{\mathrm{T}}}{kT}\right) \tag{6.24}$$

By a *C–V* measurement at 300°C, trap levels up to 1.0 eV can be characterized. Another quick and effective method is to use photoassisted *C–V*, as shown in Fig. 6.12 [159]. At first a *C–V* measurement at high frequency (such as 1 MHz) is carried out with the gate bias swept from a positive voltage (accumulation region) to a negative bias (depletion region); then the gate bias is kept at this negative voltage and the sample is exposed to UV illustration for some time, for example, 10 min. After the UV light is turned off, the gate bias is swept back to the positive voltage, that is, the accumulation region with capacitance being measured. At the first scan, a lot of electrons are captured at the deep-level states at the interface between the gate oxide and the nitride and cannot be released at the end of the sweeping as the time constants are too long at room temperature. Due to the extremely long generation time of the minority carriers in GaN, an inversion layer will not be formed. After UV light illustrates the sample for some time, holes are generated in the depletion region and the depletion region is shrunk; thus the measured gate capacitance increases. And these holes are captured by the interface traps, which are then positively charged. After the UV light is turned off, the remeasured *C–V* curve will have a parallel shift compared to the first *C–V* curve. This shift corresponds to the trapped holes so that the trap density can be expressed as

$$N_{\mathrm{it}} = \Delta V_{\mathrm{th}} \cdot C_{\mathrm{ox}}/q. \tag{6.25}$$

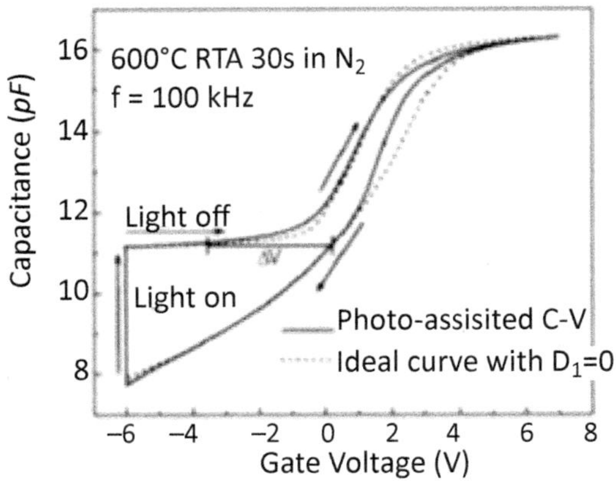

Figure 6.12 Photoassisted *C–V* measured for a GaN MOS capacitor. Reproduced from Ref. [159], with the permission of AIP Publishing.

This method can measure the deep trap levels up to 2 eV below the conduction band of GaN.

6.4 Summary

Surface states are critical because they are not only related with the 2DEG formation but also related with the gate leakage current and current collapse. Therefore, surface states' passivation becomes important in device fabrication. Using a dielectric to passivate the surface states improves the device's performance. On the basis of this, metal-insulator-semiconductor heterojunction field-effect transistors MIS-HFETs are proposed to optimize the high-power devices' performance. To measure the interface states, *C–V* measurement and other approaches are elaborated.

References

1. Ibbetson, J. P., Fini, P. T., Ness, K. D., DenBaars, S. P., Speck, J. S., and Mishra, U. K. (2000). Polarization effects, surface states, and the source of electrons in AlGaN/GaN heterostructure field effect transistors, *Appl. Phys. Lett.*, **77**, 250–252.
2. Ambacher, O., Smart, J., Shealy, J. R., Weimann, N. G., Chu, K., Murphy, M., Schaff, W. J., Eastman, L. F., Dimitrov, R., Wittmer, L., Stutzmann, M.,

Rieger, W., and Hilsenbeck, J. (1999). Two-dimensional electron gases induced by spontaneous and piezoelectric polarization charges in N- and Ga-face AlGaN/GaN heterostructures, *J. Appl. Phys.*, **85**, 3222–3233.

3. Xu, D., Chu, K., Diaz, J., Zhu, W., Roy, R., Pleasant, L. M., Nichols, K., Chao, P. C., Xu, M., and Ye, P. D. (2013). 0.2-μm AlGaN/GaN high electron-mobility transistors with atomic layer deposition Al_2O_3 passivation, *IEEE Electron Device Lett.*, **34**, 744–746.
4. Oyama, S., Hashizume, T., and Hasegawa, H. (2002). Mechanism of current leakage through metal/n-GaN interfaces, *Appl. Surf. Sci.*, **190**, 322–325.
5. Khan, M. A., Shur, M. S., Chen, Q. C., and Kuznia, J. N. (1994). Current/voltage characteristic collapse in AlGaN/GaN heterostructure insulated gate field effect transistors at high drain bias, *Electron. Lett.*, **30**, 2175–2176.
6. Dora, Y., Han, S., Klenov, D., Hansen, P. J., No, K., Mishra, U. K., Stemmer, S., and Speck, J. S. (2006). ZrO_2 gate dielectrics produced by ultraviolet ozone oxidation for GaN and AlGaN/GaN transistors, *J. Vac. Sci. Technol. B*, **24**, 575–581.
7. Vetury, R., Zhang, N. Q., Keller, S., and Mishra, U. K. (2001). The impact of surface states on the DC and RF characteristics of AlGaN/GaN HFETs, *IEEE Trans. Electron Devices*, **48**, 560–566.
8. Miao, M. S., Weber, J. R., and Van de Walle, C. G. (2010). Oxidation and the origin of the two-dimensional electron gas in AlGaN/GaN heterostructures, *J. Appl. Phys.*, **107**, 123713–123711.
9. Hao, Y., Yang, L., Ma, X., Ma, J. G., Cao, M., Pan, C., Wang, C., and Zhang, J. (2011). High-performance microwave gate-recessed AlGaN/AlN/GaN MOS-HEMT with 73% power-added efficiency, *IEEE Electron Device Lett.*, **32**, 626–628.
10. Maeda, N., Hiroki, M., Watanabe, N., Oda, Y., Yokoyama, H., Yagi, T., Makimoto, T., Enoki, T., and Kobayashi, T. (2007). Systematic study of insulator deposition effect (Si_3N_4, SiO_2, AlN, and Al_2O_3) on electrical properties in AlGaN/GaN heterostructures, *Jpn. J. Appl. Phys., Part 1*, **46**, 547–554.
11. Liu, C., Chor, E. F., and Tan, L. S. (2006). Investigations of HfO_2/AlGaN/GaN metal-oxide-semiconductor high electron mobility transistors, *Appl. Phys. Lett.*, **88**, 173504–173503.
12. Johnson, J. W., Luo, B., Ren, F., Gila, B. P., Krishnamoorthy, W., Abernathy, C. R., Pearton, S. J., Chyi, J. I., Nee, T. E., Lee, C. M., and Chuo, C. C. (2000).

Gd2O3/GaN metal-oxide-semiconductor field-effect transistor, *Appl. Phys. Lett.*, **77**, 3230–3232.

13. Tsai, C. Y., Wu, T. L., and Chin, A. (2012). High-performance GaN MOSFET with high-k $LaAlO_3/SiO_2$ gate dielectric, *IEEE Electron Device Lett.*, **33**, 35–37.
14. Lee, J.-H., Park, C., Kim, K.-W., Kim, D.-S., and Lee, J.-H. (2013). Performance of fully recessed AlGaN/GaN MOSFET prepared on GaN buffer layer grown with AlSiC precoverage on silicon substrate, *IEEE Electron Device Lett.*, **34**, 975–977.
15. Eller, B. S., Yang, J., and Nemanich, R. J. (2013). Electronic surface and dielectric interface states on GaN and AlGaN, *J. Vac. Sci. Technol. A*, **31**, 050807-050801–050807-050829.
16. Higashiwaki, M., Onojima, N., Matsui, T., and Mimura, T. (2006). Effects of SiN passivation by catalytic chemical vapor deposition on electrical properties of AlGaN/GaN heterostructure field-effect transistors, *J. Appl. Phys.*, **100**, 033714-033714–033716.
17. Khan, M. Asif, Hu, X., Tarakji, A., Simin, G., Yang, J., Gaska, R., and Shur, M. S. (2000). AlGaN/GaN metal-oxide-semiconductor heterostructure field-effect transistors on SiC substrates, *Appl. Phys. Lett.*, **77**, 1339–1341.
18. Cook, T. E., Fulton, C. C., Mecouch, W. J., Davis, R. F., Lucovsky, G., and Nemanich, R. J. (2003). Band offset measurements of the GaN (0001)/ HfO_2 interface, *J. Appl. Phys.*, **94**, 7155–7158.
19. Liu, Z. H., Ng, G. I., Arulkumaran, S., Maung, Y. K. T., Teo, K. L., Foo, S. C., and Sahmuganathan, V. (2009). Improved two-dimensional electron gas transport characteristics in AlGaN/GaN metal-insulator-semiconductor high electron mobility transistor with atomic layer-deposited Al_2O_3 as gate insulator, *Appl. Phys. Lett.*, **95**, 223501–223503.
20. Palacios, T., Rajan, S., Chakraborty, A., Heikman, S., Keller, S., DenBaars, S. P., and Mishra, U. K. (2005). Influence of the dynamic access resistance in the g_m and f_T linearity of AlGaN/GaN HEMTs, *IEEE Trans. Electron Devices*, **52**, 2117–2123.
21. Binari, S. C., Ikossi, K., Roussos, J. A., Kruppa, W., Doewon, P., Dietrich, H. B., Koleske, D. D., Wickenden, A. E., and Henry, R. L. (2001). Trapping effects and microwave power performance in AlGaN/GaN HEMTs, *IEEE Trans. Electron Devices*, **48**, 465–471.
22. Green, B. M., Chu, K. K., Chumbes, E. M., Smart, J. A., Shealy, J. R., and Eastman, L. F. (2000). The effect of surface passivation on the microwave characteristics of undoped AlGaN/GaN HEMTs, *IEEE Electron Device Lett.*, **21**, 268–270.

23. Arulkumaran, S., Ng, G. I., and Liu, Z. H. (2007). Effect of gate-source and gate-drain Si_3N_4 passivation on current collapse in AlGaN/GaN high-electron-mobility transistors on silicon, *Appl. Phys. Lett.*, **90**, 173504-173501–173504-173503.

24. Binari, S. C., Klein, P. B., and Kazior, T. E. (2002). Trapping effects in GaN and SiC microwave FETs, *Proc. IEEE*, **90**, 1048–1058.

25. Harrison, I., Clayton, N. W., and Jeffs, N. J. (2001). High temperature RF characterisation of SiN passivated and unpassivated AlGaN/GaN HFETs, *Phys. Status Solidi A*, **188**, 275–278.

26. Lu, W., Kumar, V., Schwindt, R., Piner, E., and Adesida, I. (2002). A comparative study of surface passivation on AlGaN/GaN HEMTs, *Solid-State Electron.*, **46**, 1441–1444.

27. Lee, D. S., Chung, J. W., Wang, H., Gao, X., Guo, S., Fay, P., and Palacios, T. (2011). 245-GHz InAlN/GaN HEMTs with oxygen plasma treatment, *IEEE Electron Device Lett.*, **32**, 755–757.

28. Wang, R., Li, G., Laboutin, O., Cao, Y., Johnson, W., Snider, G., Fay, P., Debdeep, J., and Xing, H. (2011). 210-GHz InAlN/GaN HEMTs with dielectric-free passivation, *IEEE Electron Device Lett.*, **32**, 892–894.

29. Liu, H. Y., Lee, C. S., Hsu, W. C., Tseng, L. Y., Chou, B. Y., Ho, C. S., and Wu, C. L. (2013). Investigations of AlGaN/AlN/GaN MOS-HEMTs on Si substrate by ozone water oxidation method, *IEEE Trans. Electron Devices*, **60**, 2231–2237.

30. Liu, H. Y., Chou, B. Y., Hsu, W. C., Lee, C. S., Sheu, J. K., and Ho, C. S. (2013). Enhanced AlGaN/GaN MOS-HEMT performance by using hydrogen peroxide oxidation technique, *IEEE Trans. Electron Devices*, **60**, 213–220.

31. Liu, X., Kim Fong Low, E., Pan, J., Liu, W., Leong Teo, K., Tan, L.-S., and Yeo, Y.-C. (2011). Impact of In situ vacuum anneal and SiH4 treatment on electrical characteristics of AlGaN/GaN metal-oxide-semiconductor high-electron mobility transistors, *Appl. Phys. Lett.*, **99**, 093504.

32. Dong, Y., Feenstra, R. M., and Northrup, J. E. (2006). Electronic states of oxidized GaN(0001) surfaces, *Appl. Phys. Lett.*, **89**, 171920–171923.

33. Bae, C., and Lucovsky, G. (2004). Low-temperature preparation of GaN-SiO_2 interfaces with low defect density. I. Two-step remote plasma-assisted oxidation-deposition process, *J. Vac. Sci. Technol. A*, **22**, 2402.

34. Therrien, R., Lucovsky, G., and Davis, R. (2000). Charge redistribution at GaN–Ga_2O_3 interfaces: a microscopic mechanism for low defect density interfaces in remote-plasma-processed MOS devices prepared on polar GaN faces, *Appl. Surf. Sci.*, **166**, 513–519.

35. Moser, N. A., Gillespie, J. K., Via, G. D., Crespo, A., Yannuzzi, M. J., Jessen, G. H., Fitch, R. C., Luo, B., Ren, F., Gila, B. P., Onstine, A. H., Abernathy, C. R., and Pearton, S. J. (2003). Effects of surface treatments on isolation currents in AlGaN/GaN high-electron-mobility transistors, *Appl. Phys. Lett.*, **83**, 4178–4180.

36. Lee, K. B., Green, R. T., Houston, P. A., Tan, W. S., Uren, M. J., Wallis, D. J., and Martin, T. (2010). Bi-layer SixNy passivation on AlGaN/GaN HEMTs to suppress current collapse and improve breakdown, *Semicond. Sci. Technol.*, **25**, 1244–1247.

37. Arulkumaran, S., Egawa, T., Ishikawa, H., Jimbo, T., and Sano, Y. (2004). Surface passivation effects on AlGaN/GaN high-electron-mobility transistors with SiO_2, Si_3N_4, and silicon oxynitride, *Appl. Phys. Lett.*, **84**, 613–615.

38. Chen, K. J., and Huang, S. (2013). AlN passivation by plasma-enhanced atomic layer deposition for GaN-based power switches and power amplifiers, *Semicond. Sci. Technol.*, **28**, 074015-074011–074015-074018.

39. Uemoto, Y., Shibata, D., Yanagihara, M., Ishida, H., Matsuo, H., Nagai, S., Batta, N., Li, M., Ueda, T., Tanaka, T., and Ueda, D. (2007). 8300V blocking voltage AlGaN/GaN power HFET with thick poly-AlN passivation, *Int. Electron Device Meeting*, 861–864.

40. Kumar, V., Selvanathan, D., Kuliev, A., Kim, S., Flynn, J., and Adesida, I. (2003). Characterisation of iridium Schottky contacts on n-$Al_xGa_{1-x}N$, *Electron. Lett.*, **39**, 747–748.

41. Seok, J. H., Choi, S., Lim, H. K., Lee, S. H., Kim, I., and Ham, B. J. (2013). Effect of the COMT val158met polymorphism on white matter connectivity in patients with major depressive disorder, *Neurosci. Lett.*, **545**, 35–39.

42. Ha, M.-W., Lee, S.-C., Her, J.-C., Seo, K.-S., and Han, M.-K. (2006). New inductively coupled plasma-chemical vapor deposition SiO_2 passivation for high-voltage switching AlGaN/GaN heterostructure field-effect transistors, *Jpn. J. Appl. Phys.*, **45**, 3391–3394.

43. Hanawa, H., Onodera, H., Nakajima, A., and Horio, K. (2014). Numerical analysis of breakdown voltage enhancement in AlGaN/GaN HEMTs with a high-k passivation layer, *IEEE Trans. Electron Devices*, **61**, 769–775.

44. Menozzi, R., Cova, P., Canali, C., and Fantini, F. (1996). Breakdown walkout in pseudomorphic HEMT's, *IEEE Trans. Electron Devices*, **43**, 543–546.

45. Harrison, I., Clayton, N. W., and Jeffs, N. J. (2001). High temperature RF characterisation of SiN passivated and unpassivated AlGaN/GaN HFETs, *Phys. Status Solidi A*, **188**, 275–278.

46. Moser, N., Fitch, R. C., Crespo, A., Gillespie, J. K., Jessen, G. H., Via, G. D., Luo, B., Ren, F., Gila, B. P., Abernathy, C. R., and Pearton, S. J. (2004). Dramatic improvements in AlGaN/GaN HEMT device isolation characteristics after UV-ozone pretreatment, *J. Electrochem. Soc.*, **151**, G915–G918.

47. Khan, M. A., Hu, X., Sumin, G., Lunev, A., Yang, J., Gaska, R., and Shur, M. S. (2000). AlGaN/GaN metal oxide semiconductor heterostructure field effect transistor, *IEEE Electron Device Lett.*, **21**, 63–65.

48. Kambayashi, H., Satoh, Y., Ootomo, S., Kokawa, T., Nomura, T., Kato, S., and Chow, T. P. (2010). Over 100A operation normally-off AlGaN/GaN hybrid MOS-HFET on Si substrate with high-breakdown voltage, *Solid-State Electron.*, **54**, 660–664.

49. Hu, C.-C., Lin, M.-S., Wu, T.-Y., Adriyanto, F., Sze, P.-W., Wu, C.-L., and Wang, Y.-H. (2012). AlGaN/GaN metal–oxide–semiconductor high-electron mobility transistor with liquid-phase-deposited barium-doped as a gate dielectric, *IEEE Trans. Electron Devices*, **59**, 121–127.

50. Lee, H.-S., Piedra, D., Sun, M., Gao, X., Guo, S., and Palacios, T. (2012). 3000-V 4.3-m$\Omega\cdot$cm^2 InAlN/GaN MOSHEMTs with AlGaN back barrier, *IEEE Electron Device Lett.*, **33**, 982–984.

51. Hu, X., Koudymov, A., Simin, G., Yang, J., Khan, M. A., Tarakji, A., Shur, M. S., and Gaska, R. (2001). Si_3N_4/AlGaN/GaN–metal–insulator–semiconductor heterostructure field–effect transistors, *Appl. Phys. Lett.*, **79**, 2832–2834.

52. Ikeda, N., Kaya, S., Li, J., Kokawa, T., Masuda, M., and Katoh, S. (2009). High-power AlGaN/GaN MIS-HFETs with field-plates on Si substrates, in *21st International Symposium on Power Semiconductor Devices & IC's*, Barcelona, pp. 251–254.

53. Moens, P., Liu, C., Banerjee, A., Vanmeerbeek, P., Coppens, P., Ziad, H., Constant, A., Li, Z., De Vleeschouwer, H., Roig-Guitart, J., Gassot, P., Bauwens, F., De Backer, E., Padmanabhan, B., Salih, A., Parsey, J., and Tack, M. (2014). An industrial process for 650V rated GaN-on-Si power devices using in-situ SiN as a gate dielectric, in *IEEE 26th International Symposium on Power Semiconductor Devices & IC's (ISPSD)*, Waikoloa, HI, pp. 374–377.

54. Oka, T., and Nozawa, T. (2008). AlGaN/GaN recessed MIS-Gate HFET with high-threshold-voltage normally-off operation for power electronics applications, *IEEE Electron Device Lett.*, **29**, 668–670.

55. Van Hove, M., Boulay, S., Bahl, S. R., Stoffels, S., Xuanwu, K., Wellekens, D., Geens, K., Delabie, A., and Decoutere, S. (2012). CMOS process-compatible high-power low-leakage AlGaN/GaN MISHEMT on silicon, *IEEE Electron Device Lett.*, **33**, 667–669.

56. Freedsman, J. J., Kubo, T., and Egawa, T. (2013). High drain current density E-mode Al_2O_3/AlGaN/GaN MOS-HEMT on Si with enhanced power device figure-of-merit (4 x 10(8) V-2 Omega(-1)cm(-2)), *IEEE Trans. Electron Devices*, **60**, 3079–3083.

57. Shi, J., Eastman, L. F., Xin, X., and Pophristic, M. (2009). High performance AlGaN/GaN power switch with HfO_2 insulation, *Appl. Phys. Lett.*, **95**, 042103-042101-042103.

58. Seok, O., Ahn, W., Han, M. K., and Ha, M. W. (2013). High-breakdown voltage and low onresistance AlGaN/GaN on Si MOS-HEMTS employing an extended tan gate on HfO_2 gate insulator, *Electron. Lett.*, **49**, 425–427.

59. Huang, L.-H., Yeh, S.-H., Lee, C.-T., Tang, H., Bardwell, J., and Webb, J. B. (2008). AlGaN/GaN metal-oxide-semiconductor high-electron mobility transistors using oxide insulator grown by photoelectrochemical oxidation method, *IEEE Electron Device Lett.*, **29**, 284–286.

60. Alomari, M., Medjdoub, F., Carlin, J. F., Feltin, E., Grandjean, N., Chuvilin, A., Kaiser, U., Gaquiere, C., and Kohn, E. (2009). InAlN/GaN MOSHEMT with self-aligned thermally generated oxide recess, *IEEE Electron Device Lett.*, **30**, 1131–1133.

61. Luo, B., Mehandru, R., Kim, Jihyun, Ren, F., Gila, B. P., Onstine, A. H., Abernathy, C. R., Pearton, S. J., Gotthold, D., Birkhahn, R., Peres, B., Fitch, R. C., Moser, N., Gillespie, J. K., Jenkins, T., Sewell, J., Via, D., and Crespo, A. (2003). High three-terminal breakdown voltage and output power of Sc_2O_3 passivated AlGaN/GaN high electron mobility transistors, *Electron. Lett.*, **39**, 809–810.

62. Kim, J., Mehandru, R., Luo, B., Ren, F., Gila, B. P., Onstine, A. H., Abernathy, C. R., Pearton, S. J., and Irokawa, Y. (2002). Charge pumping in Sc_2O_3/GaN gated mos diodes, *Electron. Lett.*, **38**, 920.

63. Wu, T.-Y., Lin, S.-K., Sze, P.-W., Huang, J.-J., Chien, W.-C., Hu, C.-C., Tsai, M.-J., and Wang, Y.-H. (2009). AlGaN/GaN MOSHEMTs with liquid-phase-deposited TiO_2 as gate dielectric, *IEEE Trans. Electron Devices*, **56**, 2911–2916.

64. Hu, C. C., Lin, M. S., Wu, T. Y., Adriyanto, F., Sze, P. W., Wu, C. L., and Wang, Y. H. (2012). AlGaN/GaN metal-oxide-semiconductor high-electron mobility transistor with liquid-phase-deposited barium-doped TiO_2 as a gate dielectric, *IEEE Trans. Electron Devices*, **59**, 121–127.

65. Lee, C.-T., Chiou, Y.-L., and Lee, C.-S. (2010). AlGaN/GaN MOS-HEMTs with gate ZnO dielectric layer, *IEEE Electron Device Lett.*, **31**, 1220–1223.

66. Kaneko, N., Machida, O., Yanagihara, M., Iwakami, S., Baba, R., Goto, H., and Iwabuchi, A. (2009). Normally-off AlGaN/GaN HFETs using NiO_x gate with recess, in *21st International Symposium on Power Semiconductor Devices & IC's*, Barcelona, pp. 25–28.

67. Oh, C. S., Youn, C. J., Yang, G. M., Lim, K. Y., and Yang, J. W. (2004). AlGaN/GaN metal-oxide-semiconductor heterostructure field-effect transistor with oxidized Ni as a gate insulator, *Appl. Phys. Lett.*, **85**, 4214–4216.

68. Kuzmik, J., Pozzovivo, G., Abermann, S., Carlin, J.-F., Gonschorek, M., Feltin, E., Grandjean, N., Bertagnolli, E., Strasser, G., and Pogany, D. (2008). Technology and performance of InAlN/AlN/GaN HEMTs with gate insulation and current collapse suppression using ZrO_2 or HfO_2, *IEEE Trans. Electron Devices*, **55**, 937–941.

69. Hatano, M., Taniguchi, Y., Kodama, S., Tokuda, H., and Kuzuhara, M. (2014). Reduced gate leakage and high thermal stability of AlGaN/GaN MIS-HEMTs using ZrO_2/Al_2O_3 gate dielectric stack, *Appl. Phys. Express*, **7**, 044101.

70. Kanamura, M., Ohki, T., Imanishi, K., Makiyama, K., Okamoto, N., Kikkawa, T., Hara, N., and Joshin, K. (2008). High power and high gain AlGaN/GaN MIS-HEMTs with high-k dielectric layer, *Phys. Status Solidi C*, **5**, 2037–2040.

71. Chu, R., Hughes, B., Chen, M., Brown, D., Li, R., Khalil, S., Zehnder, D., Chen, S., Williams, A., Garrido, A., Musni, M., and Boutros, K. (2013). Normally-off GaN-on-Si transistors enabling nanosecond power switching at one kilowatt, in *Proceedings of the Device Research Conference (DRC), 2013 71st Annual*, 199–200.

72. Robertson, J., and Falabretti, B. (2006). Band offsets of high K gate oxides on III-V semiconductors, *J. Appl. Phys.*, **100**, 014111–014111.

73. Yang, J., Eller, B. S., Zhu, C., England, C., and Nemanich, R. J. (2012). Comparative band alignment of plasma-enhanced atomic layer deposited high-k dielectrics on gallium nitride, *J. Appl. Phys.*, **112**, 053710.

74. Coan, M. R., Woo, J. H., Johnson, D., Gatabi, I. R., and Harris, H. R. (2012). Band offset measurements of the GaN/dielectric interfaces, *J. Appl. Phys.*, **112**, 024508–024506.

75. Suri, R. (2010). Investigation of MOS *Interfaces with Atomic-Layer-Deposited High-k Gate Dielectrics on III-V Semiconductors*, PhD thesis, North Carolina State University.

76. Esposto, M., Krishnamoorthy, S., Nath, D. N., Bajaj, S., Hung, T.-H., and Rajan, S. (2011). Electrical properties of atomic layer deposited aluminum oxide on gallium nitride, *Appl. Phys. Lett.*, **99**, 133503–133503.

77. Hashizume, T., Ootomo, S., Inagaki, T., and Hasegawa, H. (2003). Surface passivation of GaN and GaN/AlGaN heterostructures by dielectric films and its application to insulated-gate heterostructure transistors, *J. Vac. Sci. Technol. B*, **21**, 1828–1838.

78. Mizue, C., Hori, Y., Miczek, M., and Hashizume, T. (2011). Capacitance–voltage characteristics of Al_2O_3/AlGaN/GaN structures and state density distribution at Al_2O_3/AlGaN interface, *Jpn. J. Appl. Phys.*, **50**, 021001-021001–021001-021007.

79. Qin, X., Cheng, L., McDonnell, S., Azcatl, A., Zhu, H., Kim, J., and Wallace, R. M. (2015). A comparative study of atomic layer deposition of Al_2O_3 and HfO_2 on AlGaN/GaN, *J. Mater. Sci.: Mater. Electron.*, **26**, 4638–4643.

80. Martin, G., Strite, S., Botchkarev, A., Agarwal, A., Rockett, A., Morkoç, H., Lambrecht, W. R. L., and Segall, B. (1994). Valence-band discontinuity between GaN and AlN measured by x-ray photoemission spectroscopy, *Appl. Phys. Lett.*, **65**, 610–612.

81. Waldrop, J. R. and Grant, R. W. (1996). Measurement of AlN/GaN (0001) heterojunction band offsets by x-ray photoemission spectroscopy, *Appl. Phys. Lett.*, **68**, 2879–2881.

82. Craft, H. S., Collazo, R., Losego, M. D., Mita, S., Sitar, Z., and Maria, J.-P. (2008). Spectroscopic analysis of the epitaxial CaO (111)–GaN (0002) interface, *Appl. Phys. Lett.*, **92**, 082907.

83. Wei, W., Qin, Z., Fan, S., Li, Z., Kai, S., Zhu, Q., and Zhang, G. (2012). Valence band offset of β-Ga_2O_3 /wurtzite GaN heterostructure measured by X-ray photoelectron spectroscopy, *Nanoscale Res. Lett.*, **7**, 1–5.

84. Lay, T. S., Huang, K. H., Hung, W. H., Hong, M., Kwo, J., and Mannaerts, J. P. (2001). Probing the microscopic compositions at Ga_2O_3 (Gd_2O_3)/GaAs interface by core level photoelectron spectroscopy, *Solid-State Electron.*, **45**, 423–426.

85. Craft, H. S. (2009). *Spectroscopy of Oxide-GaN Interfaces*, PhD dissertation, University of North Carolina.

86. Wheeler, V. D. (2010). *Structure and Properties of Epitaxial Dielectrics on GaN*, PhD dissertation, North Carolina State University.

87. Gao, L. G., Xu, B., Guo, H. X., Xia, Y. D., Yin, J., and Liu, Z. G. (2009). Band alignments and improved leakage properties of (La2O3)0.5(SiO2)0.5/SiO2/GaN stacks for high-temperature metal-oxide-semiconductor field-effect transistor applications, *Appl. Phys. Lett.*, **94**, 252901.

88. Craft, H. S., Collazo, R., Losego, M. D., Mita, S., Sitar, Z., and Maria, J.-P. (2007). Band offsets and growth mode of molecular beam epitaxy grown MgO (111) on GaN (0002) by x-ray photoelectron spectroscopy, *J. Appl. Phys.*, **102**, 074104.

89. Chen, J.-J., Gila, B. P., Hlad, M., Gerger, A., Ren, F., Abernathy, C. R., and Pearton, S. J. (2006). Determination of MgO/GaN heterojunction band offsets by x-ray photoelectron spectroscopy, *Appl. Phys. Lett.*, **88**, 042113.

90. Wu, C.-L., Shen, C.-H., and Gwo, S. (2006). Valence band offset of wurtzite InN/AlN heterojunction determined by photoelectron spectroscopy, *Appl. Phys. Lett.*, **88**, 032105.

91. J. J. Chen, M. Hlad, A. P. Gerger, B. P. Gila, F. Ren, C. R. Abernathy, and Pearton, S. J. (2007). Band offsets in the Mg0.5Ca0.5O/GaN heterostructure system, *J. Electron. Mater.*, **36**, 368–372.

92. Chen, J.-J., Gila, B. P., Hlad, M., Gerger, A., Ren, F., Abernathy, C. R., and Pearton, S. J. (2006). Band offsets in the Sc_2O_3/GaN heterojunction system, *Appl. Phys. Lett.*, **88**, 142115.

93. Cook, T. E., Fulton, C. C., Mecouch, W. J., Tracy, K. M., Davis, R. F., Hurt, E. H., Lucovsky, G., and Nemanich, R. J. (2003). Measurement of the band offsets of SiO_2 on clean n- and p-type GaN(0001), *J. Appl. Phys.*, **93**, 3995–4004.

94. Cook, T. E., Fulton, C. C., Mecouch, W. J., Davis, R. F., Lucovsky, G., and Nemanich, R. J. (2003). Band offset measurements of the Si_3N_4/GaN (0001) interface, *J. Appl. Phys.*, **94**, 3949–3954.

95. Nakasaki, R., Hashizume, T., and Hasegawa, H. (2000). Insulator-GaN interface structures formed by plasma-assisted chemical vapor deposition, *Physica E*, **7**, 953–957.

96. Heidelberger, G. (2009). PhD dissertation, RWTH Aachen University.

97. Hansen, P. J., Vaithyanathan, V., Wu, Y., Mates, T., Heikman, S., Mishra, U. K., York, R. A., Schlom, D. G., and Speck, J. S. (2005). Rutile films grown by molecular beam epitaxy on GaN and AlGaN/GaN, *J. Vac. Sci. Technol. B*, **23**, 499–506.

98. Hong, S.-K., Hanada, T., Makino, H., Chen, Y., Ko, H.-J., Yao, T., Tanaka, A., Sasaki, H., and Sato, S. (2001). Band alignment at a ZnO/GaN (0001) heterointerface, *Appl. Phys. Lett.*, **78**, 3349–3351.

99. Liu, H. F., Hu, G. X., Gong, H., Zang, K. Y., and Chua, S. J. (2008). Effects of oxygen on low-temperature growth and band alignment of ZnO/GaNZnO/GaN heterostructures, *J. Vac. Sci.Technol. A*, **26**, 1462–1468.

100. Alivov, Y. I., Xiao, B., Akarcabiyikli, S., Fan, Q., Morkoç, H., Johnstone, D., Lopatiuktirpak, O., Chernyak, L., and Litton, W. (2008). Properties of isotype n-ZnO/n-GaN heterostructures studied by I V T and electron beam induced current methods, *J. Phys.: Condens. Matter*, **20**, 2905–2910.

101. Kramm, B., Laufer, A., Reppin, D., Kronenberger, A., Hering, P., Polity, A., and Meyer, B. K. (2012). The band alignment of Cu2O/ZnO and Cu2O/GaN heterostructures, *Appl. Phys. Lett.*, **100**, 094102.

102. Chiou, Y. L., and Lee, C. T. (2011). Band alignment and performance improvement mechanisms of chlorine-treated ZnO-gate AlGaN/GaN metal-oxide-semiconductor high-electron mobility transistors, *IEEE Trans. Electron Devices*, **58**, 3869–3875.

103. Chiou, Y.-L., Lee, C.-S., and Lee, C.-T. (2010). AlGaN/GaN metal-oxide-semiconductor high-electron mobility transistors with ZnO gate layer and (NH4)2Sx surface treatment, *Appl. Phys. Lett.*, **97**, 032107.

104. Veal, T. D., King, P. D. C., Hatfield, S. A., Bailey, L. R., McConville, C. F., Martel, B., Moreno, J. C., Frayssinet, E., Semond, F., and Zúñiga-Pérez, J. (2008). Valence band offset of the ZnO/AlN heterojunction determined by x-ray photoemission spectroscopy, *Appl. Phys. Lett.*, **93**, 202108.

105. Ye, G., Wang, H., Arulkumaran, S., Ng, G. I., Li, Y., Liu, Z. H., and Ang, K. S. (2014). Influence of post-deposition annealing on interfacial properties between GaN and ZrO_2 grown by atomic layer deposition, *Appl. Phys. Lett.*, **105**, 152104.

106. Robertson, J. (2004). High dielectric constant oxides, *Eur. Phys. J. Appl. Phys.*, **28**, 265–291.

107. Nepal, N., Garces, N. Y., Meyer, D. J., Hite, J. K., Mastro, M. A., and Eddy, J. C. R. (2011). Assessment of GaN surface pretreatment for atomic layer deposited high-k dielectrics, *Appl. Phys. Express*, **4**, 055802–055802.

108. Wilk, G. D., Wallace, R. M., and Anthony, J. M. (2001). High-k gate dielectrics: current status and materials properties considerations, *J. Appl. Phys.*, **89**, 5243–5275.

109. Maeda, N., Wang, C., Enoki, T., Makimoto, T., and Tawara, T. (2005). High drain current density and reduced gate leakage current in channel-doped AlGaN/GaN heterostructure field-effect transistors with Al_2O_3/Si_3N_4 gate insulator, *Appl. Phys. Lett.*, **87**, 073504.

110. Lin, H. C., Ye, P. D., and Wilk, G. D. (2005). Leakage current and breakdown electric-field studies on ultrathin atomic-layer-deposited Al_2O_3 on GaAs, *Appl. Phys. Lett.*, **87**, 182904.

111. Ye, P. D., Yang, B., Ng, K. K., Bude, J., Wilk, G. D., Halder, S., and Hwang, J. C. M. (2005). GaN metal-oxide-semiconductor high-electron-mobility-transistor with atomic layer deposited Al_2O_3 as gate dielectric, *Appl. Phys. Lett.*, **86**, 063501–063503.

112. Chang, Y. C., Chang, W. H., Chiu, H. C., Tung, L. T., Lee, C. H., Shiu, K. H., Hong, M., Kwo, J., Hong, J. M., and Tsai, C. C. (2008). Inversion-channel GaN metal-oxide-semiconductor field-effect transistor with atomic-layer-deposited Al_2O_3 as gate dielectric, *Appl. Phys. Lett.*, **93**, 053504.

113. Gusev, E. P., Narayanan, V., and Frank, M. M. (2006). Advanced high-k dielectric stacks with polySi and metal gates: recent progress and current challenges, *IBM J. Res. Dev.*, **50**, 387–410.

114. Liu, C., Chor, E. F., and Tan, L. S. (2007). Enhanced device performance of AlGaN/GaN HEMTs using HfO_2 high-k dielectric for surface passivation and gate oxide, *Semicond. Sci. Technol.*, **22**, 522–527.

115. Yuanzheng, Y., Hao, Y., Zhang, J. C., Ni, J., Mao, W., Qian, F., and Linjie, L. (2008). AlGaN/GaN MOS-HEMT with HfO_2 dielectric and Al_2O_3 interfacial passivation layer grown by atomic layer deposition, *IEEE Electron Device Lett.*, **29**, 838–840.

116. Tokranov, V., Rumyantsev, S. L., Shur, M. S., Gaska, R., Oktyabrsky, S., Jain, R., and Pala, N. (2007). HfO_2/AlGaN/GaN structures with HfO_2 deposited at ultra low pressure using an e-beam, *Phys. Status Solidi RRL*, **1**, 199–201.

117. Abermann, S., Pozzovivo, G., Kuzmik, J., Strasser, G., Pogany, D., Carlin, J.-F., Grandjean., N., and Bertagnolli, E. (2007). MOCVD of HfO2 and ZrO2 high-k gate dielectrics for InAlN/AlN/GaN MOS-HEMTs, *Semicond. Sci. Technol.*, **22**, 1272–1275.

118. Sire, C., Blonkowski, S., Gordon, M. J., and Baron, T. (2007). Statistics of electrical breakdown field in HfO_2 and SiO_2 films from millimeter to nanometer length scales, *Appl. Phys. Lett.*, **91**, 242905.

119. Wilk, G. D., Wallace, R. M., and Anthony, J. M. (2000). Hafnium and zirconium silicates for advanced gate dielectrics, *J. Appl. Phys.*, **87**, 484–492.

120. Goel, N., Majhi, P., Tsai, W., Warusawithana, M., Schlom, D. G., Santos, M. B., Harris, J. S., and Nishi, Y. (2007). High-indium-content InGaAs metal-oxide-semiconductor capacitor with amorphous LaAlO3 gate dielectric, *Appl. Phys. Lett.*, **91**, 093509.

121. Lim, B. S., Rahtu, A., de Rouffignac, P., and Gordon, R. G. (2004). Atomic layer deposition of lanthanum aluminum oxide nano-laminates for electrical applications, *Appl. Phys. Lett.*, **84**, 3957–3959.

122. Chiu, H.-C., Lin, C.-W., Chen, C.-H., Yang, C.-W., Lin, C.-K., Fu, J. S., Chang, L.-B., Lin, R.-M., and Hsueh, K.-P. (2010). Low hysteresis dispersion La_2O_3 AlGaN/GaN MOS-HEMTs, *J. Electrochem. Soc.*, **157**, H160

123. Wu, Y. H., Yang, M. Y., Chin, A., Chen, W. J., and Kwei, C. M. (2000). Electrical characteristics of high quality La_2O_3 gate dielectric with equivalent oxide thickness of 5 /spl aring, *IEEE Electron Device Lett.*, **21**, 341–343.

124. Qian, F., Qian, L., Tao, X., Qiang, W., Cheng, Z. J., and Yue, H. (2012). Performance of La_2O_3/InAlN/GaN metal oxide semiconductor high electron mobility transistors, *Chin. Phys. B*, **21**, 067305-067301–067305-067306.

125. Oon, H. S., and Cheong, K. Y. (2013). Recent development of gallium oxide thin film on GaN, *Mater. Sci. Semicond. Process.*, **16**, 1217.

126. Lee, C.-T., Chen, H.-W., and Lee, H.-Y. (2003). Metal–oxide–semiconductor devices using Ga2O3 dielectrics on n-type GaN, *Appl. Phys. Lett.*, **82**, 4304–4306.

127. Kim, H., Park, S.-J., and Hwang, H. (2001). Thermally oxidized GaN film for use as gate insulators, *J. Vac. Sci. Technol. B*, **19**, 579–581.

128. Chang, W. H., Chang, P., Lai, T. Y., Lee, Y. J., Kwo, J., Hsu, C.-H., and Hong, M. (2010). Structural characteristics of nanometer thick Gd_2O_3 films grown on GaN (0001), *Cryst. Growth Des.*, **10**, 5117–5122.

129. Chang, W. H., Lee, C. H., Chang, P., Chang, Y. C., Lee, Y. J., Kwo, J., Tsai, C. C., Hong, J. M., Hsu, C.-H., and Hong, M. (2009). High κ dielectric single-crystal monoclinic Gd2O3 on GaN with excellent thermal, structural, and electrical properties, *J. Cryst. Growth*, **311**, 2183–2186.

130. Ren, F., Abernathy, C. R., MacKenzie, J. D., Gila, B. P., Pearton, S. J., Hong, M., Marcus, M. A., Schurman, M. J., Baca, A.G., and Shul, R. J. (1998). Demonstration of GaN MIS diodes by using AlN and $Ga_2O_3(Gd_2O_3)$ as dielectrics, *Solid-State Electron.*, **42**, 2177–2181.

131. Robertson, J. (2000). Band offsets of wide-band-gap oxides and implications for future electronic devices, *J. Vac. Sci. Technol. B*, **18**, 1785–1791.

132. Chiu, F.-C., Wang, J.-J., Lee, J. Y., and Wu, S. C. (1997). Leakage currents in amorphous Ta_2O_5 thin films, *J. Appl. Phys.*, **81**, 6911–6915.

133. Kim, H.-S., Campbell, S. A., and Gilmer, D. C. (1997). Charge trapping and degradation in high-permittivity TiO_2 dielectric films, *IEEE Electron Device Lett.*, **18**, 465–467.

134. Rausch, N., and Burte, E. P. (1993). Thin TiO_2 films prepared by low pressure chemical vapor deposition, *J. Electrochem. Soc.*, **140**, 145–149.

135. Kirkpatrick, C. J., Bongmook, L., Suri, R., Xiangyu, Y., and Misra, V. (2012). Atomic layer deposition of SiO_2 for AlGaN/GaN MOS-HFETs, *IEEE Electron Device Lett.*, **33**, 1240–1242.

136. Pang, L., Lian, Y., Kim, D.-S., Lee, J.-H., and Kim, K. (2013). Sputtered-gate-SiO_2/AiGaN/GaN MOSHEMT for high breakdown voltage achievement, in *Power and Energy Conference at Illinois*, pp. 13–17.

137. Chang, K.-M., Cheng, C.-C., and Lang, C.-C. (2002). Electrical properties of SiN/GaN MIS diodes formed by ECR-CVD, *Solid-State Electron.*, **46**, 1399–1403.

138. Chen, C. P., Hong, M., Kwo, J., Cheng, H. M., Huang, Y. L., Lin, S. Y., Chi, J., Lee, H. Y., Hsieh, Y. F., and Mannaerts, J. P. (2005). Thin single-crystal Sc_2O_3 films epitaxially grown on Si (111)-structure and electrical properties, *J. Cryst. Growth*, **278**, 638–642.

139. Gregušová, D., Hušeková, K., Stoklas, R., Blaho, M. l., Jurkovič, M., Carlin, J.-F., Grandjean, N., and Kordoš, P. (2013). ZrO_2/InAlN/GaN metal-oxide–semiconductor heterostructure field-effect transistors with InAlN barrier of different compositions, *Jpn. J. Appl. Phys.*, **52**, 08JN07.

140. Chen, C.-H., Chiu, H.-C., Chien, F.-T., Chuang, H.-W., Chang, K.-J., and Gau, Y.-T. (2012). High thermal stability and low hysteresis dispersion AlGaN/GaN MOS-HEMTs with zirconia film design, *Microelectron. Reliab.*, **52**, 2551–2555.

141. Abermann, S., Pozzovivo, G., Kuzmik, J., Strasser, G., Pogany, D., Carlin, J.-F., Grandjean, N., and Bertagnolli, E. (2007). MOCVD of HfO_2 and ZrO_2 high-k gate dielectrics for InAlN/AlN/GaN MOS-HEMTs, *Semicond. Sci. Technol.*, **22**, 1272–1275.

142. Kim, J., Mehandru, R., Luo, B., Ren, F., Gila, B. P., Onstine, A. H., Abernathy, C. R., Pearton, S. J., and Irokawa, Y. (2002). Characteristics of MgO/GaN gate-controlled metal–oxide– semiconductor diodes, *Appl. Phys. Lett.*, **80**, 4555–4557.

143. Gila, B. P., Onstine, A. H., Kim, J., Allums, K. K., Ren, F., Abernathy, C. R., and Pearton, S. J. (2003). Magnesium oxide gate dielectrics grown on GaN using an electron cyclotron resonance plasma, *J. Vac. Sci. Technol. B*, **21**, 2368–2370.

144. Zhou, H., Ng, G. I., Liu, Z. H., and Arulkumaran, S. (2011). Improved device performance by post-oxide annealing in atomic-layer-deposited Al_2O_3/AlGaN/GaN metal-insulator-semiconductor high electron mobility transistor on Si, *Appl. Phys. Express*, **4**, 104102.

145. Ramanan, N., Lee, B., and Misra, V. (2015). ALD gate dielectrics for improved threshold voltage stability in AlGaN/GaN MOS-HFETs for power applications, *Semicond. Sci. Technol.*, **30**, 125017.

146. Ramanan, N., Lee, B., and Misra, V. (2015). Accurate characterization and understanding of interface trap density trends between atomic layer deposited dielectrics and AlGaN/GaN with bonding constraint theory, *Appl. Phys. Lett.*, **106**, 243503.

147. Long, R. D., Jackson, C. M., Yang, J., Hazeghi, A., Hitzman, C., Majety, S., Arehart, A. R., Nishi, Y., Ma, T. P., Ringel, S. A., and McIntyre, P. C. (2013). Interface trap evaluation of Pd/Al_2O_3/GaN metal oxide semiconductor capacitors and the influence of near-interface hydrogen, *Appl. Phys. Lett.*, **103**, 201607.

148. Ramanan, N., Lee, B., Kirkpatrick, C., Suri, R., and Misra, V. (2013). Properties of atomic layer deposited dielectrics for AlGaN/GaN device passivation, *Semicond. Sci. Technol.*, **28**, 074004.

149. Marron, T., Takashima, S., Li, Z., and Chow, T. P. (2012). Impact of annealing on ALD Al_2O_3 gate dielectric for GaN MOS devices, *Phys. Status Solidi C*, **9**, 907.

150. Terman, L. M. (1962). An investigation of surface states at a silicon/silicon oxide interface employing metal-oxide-silicon diodes, *Solid-State Electron.*, **5**, 285–299.

151. Witczak, S. C., Suehle, J. S., and Gaitan, M. (1992). An experimental comparison of measurement techniques to extract Si-SiO_2 interface trap density, *Solid-State Electron.*, **35**, 345–355.

152. Ma, T. P., and Barker, R. C. (1974). Surface-state spectra from thick-oxide MOS tunnel junctions, *Solid-State Electron.*, **17**, 913–929.

153. Kordoš, P., Stoklas, R., Gregušová, D., and Novák, J. (2009). Characterization of AlGaN/GaN metal-oxide-semiconductor field-effect transistors by frequency dependent conductance analysis, *Appl. Phys. Lett.*, **94**, 223512.

154. Laroche, J. R. (2004). Comparison of interface state density characterization methods for SiO_2/4H SiC MOS diodes, *Electrochem. Solid-State Lett.*, **7**, G21–G24.

155. ManjulaRani, K. N., Rao, V. R., and Vasi, J. (2003). A new method to characterize border traps in submicron transistors using hysteresis in the drain current, *IEEE Trans. Electron Devices*, **50**, 973–979.

156. Lu, B., Matioli, E., and Palacios, T. (2012). Tri-gate normally-off GaN power MISFET, *IEEE Electron Device Lett.*, **33**(3), 360–362.

157. Sun, X., Saadat, O. I., Chang-Liao, K. S., Palacios, T., Cui, S., and Ma, T. P. (2013). Study of gate oxide traps in HfO_2/AlGaN/GaN metal-oxide-semiconductor high-electron-mobility transistors by use of ac transconductance method, *Appl. Phys. Lett.*, **102**, 103504.

158. Miczek, M., Mizue, C., Hashizume, T., and Adamowicz, B. (2008). Effects of interface states and temperature on the C-V behavior of metal/insulator/AlGaN/GaN heterostructure capacitors, *J. Appl. Phys.*, **103**, 104510.

159. Wu, Y. Q., Shen, T., Ye, P. D., and Wilk, G. D. (2007). Photo-assisted capacitance-voltage characterization of high-quality atomic-layer-deposited Al_2O_3/GaN metal-oxide-semiconductor structures, *Appl. Phys. Lett.*, **90**, 143504.

Chapter 7

GaN Vertical Power Devices

Liu Zhihong
Singapore-MIT Alliance for Research and Technology Center, 1 CREATE Way, Innovation Wing, #05-09, 138602, Singapore
liuzhihong@ieee.org

7.1 Introduction

Semiconductor power devices can be categorized into two groups on the basis of the output current direction, lateral power devices and vertical power devices. For the former, all the electrodes, including the anode and cathode for diodes and the source, drain, and gate for field-effect transistors (FETs), are fabricated coplanarly and the current is laterally flowing. Oppositely for the latter, the electrodes (anode/cathode or source/drain) are fabricated on two sides of the wafers separately and the current is vertically flowing. Lateral power devices have the advantages that their performance is mainly determined by the quality of the materials in the upper parts, so the active layers can be grown on a foreign substrate. Coplanar electrodes will also render the power devices easy to integrate monolithically with other circuits, such as drivers,

Gallium Nitride Power Devices
Edited by Hongyu Yu and Tianli Duan

ISBN 978-981-4774-09-3 (Hardcover), 978-1-315-19662-6 (eBook)
www.panstanford.com

modulators, or signal-processing circuits. Vertical power devices have a couple of advantages over the lateral counterparts, as follows: First, vertical devices are size effective, and their current values are almost proportional to the electrode sizes. Second, the current is flowing vertically, and the device does not suffer from depletion and scattering from the surface states, so a trivial current collapse or frequency dispersion is expected. Third, the wire connection to integrate the power device into a power system is much easier for vertical devices. And moreover, the cooling system can be designed on both sides of the wafer, which can improve the cooling efficiency and the device's thermal performance.

It is generally thought that GaN lateral devices, including Schottky barrier diodes (SBDs) and high-electron-mobility transistors (HEMTs), are suitable for low-breakdown-level applications, such as 600 V. When the breakdown requirement is higher, we have to increase the distance between electrodes, which requires a large size for the devices. For example, the reported GaN HEMT with the highest breakdown voltage (V_{BR}) of 10.4 kV has a long gate-to-drain distance of 125 μm [1]. So vertical devices are preferred for high-power applications of 1200 V and above.

The hurdle in developing a GaN vertical device with a high breakdown voltage of more than 1200 V is the lack of high-quality, large-size, and cost-effective stand-free bulk GaN wafers. Until now most GaN materials are grown on foreign substrates such as SiC, sapphire, or Si, and the lattice mismatch and thermal mismatch between the GaN epilayers and the substrates cause a high threading dislocation density (TDD) up to 10^9–10^{10} cm^{-2} and a high background doping level of more than 10^{16} cm^{-3}. Such a high TDD level brings down the critical electrical field to two or three times less than the theoretical. Presently most of the commercially available bulk GaN substrates have a TDD level of 10^6 [2], which ensures the fabrication of GaN vertical diodes at the 1200 V level. The bulk GaN substrates at a lower TDD up to 10^4 cm^{-2} start to be commercially available [3] but still have quite a small size and are very expensive. Table 7.1 lists the sizes and costs of currently available substrates used for GaN power devices.

Table 7.1 Comparison of the size and cost of currently available substrates used for GaN power devices [4]

Substrate	Available size	Cost
SiC	75 mm	~$6
Sapphire	100 mm	~$2.2
Si	200 mm	~$0.08
GaN	50 mm	~$100

To date a few GaN vertical power devices have been demonstrated in both universities/institutes and industries. Most of the demonstrated devices are fabricated on freestanding bulk GaN substrates, and the breakdown level has been promoted to above 1200 V. In this chapter we will discuss the details of various GaN vertical devices, including two-terminal devices, such as PN diodes and Schottky diodes, and three-terminal devices, such as current aperture vertical electronic transistors (CAVETs), junction FETs (JFETs), and trench FETs.

7.2 GaN Vertical PN Diodes and Schottky Diodes

7.2.1 Device Structures

The schematic of the cross section of a typical GaN vertical PN diode on a bulk GaN substrate is shown in Fig. 7.1a. The two terminal electrodes, anode and cathode, are formed on the front and back sides of the wafer, respectively. From the bottom to the top, the PN diode includes a layer of N^+-GaN to assist the ohmic contact for the cathode, a layer of N^--GaN for electron drift, and a layer of P^+-GaN on the top to form a junction and assist the ohmic contact for the anode. The anode is typically made of metals of Ni/Au or Pd/Au, while the cathode is typically Ti/Al based. Annealing at higher than 500°C is required to realize low ohmic contact resistance. As for a vertical PN diode fabricated on a foreign substrate, such as SiC, sapphire, or Si(111), the cathode contacts the N^+-GaN layer at a trench by etching away the P^+-GaN and N^--GaN layers above it, as shown in Fig. 7.1b.

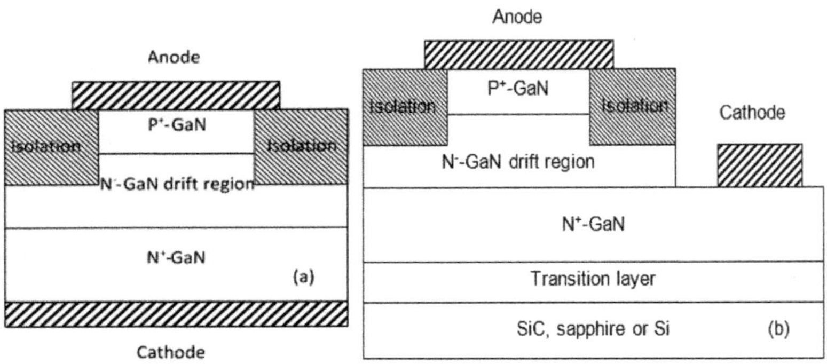

Figure 7.1 Schematic of the cross section of a typical GaN vertical PN diode (a) on a bulk GaN substrate and (b) on a foreign substrate.

It is needed for the GaN substrate to have a low dislocation density and a low background carrier doping concentration. In a sample with a high dislocation and defect density, the electron and hole mobility is low due to the scattering by the charged states. The current leakage is also high as the carrier traps formed by these defects assist the hopping current. These dislocations and defects also enhance the local electrical field and then lower the critical electrical field and limit the device's breakdown voltage. The background carrier concentration will limit the doping level of the N drift region and thereby limit the whole device's breakdown voltage.

The poor material quality of the available freestanding bulk GaN substrates is one of the key obstacles preventing the fabricated PN diodes from achieving high performance close to the theoretical limit. The traditional fabrication method for freestanding bulk GaN substrates is to grow a thick layer of GaN on a foreign substrate such as sapphire by high-vapor-pressure epitaxy (HVPE), followed by laser liftoff or a mechanical process to take off the thick GaN layer [5]. The typical growth rate can be as high as several tens of microns per hour. The TDD is typically 10^6 cm^{-2}, and the best value of 10^4 cm^{-2} is also commercially available now [2, 3]. The typical background doping is N-type with a concentration higher than 10^{16} cm^{-3}. There are also other fabrication methods for bulk GaN substrates, including N-flux liquid-phase epitaxy (LPE) and the ammonothermal method, and the results of TDD and background doping are similar to those grown by HVPE. Currently the wafer size of the bulk GaN substrate is quite limited (<2 inches) and the price is very high. With the continuous development of growth technology, the price of the bulk GaN substrate can be expected to come down and the quality will improve in the future.

A high P-doping level is required for the top P$^+$-GaN layer in order to reduce the contact resistance of the anode to GaN and increase the injected numbers of holes to obtain high on-state current. However, there is lack of P-dopants with low enough activation energy. Typically Mg is used as the P-type dopant for GaN with high activation energy of 170 meV so that the activation efficiency is low and smaller than 5% [6]. The typical hole concentration of the top P$^+$-GaN is lower than 5×10^{17} cm^{-3}, while the typical electron concentration of the bottom N$^+$-nitride semiconductor GaN is larger than 1×10^{19} cm^{-3}. The hole mobility in the P-GaN is also low, with a typical value of 10–50 cm^{-2}/Vs. The low hole concentration and low mobility limit the current level that a GaN PN diode can achieve.

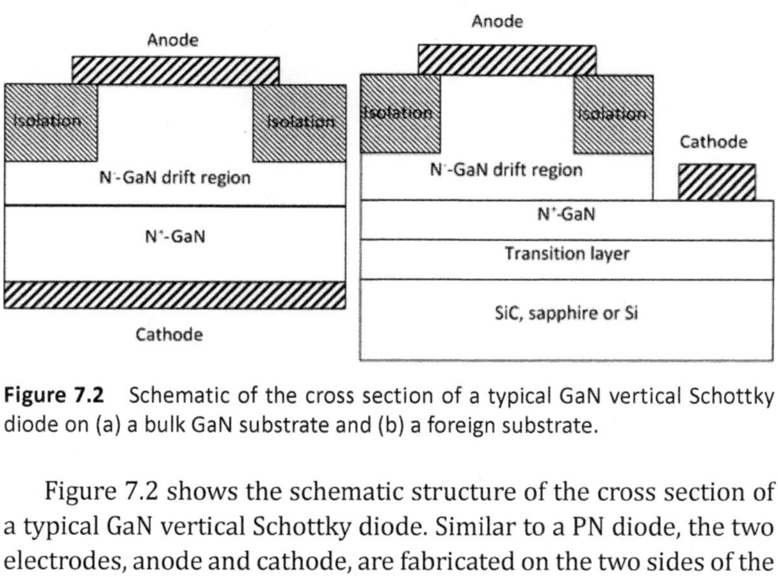

Figure 7.2 Schematic of the cross section of a typical GaN vertical Schottky diode on (a) a bulk GaN substrate and (b) a foreign substrate.

Figure 7.2 shows the schematic structure of the cross section of a typical GaN vertical Schottky diode. Similar to a PN diode, the two electrodes, anode and cathode, are fabricated on the two sides of the wafer. From the bottom to the top, there are a cathode, a layer of N$^+$-GaN to help the ohmic contact for cathode, a layer of N$^-$-GaN for electron drift, and an anode directly fabricated on the N$^-$-GaN layer to form a Schottky contact. The metals for the cathode and anode are similar to those for the PN diode, which are Ti/Al for the cathode and Ni/Au or Pd/Au for the anode. A different point is that the anode metal forms a Schottky contact with the GaN layer underneath it.

The Schottky barrier is typically ~1.0 eV for Ni and slightly higher for Pd. The turn-on voltage of a GaN Schottky diode is limited by the Schottky barrier height (SBH) between the anode and the N-GaN. The requirements for the different layers of GaN are similar to the case of a PN diode.

7.2.2 Design of the Vertical Diode Structures

Ideally, at off-state breakdown a GaN vertical PN diode or a Schottky diode should be working in the punch-through mode. In this case the breakdown voltage is mainly dependent on the doping level and the thickness of the N drift layer. For a PN diode with a structure shown in Fig. 7.1, at a reverse bias the electrical field distribution is as shown in Fig. 7.3. The y axis is set to originate from the junction of P^+-GaN/N^--GaN and directs from the anode to the cathode. t_p denotes the thickness of the P^+-GaN region, and t_n denotes the thickness of the N^--GaN region. W_p and W_n denote the depletion width of the P^+-GaN region and the N^--GaN region, respectively. The electrical field peak E_{max} is located at the P^+-/N^--GaN junction (y = 0).

According to Poisson's equation, the electrostatic potential ϕ in the PN diode at a location of y is given by

$$\frac{d^2\phi}{dy^2} = -\frac{\rho}{\varepsilon_S} = -\frac{q}{\varepsilon_S}(p - n + N_D^+ - N_A^-), \tag{7.1}$$

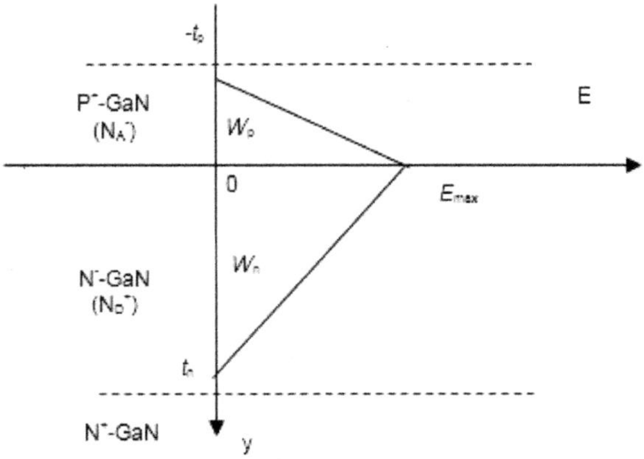

Figure 7.3 Electrical field distribution in a GaN vertical PN diode under a reverse bias.

Where ρ is the net charge density at the location of y, ε_s is the dielectric constant of GaN, and p, n, N_D^+, and N_A^- are the hole density, electron density, ionized donor density, and ionized acceptor density, respectively, at the location y. So the electrical field E can be expressed as

$$\frac{dE}{dy}=\frac{q}{\varepsilon_S}(p-n+N_D^+-N_A^-). \tag{7.2}$$

Under full-depletion approximation, $p = n = 0$; then

$$\frac{dE}{dy}=\frac{q}{\varepsilon_S}(N_D^+-N_A^-). \tag{7.3}$$

The electrical field E at the borders of the depletion regions will be

$$E(y = -W_p) = 0 \tag{7.4}$$

$$E(y = 0) = E_{max} \tag{7.5}$$

$$E(y = W_n) = 0 \tag{7.6}$$

And the electrical field E at any location y can be obtained by the integral of Eq. 7.3 using the above three border conditions:

$$E(-W_p < y < 0)=\frac{qN_{Ap}^-}{\varepsilon_S}(y+W_p) \tag{7.7}$$

and

$$E(0 < y < W_n)=\frac{qN_{Dn}^+}{\varepsilon_S}(W_n-y), \tag{7.8}$$

where N_{Ap}^- is the ionized acceptor density in the P⁻-GaN region and N_{Dn}^+ is the ionized donor density in the N⁻-GaN drift region. The peak electrical field E_{max} at the PN junction can be calculated by the integral of dE/dy across the two depletion regions:

$$E_{max}=\frac{qN_{Ap}^-W_p}{\varepsilon_S}=\frac{qN_{Dn}^+W_n}{\varepsilon_S} \tag{7.9}$$

Then we can get

$$N_{Ap}^-W_p=N_{Dn}^+W_n. \tag{7.10}$$

And the potential drop V_{pn} across the two depletion regions can be calculated through

$$V_{\text{pn}} = \int_{-W_p}^{0} E(y)dy + \int_{0}^{W_n} E(y)dy. \tag{7.11}$$

And V_{pn} can be expressed as

$$V_{\text{pn}} = \frac{\varepsilon_{\text{S}} E_{\text{max}}^2}{2q}\left(\frac{1}{N_{\text{Ap}}^-} + \frac{1}{N_{\text{Dn}}^+}\right), \tag{7.12}$$

where E_{max} is the peak electrical field located at the PN junction (see Fig. 7.3). For a power PN diode, generally the electron concentration in the N^- drift region is much lower than the hole concentration in the P^+-GaN region, so the above equation can be simplified as

$$V_{\text{pn}} = \frac{\varepsilon_{\text{S}} E_{\text{max}}^2}{2qN_{\text{Dn}}^+}. \tag{7.13}$$

In an ideal PN diode the avalanche breakdown occurs when the peak electrical field E_{max} reaches the critical electrical field E_{C} of the material. In this case at the junction of PN, the impact ionization occurs and a lot of electrons and holes are generated in a short time and the reverse current increases sharply. Neglecting the incomplete ionization for the n-dopants in N^--GaN, that is, $N_{\text{d}} = N_{\text{Dn}}^+$, the relationship between the reverse breakdown voltage (called the reverse blocking voltage) V_{BR} and the critical electrical field E_{C} can be expressed as

$$V_{\text{BR}} = \frac{\varepsilon_{\text{S}} E_{\text{C}}^2}{2qN_{\text{D}}}. \tag{7.14}$$

The depletion width of N^--GaN drifting region W_{n} can be expressed as

$$W_{\text{n}} = \frac{\varepsilon_{\text{S}} E_{\text{C}}}{qN_{\text{D}}}. \tag{7.15}$$

The depletion width of the P^+-GaN region can be expressed as

$$W_{\text{p}} = \frac{\varepsilon_{\text{S}} E_{\text{C}}}{qN_{\text{Ap}}^-}. \tag{7.16}$$

The relationship between the critical electric field E_{C} and the doping level in the drift region, N_{D}, can be further calculated on the basis of the theory of impact ionization. From the impact ionization theory, the maximum voltage that a semiconductor device can sustain, that is, the breakdown voltage, is limited by the onset of the avalanche breakdown triggered by the impact ionization process. The impact ionization coefficients, generally different for electrons and holes, are used to describe the impact ionization process. The impact ionization coefficient is defined as the number of electron–hole pairs generated by a high-energy electron or hole traveling over a unit distance (1 cm here) along the direction of the electrical field. From Chynoweth's equations [7], the impact ionization coefficient α can be expressed as

$$\alpha = ae^{-\frac{b}{E}}, \quad (7.17)$$

where a and b are two coefficients different for electrons and holes. The avalanche breakdown will happen when the ionization integral across the depletion regions becomes equal to unity, that is,

$$\int_0^{W_n+W_p} \alpha = 1. \quad (7.18)$$

To simplify the calculation, we use an Ozbek-Baliga power law as an approximation of the Eq. 7.17 [8]:

$$\alpha = A \cdot E^7, \quad (7.19)$$

where $A = 1.5 \times 10^{-42}$ for GaN. Combine Eqs. 7.7, 7.8, and 7.19 and calculat e the integral in Eq. 7.18 and we can get

$$\frac{1}{8}A\left(\frac{qN_\mathrm{D}W_\mathrm{n}}{\varepsilon_\mathrm{S}}\right)^7(W_\mathrm{n}+W_\mathrm{p})=1. \quad (7.20)$$

Considering $N_{\mathrm{Ap}}^- >> N_\mathrm{D}$ and thereby $W_\mathrm{n} \gg W_\mathrm{n}$, after combining Eq. 7.15, the relationship between E_C and N_D can be expressed as

$$E_\mathrm{C} = \left(\frac{8q}{A\varepsilon_\mathrm{S}}\right)^{\frac{1}{8}} N_\mathrm{D}^{\frac{1}{8}} = 3.14 \times 10^4 N_\mathrm{D}^{\frac{1}{8}}. \quad (7.21)$$

Substituting Eq. 7.21 into Eq. 7.14, the relationship between V_BR and N_D can be expressed as

$$V_{\text{BR}}=\left(\frac{\varepsilon_{\text{S}}}{q}\right)^{\frac{3}{4}}(2A)^{-\frac{1}{4}}N_{\text{D}}^{-\frac{3}{4}}=2.51\times10^{15}N_{\text{D}}^{-\frac{3}{4}}. \tag{7.22}$$

Substituting Eq. 7.21 into Eq. 7.15, the relationship between depletion width W_{n} and N_{D} at breakdown can be expressed as

$$W_{\text{n}}=\left(\frac{8}{A}\right)^{\frac{1}{8}}\left(\frac{\varepsilon_{\text{S}}}{q}\right)^{\frac{7}{8}}N_{\text{D}}^{-\frac{7}{8}}=1.57\times10^{11}N_{\text{D}}^{-\frac{7}{8}}. \tag{7.23}$$

Equations 7.22, 7.23, and 7.16 can be used to design the PN diode structure, including the doping level of the N$^-$-GaN drift region, and the minimum thickness of these two regions, targeted at a different breakdown voltage level.

The specific on-resistance $R_{\text{on,sp}}$ of a power device is defined as the resistance per surface area when the device is working in the on state. In a GaN vertical PN diode, the $R_{\text{on,sp}}$ is determined by the resistance of the drift region for the unit cross-sectional area. So the $R_{\text{on,sp}}$ can be calculated as

$$R_{\text{on,sp}}=\rho_{\text{dr}}W_{\text{n}}=\frac{1}{q\mu_{\text{n}}N_{\text{D}}}W_{\text{n}}. \tag{7.24}$$

Combine Eqs. 7.24 and 7.23, and we can get

$$R_{\text{on,sp}}=\frac{8^{\frac{1}{8}}q^{-\frac{7}{8}}\varepsilon_{\text{S}}^{\frac{7}{8}}}{\mu_{\text{n}}A^{\frac{1}{8}}}N_{\text{D}}^{-\frac{15}{8}}=1.41\times10^{-3}N_{\text{D}}^{-\frac{15}{8}}. \tag{7.25}$$

And the relationship between $R_{\text{on,sp}}$ and V_{BR} can be obtained by the combination of Eqs. 7.24, 7.14, and 7.15:

$$R_{\text{on,sp}}=\frac{4BV^{2}}{\varepsilon_{\text{S}}\mu_{\text{n}}E_{\text{C}}^{3}} \tag{7.26}$$

The above equations are also true for the diodes made with other materials. The critical electrical field E_{C}, the V_{BR}, and the depletion width W_{n} at breakdown for different materials, including Si, SiC, and GaN, are plotted as a function of the doping concentration in the drift region in Figs. 7.4 and 7.6 [8]. It can be seen from Fig. 7.5 that at the same doping concentration level in the N drift region, the breakdown voltage of a PN diode made of GaN is more than 30 times higher than that made of Si and also higher than SiC.

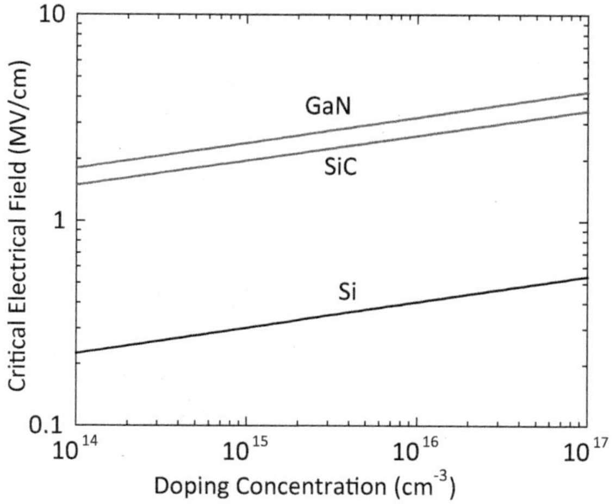

Figure 7.4 The critical electrical field as a function of the doping concentration in the drift region for different materials: Si, SiC, and GaN.

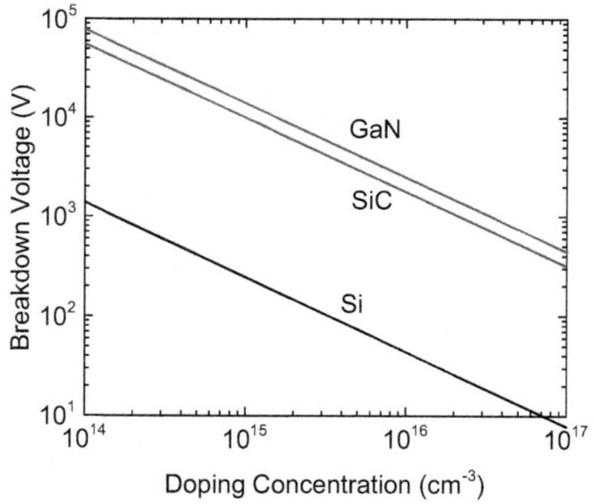

Figure 7.5 The breakdown voltage as a function of the doping concentration in the drift region for different materials: Si, SiC, and GaN.

The comparison of the limits considering V_{BR} and the specific on-resistance ($R_{on,sp}$) for various materials, including Si, SiC, and GaN, is shown in Fig. 7.7. It can be seen that at the same breakdown voltage level, the specific on-resistance of a GaN PN diode can be 3 orders of magnitude lower than the Si counterpart. This indicates the great potential for the GaN power diode to have much lower power loss and higher efficiency compared to Si diodes.

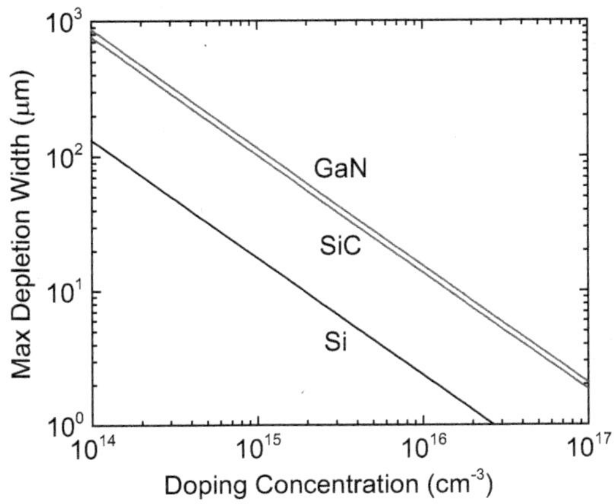

Figure 7.6 The depletion width at breakdown as a function of the doping concentration in the drift region for different materials: Si, SiC, and GaN.

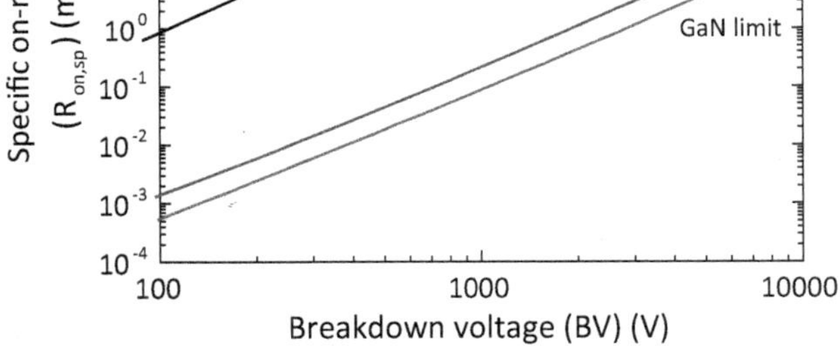

Figure 7.7 The calculated material limits of breakdown voltage and specific on-resistance for various power semiconductors, such as Si, SiC, and GaN.

It is worth noting that Eqs. 7.14, 7.15, and 7.21–7.26 are deduced from the theory of a GaN vertical PN diode but they are also correct for a vertical Schottky diode. Figure 7.8 shows the schematic of the vertical distributions of the electrical field in a GaN vertical Schottky diode (Fig. 7.2). The maximum electrical field is located at the interface between the anode and the N^--GaN drift region.

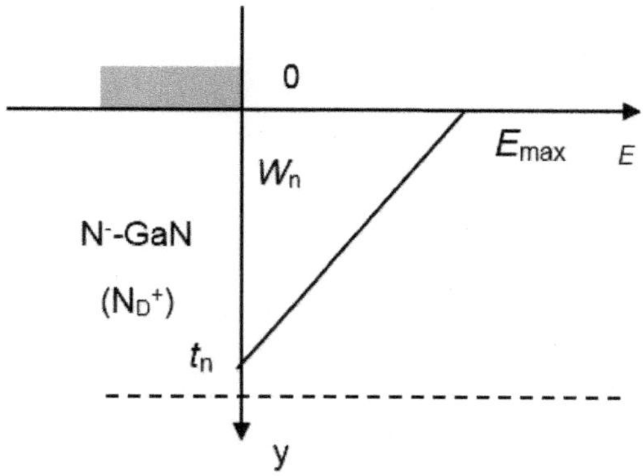

Figure 7.8 Electrical field distribution in a GaN Schottky diode at a reverse bias.

Another important parameter for a power PN diode is its turn-on voltage (V_{on}), also called the forward voltage drop (V_F). The turn-on voltage V_{on} is defined as the forward voltage biased, above which the forward current starts to sharply increase in the diode. Generally V_{on} is taken as the voltage when the forward current is equal to 100 A/cm^2 [9].

The V_{on} of a PN diode consists of the voltage drop across the middle part of the N^--GaN drift region (V_m) and the drop across the two junctions between P^+-/N^- drift GaN and N^- drift/N^+-GaN. V_{on} can be expressed as [9]

$$V_{on} = \frac{kT}{q}\ln\left(\frac{n(y=0)\cdot n(y=W_n)}{n_i^2}\right) + V_m, \tag{7.27}$$

where k is the Boltzmann constant, T is the absolute temperature, q is the electron charge, $n(y = 0)$ and $n(y = W_n)$ are the electron concentrations at the two surfaces of the N^--GaN drift region (See Fig. 7.3), and n_i is the intrinsic carrier centration in the GaN at the temperature of T. n_i can be calculated by

$$n_i = (N_C N_V)^{\frac{1}{2}} \exp^{-\frac{E_g}{2kT}}, \tag{7.28}$$

where N_C is the effective density of states for electrons in the conduction band for GaN ($N_C = 2.3 \times 10^{18}$ cm^{-3} at room temperature), N_V is the density of states for holes in the valence band for GaN ($N_V = 4.6 \times 10^{19}$ cm^{-3} at room temperature), and E_g is the bandgap of GaN (3.42 eV). Due to the wide bandgap of GaN, the value of n_i is very small and at a magnitude of order of 10^{-10} cm^{-3} at room temperature [10]. V_m in Eq. 7.27 can be calculated by [11]

$$V_m = \frac{4d^2}{(\mu_n + \mu_p)\tau_{HL}}, \tag{7.29}$$

where d is half width of the N$^-$-GaN drift region ($d = t_n/2$), μ_n and μ_p are the electron mobility and hole mobility, respectively, and τ_{HL} is the high-level minority carrier lift time in the N$^-$-GaN drift region. Substituting Eq. 7.29 into Eq. 7.27 and solving the current continuity equations to get $n(y = 0)$ and $n(y = W_n)$, we can obtain [11]

$$V_{on} = \frac{2kT}{q}\ln\left[\frac{J_{on}d}{2qD_a n_i F\left(d/L_a\right)}\right], \tag{7.30}$$

where J_{on} is the forward current value at which we define the turn-on voltage (or forward voltage drop) V_{on} and usually $J_{on} = 100$ A/cm^2. D_a is the ambipolar diffusion coefficient, and based on the Einstein relationship, we can get

$$D_a = \frac{kT}{q}\mu_a, \tag{7.31}$$

where μ_a is the ambipolar mobility. L_a in Eq. 7.30 is the ambipolar diffusion length, and it can be expressed as

$$L_a = \sqrt{D_a \tau_{HL}}. \tag{7.32}$$

$F(d/L_a)$ in Eq. 7.30 is a function of d/L_a and can be simplified as

$$F\left(d/L_a\right) = \left(\frac{d}{L_a}\right)^2. \tag{7.33}$$

Combining Eq. 7.28 and 7.30–7.33, V_{on} can be simplified as

$$V_{on} = \frac{2kT}{q}\ln\left(\frac{J_{on}\tau_{HL}}{2qd}\right) - \frac{kT}{q}\ln(N_C N_V) + \frac{E_g}{q}. \tag{7.34}$$

It can be clearly seen from Eq. 7.34 that the turn-on voltage V_{on} of a vertical PN diode is determined by the bandgap of the material E_g, which means that the V_{on} of a GaN PN diode will roughly be 2 V larger than that of the Si PN diode. Considering a group of typical values, including J_{on} = 100 A/cm^2, t_{HL} = 5 ns, d = 5 μm, T = 300 K, N_C = 2.3 × 10^{18} cm^{-3}, and N_V = 4.6 × 10^{19} cm^{-3}, we can calculate V_{on} = ~3.0 V. It is clear from Eq. 7.34 that V_{on} will increase with the increase of t_{HL} and/or decrease of the width of the N$^-$-GaN drift region t_n = 2d.

For a vertical power Schottky diode, from the theory of the thermionic emission, V_{on} can be expressed as [9]

$$V_{on} = \frac{nkT}{q}\ln\left(\frac{J_{on}}{A^{**}T^2}\right) + n\phi_B + R_{on}J_{on}, \qquad (7.35)$$

where n is the ideality factor of the Schottky contact, A^{**} is the effective Richardson's constant, ϕ_B is the SBH, and R_{on} is the on-resistance. Richardson's constant is expressed as

$$A^{**} = \frac{4\pi q m_n^* k^2}{h^3}, \qquad (7.36)$$

where m_n^* is the effective electron quality and h is Planck's constant. The theoretically calculated value of A^{**} for GaN is 26.4 Acm^{-2}K^{-2}. From Eq. 7.35, V_{on} is determined mainly by the SBH between the anode and the N$^-$-GaN drift region. Similar to the PN power diode, normally V_{on} is measured at J_F = 100 A/cm^2 and is around 1.0 V.

Generally one advantage of a Schottky power diode over a PN power diode is that it is a unipolar device and does not suffer from the minority carrier storage effect. So a Schottky diode is expected to have a higher speed and lower power loss than a PN diode. The minority carrier storage effect in a PN diode mainly manifests itself as the negative current overshoot and long reverse recovery time after the device is switched from a forward-high-current on state to a high-reverse-voltage off state. A typical reverse recovery current transient waveform of a SiC PN diode is shown in Fig. 7.9 [9], which recorded the current as a function of the time when the diode was switched from a forward current density of 1250 A/cm^2 to a reverse voltage of 150 V.

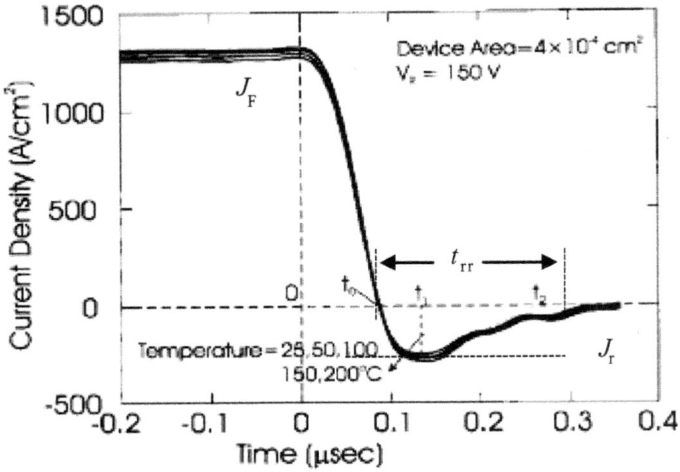

Figure 7.9 A typical reverse recovery current transient waveform of a SiC PN diode, which shows the behavior caused by the minority carrier storage effect. Reprinted from Ref. [9], Copyright (1999), with permission from Elsevier.

After the forward bias is turned off, the number of holes (per area in a cross section) stored in the N^--GaN drift region, Q_{SC}, is equal to the enclosed area by the current transient below zero in Fig. 7.9. Q_{SC} can be calculated using a triangle-shape approximation as below:

$$Q_{SC} = \frac{t_{rr} J_{PR}}{2} = \tau_{HL} J_F, \tag{7.37}$$

where τ_{rr} is the reverse recovery time and $t_{rr} = t_2 - t_0$, J_{PR} is the maximum reverse current, and J_F is the on-state forward current. It can be seen that the reverse recovery of a PN diode depends on the high-level minority lifetime τ_{HL}. From Eqs. 7.37 and 7.34, a short τ_{HL} is required for a suppressed reverse recovery effect and a small turn-on voltage. There are mainly three factors, that is, the Shockley–Read–Hall (SRH) recombination lifetime τ_{SRH}, the radiative recombination lifetime τ_{rad}, and the Aug recombination lifetime τ_{Aug}, contributing to the τ_{HL}:

$$\frac{1}{\tau_{HL}} = \frac{1}{\tau_{SRH}} + \frac{1}{\tau_{rad}} + \frac{1}{\tau_{Aug}} \tag{7.38}$$

All the three lifetimes, τ_{SRH}, τ_{rad}, and τ_{Aug}, are functions of the doping concentrations N_A and N_D. An advantage for a GaN PN diode is that it has a much smaller minority carrier lifetime compared to Si and SiC. Typically the t_{HL} in a Si or SiC PN diode is more than 0.1 μs, but the t_{HL} in a GaN diode is 1–10 ns. The reverse recovery time of a GaN PN diode is determined by the capacitance rather than the minority lifetime. So GaN has a much lessened minority carrier storage effect and faster reverse recovery compared to Si and SiC PN diodes. Figure 7.10 shows a comparison of the switching characteristics from forward current to reverse off state (a) between a GaN PN diode and a fast Si diode and (b) between a GaN diode and a SiC Schottky diode [12, 27]. The compared device has the same current and blocking voltage levels. It can be clearly seen that the GaN PN power diodes have superior switching performance compared to Si PN diodes and SiC diodes.

On the other side, the space region generation current J_{en} related to the SRH recombination plays an important role in the whole reverse current and the low-level forward current when the forward bias $V_F < V_{on}$ in a PN diode. The generation current is inversely proportional to the effective generation lifetime $\tau_{g,eff}$ and can be written as

$$J_{gen} = q\sqrt{\frac{2\varepsilon}{qN_B}}\left(\frac{n_i}{\tau_{g,eff}}\right)\sqrt{V}, \tag{7.39}$$

where $\tau_{g,eff} \simeq \tau_{SRH}$. Thus the SRH recombination lifetime τ_{SRH} cannot be too small in order to maintain a low off-state current and low power loss, although a small τ_{SRH} is required to suppress the reverse recovery effect and reduce the forward voltage drop.

7.2.3 Edge Termination Technology

Equations 7.14–7.39 predict the GaN vertical power diode's characteristics with specific design parameters. However, in the actually fabricated devices, the achieved V_{BR} and the extracted critical electrical field E_C are obviously smaller than the expected values. Besides the influence of the imperfect material quality of the N^--GaN drift layer, the electrical field crowding at the edges is another bottleneck. Figure 3 in Ref. [13] shows the simulated electrical field line distribution in a GaN Schottky diode without any edge termination applied [13]. It clearly illustrates the field-crowding effect at the edge of the active devices. Such field crowding leads to electrical peaks around the edges, and when these electrical peaks reach the critical electrical field the breakdown occurs at the edges instead of at the junction of the N drift region.

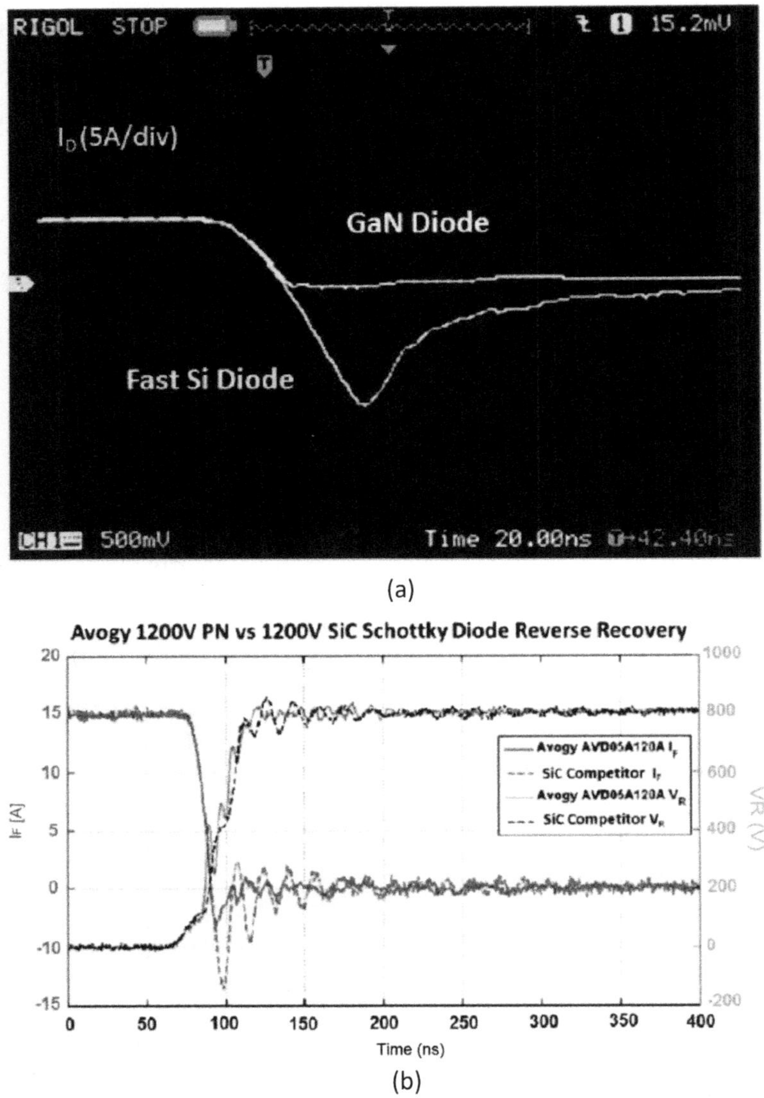

Figure 7.10 Comparison of the switching performance from forward current to a high reverse voltage between (a) a GaN PN diode and a fast Si PN diode with a level of 1200 V, 5 A at a test condition of 5 A, 600 V. From Ref. [27]. Copyright @ 2013 IEEE. (b) a GaN PN diode and a SiC Schottky diode with a level of 5 A, 1200 V and at a test condition of 15 A, 800 V. From Ref. [12]. Copyright @ 2013 IEEE.

A number of different edge termination technologies have been tried for GaN PN diodes or Schottky diodes, including mesas, P^+-GaN guard rings, high-resistivity layers created by ion implantation, and field plates, as shown in Fig. 7.11a–d. The target to introduce various edge termination treatments is to try to distribute the electrical field as evenly as possible so as to reduce the value of the electrical field peak at the edges and avoid premature breakdown.

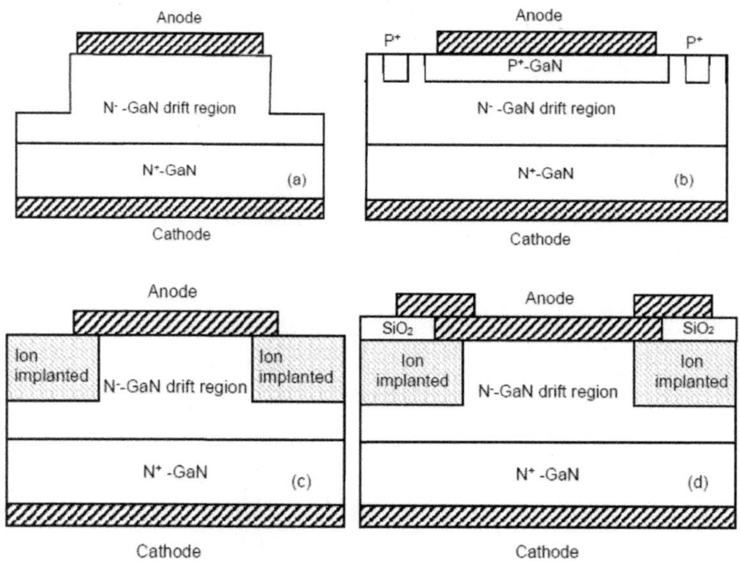

Figure 7.11 Various edge termination technologies for a power PN diode and/or Schottky diode: (a) mesa, (b) P^+-GaN guard ring, (c) ion implantation for isolation, and (d) field plate.

Laroche et al. studied the effect of P^+-GaN guard rings on the breakdown voltage of a vertical PN diode by numerical simulation [13]. The P^+ guard rings can spread the depletion region in the lateral direction, which reduces the field crowding at the edges. It is important to design carefully the guard width and spacing so as to spread the electrical field as evenly as possible at the P^+/N^- interface. Laroche et al. designed a GaN power PN diode, with a guard ring width of 5 μm and spacing of 1 μm, and found that the electrical field is evenly distributed at all the edges of the guard rings. From the simulation, the design of a single guard ring can improve the breakdown voltage up to 50%, but there is not much improvement on using multiguard rings. In addition, increasing the depth of the guard ring implantation can also efficiently increase the device's breakdown voltage.

Although the P^+ guard ring can effectively reduce the electrical field crowding and increase the breakdown voltage for GaN vertical diodes, it has a disadvantage of the need for complex Mg implantation and high-temperature annealing up to 1100°C–1200°C to activate the implanted Mg ions. Such high-temperature annealing will degrade the material quality [14]. Another effective method is to do ion implantation to form a highly resistive ring. Ozbek et al. have studied the finite-zone Argon implant edge termination for GaN power Schottky diodes [15]. The schematic of their device's structure is shown in Fig. 7.12a, which includes a layer of N^--GaN as the electron drift region with a doping concentration of 4×10^{14} cm^{-3} and a thickness of 4 μm and a 2 μm N^+-GaN layer underneath N^--GaN to assist in forming the ohmic contact. The N^- and N^+-GaN layers were grown on a sapphire substrate. A 100 nm Ni layer deposited on the top surface of the N^- drift region was used as the anode electrode. The edge termination was realized by Ar ion implantation with an implantation energy of 50 keV and a dosage of 2×10^{16} cm^{-2} at a 7° tilt. The lengths of the implanted region (L_{ET}) were varied from 5 to 200 μm. Figure 7.12b shows the simulated electrical field distribution along the surface around the Schottky gate edge for different implantation lengths. The simulation shows that the electrical field peak is located at the edge of the anode metal and at the interface between the anode/N^--GaN drift layer without implantation. With the implantation, the electrical field peak is lower and the electrical field distribution is flatter compared to the case without implantation. The longer the implantation width is, the flatter the electrical field distribution becomes, if L_{ET} is shorter than 50 μm. After L_{ET} = 50 μm, the electrical field distribution will not change with the increase of L_{ET}. The measured breakdown voltage as a function of the length of the implantation region is shown in Fig. 7.12c. Consistent with the trend of the electrical field distribution as a function of the L_{ET} by simulation, the measured breakdown voltage increases from 400 V to 1700 V with the increase of L_{ET} from 0 to 50 μm and saturates after L_{ET} is more than 50 μm.

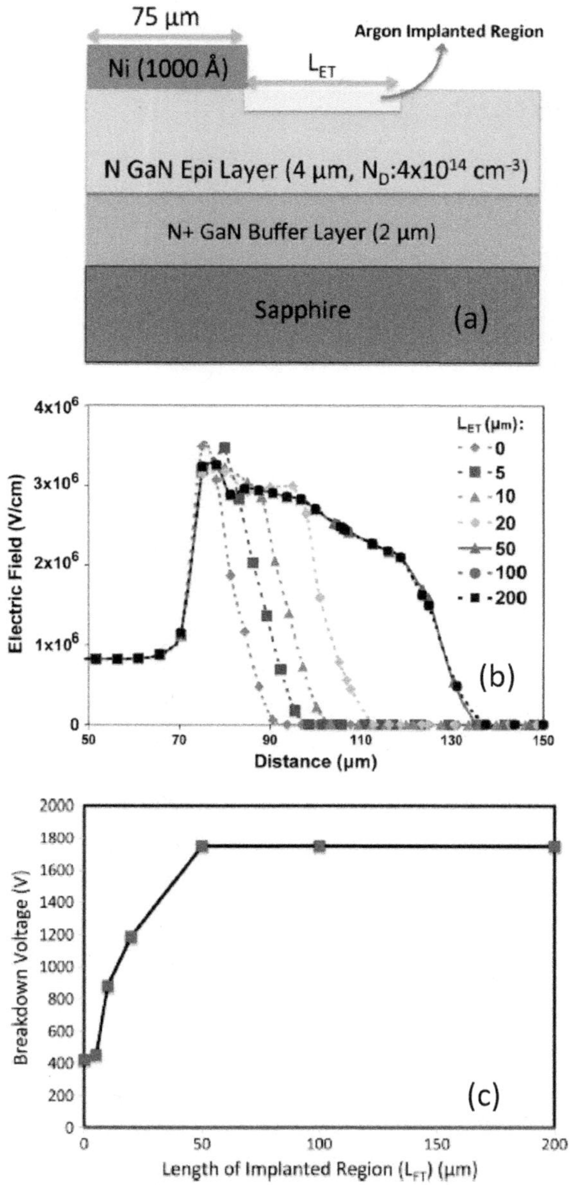

Figure 7.12 (a) Schematic of the GaN vertical Schottky diode with Ar ion implantation for edge termination. (b) Simulated lateral electrical field distribution at a high reverse voltage with different implantation lengths L_{ET}. (c) Measured breakdown voltage as a function of the length of the Ar-implanted region. From Ref. [15]. Copyright @ 2011 IEEE.

A field plate is another effective approach for edge termination in semiconductor power diodes. Sundaramoorthy et al. studied the design of various field plates for reduction of the electrical field crowding at the Schottky gate edge in GaN Schottky diodes [16]. Their Schottky diode structure includes an N^--GaN drift layer of 12 μm thickness with a doping level of 5×10^{15} cm^{-3} and an N^+-GaN layer for ohmic contact with 5 μm thickness and 1×10^{18} cm^{-3} doping concentration. The anode metal is fabricated on the top of the N^--GaN layer. At the anode edge, the anode metal extends beyond the anode edge to form a field plate, with a layer of dielectric inserted between the field plate metal and the N^--GaN. The device structure is shown in Fig. 7.13a. Sundaramoorthy et al. designed and tested two different field plates by simulation, including a bevel field plate (see Fig. 7.13b) and a multiple-step field plate (see Fig. 7.13c). The simulation shows that the optimum dimensions for the highest breakdown voltage are 18 μm for the bevel width and 1.5 μm for the oxide thickness. At a fixed bevel width, the breakdown voltage increases with the increase of the oxide thickness and reaches its maximum value at a certain point and then drops with a further increase of the oxide thickness. Such a trend is due to the fact that the breakdown voltage is dependent on the angle between the bevel field plate and the GaN surface. When the angle is smaller, the electrical field crowding becomes weaker and hence the breakdown voltage will be enhanced. The highest breakdown voltage occurs at the bevel angle of 5°.

Although the bevel-shaped field plate is very effective in reducing the edge field-crowding effect and increasing the breakdown voltage for a GaN vertical diode, it is a big challenge to do the fabricate to realize the bevel-shaped field plate with a certain designed angle. One approach to realizing the bevel shape is to do isotropic wet/dry etch and form a nonvertical etch profile [17]. However, the angle of the etch profile is dependent on the SiO_2 crystal orientation and is a fixed value. The second approach is to use photoresist patterns as a mask to do the dielectric etch, and these photoresist patterns have a nonvertical profile due to the gradually reduced exposure intensity caused by the absorption of the exposure light at the upper part of the photoresist [18]. However, there is the same limit for the second method—that usually the angle of the photoresist patterns' profile is one of photoresist's intrinsic properties and is also a fixed value. Another more flexible method is to use a gray-scale photomask during the lithography to realize a photoresist pattern with any profile [19]. The gray scale on the mask can be realized by an array of small patterns with gradually changed density. Anyway, it is complicated in the fabrication process to realize a bevel-shaped field plate.

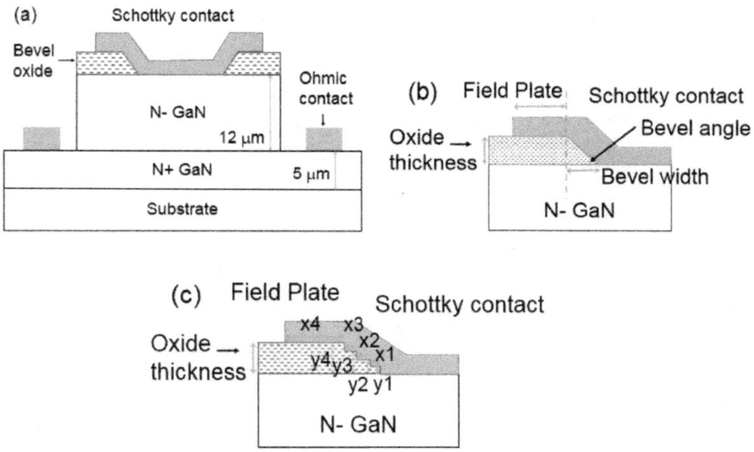

Figure 7.13 (a) Schematic of the cross section of a GaN vertical Schottky diode with a field plate for edge termination, (b) schematic of a bevel field plate, and (c) schematic of a multiple-step field plate.

Another design of a field plate to replace the bevel shape is a multistep. Figure 7.14c shows the schematic of a four-step field plate design. Sundaramoorthy et al. studied the effect of the dimension of the field plates on the device breakdown voltage by simulation [16]. It was found from the simulation that the dimensions of the first step, that is, the oxide thickness y_1 and width x_1, are the most important factors to determine the breakdown voltage. At a fixed oxide thickness, the breakdown voltage increases with the increase of x_1 from a small value to 6 μm. At x_1 = 6 μm the breakdown voltage reaches a peak and then drops with the further increase of x_1. For all the x_1 values, the thinner the oxide is, the higher the breakdown will be.

It is worth noting that compared to SiO_2, SiN is more suitable to be used in the fabrication process of GaN devices as the surface passivation effect of SiN is better than that of SiO_2. In case SiN is used as the dielectric under the metal field plate, the trend of the dependence of the breakdown voltage on the field plate shapes and dimensions is the same as SiO_2. But due to the different dielectric constant between the SiN and SiO_2, the optimum values of the dimensions for the highest breakdown are different.

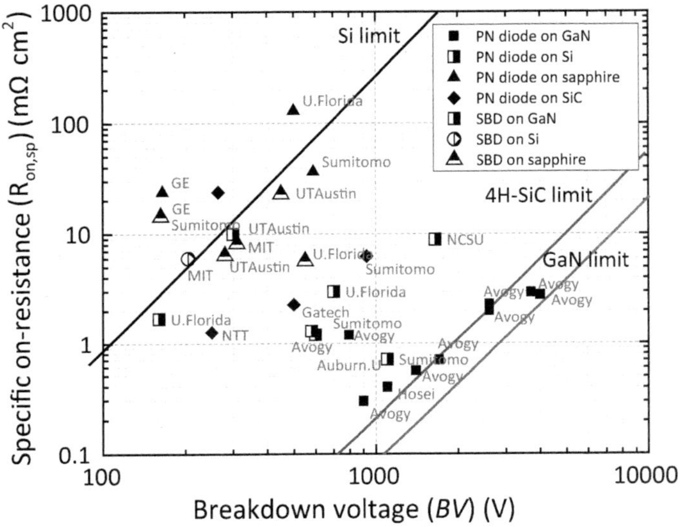

Figure 7.14 Performance of the state-of-the-art GaN vertical PN diodes and Schottky barrier diodes on various substrates, including bulk GaN, sapphire, Si, and SiC. The theoretical material limits for Si, SiC, and GaN are also drawn for comparison.

In Table 7.2 a summary of the reported GaN vertical power diodes in literature is given, along with the state-of-the-art results of the reported GaN vertical diodes, including PN diodes and Schottky barrier diodes (SBDs), on various substrates such as bulk GaN, sapphire, Si, and SiC. The main figures of merit, V_{BR} and specific on-resistance ($R_{on,sp}$) are plotted in Fig. 7.14, together with the theoretical material limits for Si, SiC, and GaN for comparison. Due to the high cost of SiC substrates, there is not much interest in developing GaN diodes on SiC, although GaN diodes on SiC have better performance due to the better material GaN epilayer quality benefitting from the lower lattice mismatch, thermal expansion mismatch, and higher thermal conductivity of SiC in comparison to the GaN epilayers grown on SiC or sapphire. Limited by the poor material quality (high threading dislocation density [TDD] values and high background doping) for the GaN epilayers, the GaN vertical diodes grown on sapphire or Si have difficulty achieving a V_{BR} higher than 600 V. On the contrary, GaN diodes grown on bulk GaN substrates have made great process during recent years, and a few devices have already obtained excellent performance beyond the theoretical material limit of SiC and close to the theoretical limit of GaN at the 4000 V level. With the further development of the growth technologies for freestanding GaN bulk materials, the GaN vertical diodes on bulk GaN substrates will be very promising to be the next generation of power diodes to replace Si and SiC counterparts.

7.3 Three-Terminal GaN Vertical Power Devices

7.3.1 Current Aperture Vertical Electronic Transistors

The schematic of the cross-sectional structure of a GaN current aperture vertical electronic transistor (CAVET) is shown in Fig. 7.15. The source and gate electrodes are fabricated on the top surface of the devices, and the drain electrode is fabricated on the bottom surface. The device usually consists of three regions from top to bottom, including an AlGaN barrier, an N^--GaN drift region, and an N^+-GaN layer for a drain ohmic contact. A discontinuous current-blocking layer (CBL) is buried in the N^--GaN drift region, and there exists a current aperture between two CBL sections. When the device is in the on state, the electron flow is directed from the source laterally through the channel formed at the AlGaN/N^--GaN interface smf then vertically through the current aperture and the N^--GaN region to the drain electrode. The two-dimensional electron gas (2DEG) existing between the AlGaN barrier and the N^--GaN channel increases the channel conductivity and lowers the device on-resistance. The N^--GaN drift region ensures a high reverse blocking voltage. The current in the 2DEG is controlled by the gate that is fabricated above the current aperture with an overlap with the 2DEG channel.

Table 7.2 Summary of the reported GaN vertical power diodes in the literature

Device type	Sub.	V_{BR} (V)	$R_{on,sp}$ (mΩcm^2)	V_{on} (V)	$N_{D,drf}$ (cm^{-3})	t_{drf} (μm)	Research group/ company	Year	Ref.
PN diode	Sapphire	500	130	5	NA	4	U.Florida	2000	[20]
PN diode	Sapphire	165	24–56	N.A	1e17	4	GE	2005	[21]
PN diode	GaN	265	24–56	N.A	1e17	4	GE	2005	[21]
PN diode	SiC	250	1.28	4.2	1e17	1.8	NTT	2006	[22]
PN diode	SiC	500	2.3	4.6	5e16	2.5	Gatech	2007	[23]
PN diode	GaN	925	6.3	3.5	3e16	7	Sumitomo	2007	[24]
PN diode	Sapphire	590	36.6	3.3	3e16	7	Sumitomo	2007	[24]
PN diode	GaN	1100	0.4	3	2e16	10	Hosei U.	2011	[25]
PN diode	GaN	900	0.3	3	1–3e16	NA	Avogy	2013	[26]
PN diode	GaN	1400	0.56	3	1–3e16	NA	Avogy	2013	[26]
PN diode	GaN	1700	0.7	3	1–3e16	NA	Avogy	2013	[26]
PN diode	GaN	2600	2.3	3	1–3e16	NA	Avogy	2013	[26]
PN diode	GaN	2600	2.0	3	1–3e16	20	Avogy	2013	[27]
PN diode	Si	300	10	3.5	2e16	1	MIT	2014	[28]
PN diode	GaN	795	1.2	3.7	2e16	6	Avogy	2014	[29]

Device type	Sub.	V_{BR} (V)	$R_{on,sp}$ (mΩcm^2)	V_{on} (V)	$N_{D,drf}$ (cm^{-3})	t_{drf} (μm)	Research group/ company	Year	Ref.
PN diode	GaN	3700	2.95	3	5e15	30	Avogy	2014	[30]
PN diode	GaN	4000	2.8	3	2–5e15	40	Avogy	2015	[3]
SBD	Sapphire	550	5.7	3.5	<10^{15}	12	U. Florida	2000	[31]
SBD	Sapphire	450	23	N.A	4e15	5	U.T. Austin	2000	[32]
SBD	Sapphire	310	8.2	N.A	7e16	2	U.T. Austin	2000	[32]
SBD	Sapphire	280	6.4	N.A	7e16	2	U.T. Austin	2000	[32]
SBD	GaN	700	3	1.8	1e17	200	U. Florida	2001	[33]
SBD	GaN	160	1.7	1.8	8e16	200	U. Florida	2002	[34]
SBD	GaN	600	1.2	2.2	7e15	460	Auburn U.	2006	[45]
SBD	Sapphire	163	14.4	1.0	1e16	7	Sumitomo	2007	[36]
SBD	GaN	580	1.3	1.2	1e16	7	Sumitomo	2007	[36]
SBD	GaN	1100	0.71	1.0	8e15	5	Sumitomo	2010	[37]
SBD	GaN	1650	8.9	0.5	1e14	4	NCSU	2011	[38]
SBD	Sapphire	1700	NA	N.A	4e14	4	NCSU	2011	[39]
SBD	GaN	600	1.2	0.9	1–3e16	NA	Avogy	2013	[26]
SBD	Si	205	6	0.5	2e16	1.5	MIT	2014	[28]

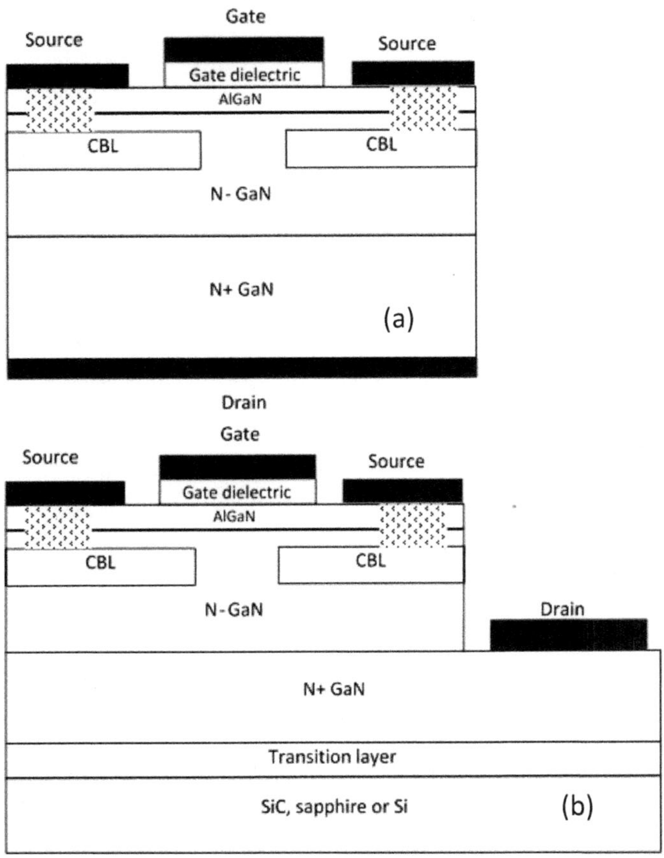

Figure 7.15 Schematic of the cross-sectional structure of a GaN current aperture vertical electron transistor (CAVET) (a) on a bulk GaN substrate and (b) on a foreign substrate.

To reduce the gate leakage current and increase the breakdown voltage, a layer of dielectric, such as SiO_2 or SiN, is usually inserted between the gate metal and the AlGaN barrier [40, 49]. The quality of the gate dielectric needs to be controlled to have a high breakdown strength and low interface-state density. The SiN grown by a metal-organic chemical vapor deposition (MOCVD) system or SiN or SiO_2 grown by a low-pressure CVD (LPCVD) system has much higher quality than the commonly used SiN or SiO_2 grown by a plasma-enhanced CVD (PECVD) system. A thin layer of P^+-GaN can also be inserted between the gate metal and the AlGaN barrier to realize enhancement-mode operation [42]. The AlGaN barrier is similar to that in a normal lateral AlGaN/GaN high-electron-mobility transistor (HEMT), with a typical Al composition of 0.2–0.3 and thickness of 15–30 nm. The design of the CBL and the N^--GaN drift region is the key to achieving high breakdown and low on-state resistance. The CBL can be realized by isolation, such as air or ion implanted, or P^+-GaN. It is difficult in process to fabricate an air CBL as there is a lack of an effective selective etch "recipe" between GaN and materials of the same family (AlGaN, etc.). Ion implantation can realize a high-insulating current-blocking region, but the ions and the defects created by the ion bombardment are prone to diffusing into the GaN channel layer above it during the high-temperature regrowth and degrade the 2DEG mobility due to the increased scattering. A P^+-GaN epilayer is a good choice for current blocking as the electron barrier height between the P^+-GaN and N^--GaN is high enough at 3.0 eV. The doping concentration and thickness of the N^--GaN layer need to be designed to meet the breakdown voltage requirement, following the similar design rule to the case of two-terminal vertical power devices discussed in the previous sections.

The total on-resistance ($R_{\text{on,t}}$) of a CAVET includes several parts:

$$R_{\text{on,t}} = R_{\text{C,S}} + R_{\text{lat}} + R_{\text{ap}} + R_{\text{drf}} + R_{\text{C,D}}, \tag{7.40}$$

where $R_{\text{C,S}}$ is the source contact resistance, R_{lat} is the resistance in the lateral 2DEG region, R_{ap} is the aperture resistance, R_{drf} is the resistance of the drift region, and $R_{\text{C,D}}$ is the drain contact resistance. The aperture resistance R_{ap} can be expressed as

$$R_{\text{ap}} = \frac{t_{\text{CBL}}}{q\mu_{\text{n}} L_{\text{ap}} N_{\text{ap}} W_{\text{g}}}, \tag{7.41}$$

where t_{CBl} is the thickness of the CBL, q is the electron charge, μ_{n} is the electron mobility in the aperture, L_{ap} is the aperture length, N_{ap} is the electron concentration in the aperture, and W_{g} is the gate width. The aperture resistance needs to be optimized by tuning the dimensions, including the aperture length L_{ap}, the overlap length L_{ov}, and the doping concentration L_{ap}. A low aperture resistance is desired, but a trade-off is needed because the adjustment of these parameters (t_{CBL}, L_{ap}, N_{ap}) will also influence the other characteristics. For example, increasing the CBL thickness t_{CBL} and aperture doping concentration N_{ap} will decrease the aperture resistance; however, the breakdown voltage will be lowered, too. Increasing the aperture length L_{ap} will weaken the control of the gate to the channel electrons and negatively shift the pinch-off voltage. Improving the material quality of N^--GaN and improving the electron mobility are an effective way to lower the aperture resistance.

A typical process flow to fabricate a CAVET is shown in Fig. 7.16, reported by Chowdhury et al. from the University of California, Santa Barbara (UCSB) in 2012 [40]. The epilayers were grown on a bulk N^+-GaN substrate. The drift layers were first grown on the substrate by MOCVD, consisting of a 3 μm thick N^--GaN layer with a Si doping level of 2×10^{16} cm^{-3} and a 0.3 μm thick GaN layer with a Si doping level of 5×10^{16} cm^{-3}. The current-blocking regions were formed by Mg ion implantation at 80 keV and with a dosage of 10^{15} cm^{-2} at selected areas. Then the unintentionally doped GaN channel and the $Al_{0.3}Ga_{0.7}N$ barrier were regrown using plasma molecule beam epitaxy (MBE) at 700°C. Compared with MOCVD regrowth, MBE can use a much lower temperature (typical MOCVD growth temperatures for GaN and AlGaN are above 1000°C) and hence reduce the Mg ion diffusion into the GaN channel above the CBL. Such diffusion will increase the scattering of the electrons in the 2DEG and reduce the electron mobility, also changing the threshold voltage. The source ohmic contact was formed by Si ion implantation and nonalloyed ohmic metals. The ion implantation ohmic contact can avoid the issue of metal spiking inside the AlGaN/GaN/CBL layers usually found in the traditional ohmic contact by Ti/Al/Ni/Au metals and high-temperature annealing above 800°C. The device isolation was realized by mesa etch. The gate electrode was formed by Ni/Au/Ni metals, and a 25 nm MOCVD-deposited SiN layer was inserted between the gate metals and the AlGaN barrier to reduce the gate leakage current. The source electrodes were short to the CBL to speed up the charging and discharging under switching operation. The measurement of the fabricated device showed that a breakdown voltage of 250 V, a specific on-resistance of around 2.2 mΩcm^2, and a threshold voltage of –6 V were achieved. There was not much difference of the output current under pulse measurement and DC measurement, which indicates the advantage of this vertical structure compared to a lateral GaN HEMT, that is, the current collapse is greatly suppressed by shifting the peak of the electrical field from the gate edge near the surface to the N^--GaN drift region underneath the gate far from the surface.

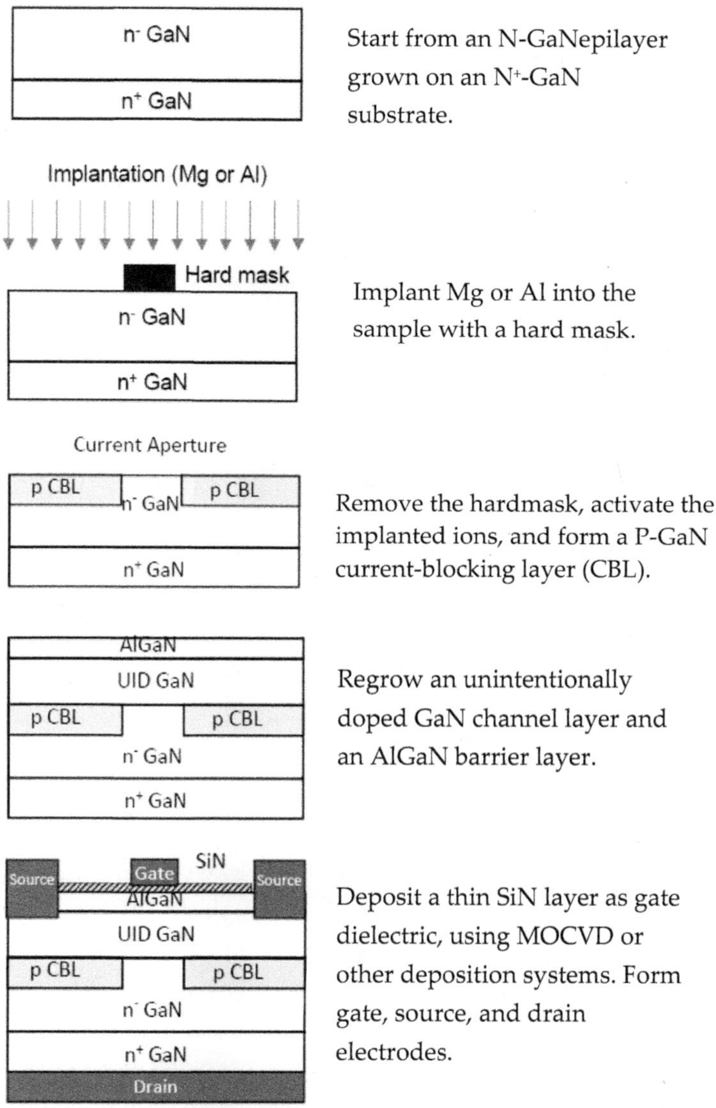

Figure 7.16 Schematic of a process flow to fabricate a CAVET [41].

More recently, Nie et al. from Avogy have reported a CAVET with excellent performance [42]. The device structure is similar to that in Fig. 7.15. The N^--GaN drift layer and the P^+-GaN CBL were grown on an N^+-GaN bulk GaN substrate by MOCVD first. Good quality of these epilayers was achieved with a TDD as low as 10^4 cm^{-2}. Different from the structure reported by Chowdhury et al. in [40], the device reported by Nie et al. has a thicker N^--GaN drift layer of 15 μm, with a net doping density of 1×10^{16} cm^{-3}. The current aperture was realized by etching off the P^+-GaN layer using a Cl_2 base plasma in an inductively coupled plasma (ICP) system. Then an N-type GaN layer was regrown to fill the etched trench and formed the lateral current channel followed by growth of an $Al_{0.2}Ga_{0.8}N$ barrier. Instead of inserting a gate dielectric between the gate metal and the AlGaN barrier, Nie et al. grew a layer of P^+-GaN between the AlGaN barrier and the gate metals of Sc/Au. The source contact was formed by Ti/Al and annealing, and the drain contact was formed by a thick layer of Al. The P^+-GaN cap below the gate metal depletes the electrons in the 2DEG and enables this device to work in enhancement mode. The device has achieved a positive threshold voltage of 0.5 V, a high breakdown voltage of 1.5 kV, and a low specific on-resistance of 2.2 mΩcm^2.

7.3.2 Junction Field-Effect Transistors

Although a current aperture vertical field-effect transistor (CAVFET) structure is promising to realize a vertical power transistor with a high reverse breakdown voltage and low on-resistance, it has a disadvantage that it is not easy to realize normally off operation. Another promising structure is a vertical junction FET (JFET), which is feasible to realize normally off operation and has successfully been utilized for Si and SiC power transistors. Figure 7.17 shows the schematic of a GaN JFET. Similar to a CAVET, the gate and source electrodes are fabricated on the front side, while the drain electrode is fabricated on the back side. The device includes an N^--GaN drift layer above the N^+-GaN substrate and a layer of N^+-GaN on the top to assist in forming the source ohmic contact. P^+-GaN regions were selectively formed on the N^--GaN layer, which is exposed after etching away the N^+-GaN layer. The gate electrode is fabricated on P^+-GaN. Opposite to the CAVET, the source electrode is fabricated on N^+-GaN at the center of the gate electrode ring. The current is directed from drain to source vertically through the current aperture below the source electrode, and the current in the aperture is controlled by the gate electrode and the P^--GaN regions. The P^--GaN depletes the electrons in the aperture so that the there is no current from drain to source when the gate is biased at zero, that is, the device is working in enhancement mode.

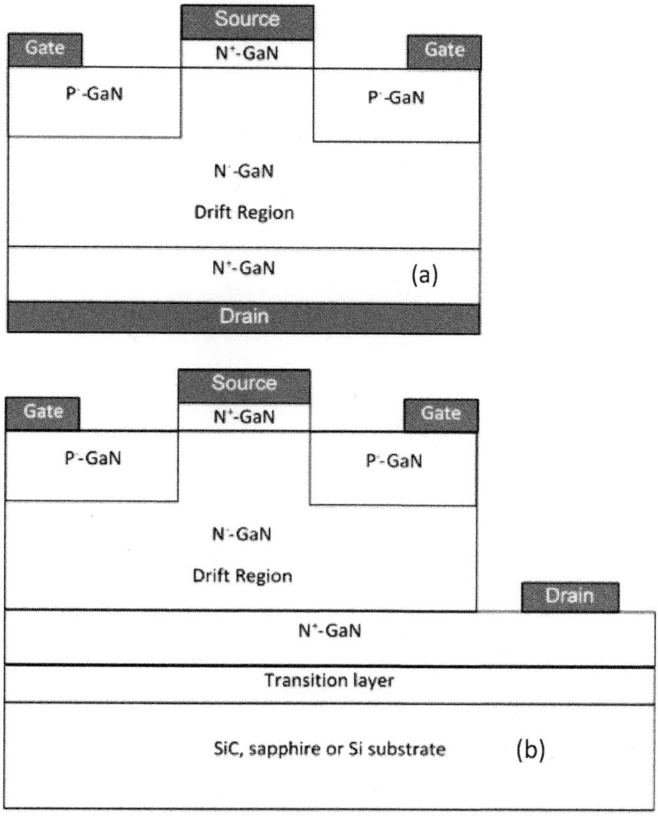

Figure 7.17 Schematic of the cross section of a GaN vertical JFET (a) on a bulk GaN substrate and (b) on a foreign substrate.

Similar to a vertical PN diode, the maximum threshold voltage (V_T) of a vertical JFET is mainly determined by the wide bandgap of the material, that is, around 3.0 eV for a GaN JFET. The breakdown voltage (V_{BR}) is also determined by the electron concentration and the thickness of the N^--GaN drift region. Besides the N^- drift layer, the dimensions of the current aperture, that is, the aperture length (L_{ap}) and the aperture thickness (t_{ap}), also play an important role in the device characteristics such as breakdown voltage, the on-resistance (R_{on}), and the threshold voltage. Ji et al. from the Arizona State University (ASU) studied the design of a GaN vertical JFET in 2015 by simulation using the Silvaco TCAD tool, and some results are shown in Fig. 7.18 [43]. The schematic of the dimensions L_{ap} and t_{ap} is shown in Fig. 7.18a. Figure 7.18b shows the simulated results of the breakdown voltage and on-resistance as a function of the thickness of the current aperture. In a JFET, the thickness of the current aperture is like the actual gate length. The breakdown voltage increases with the increase of the thickness of the current aperture due to the increase of the depletion width and saturates after 2.2 μm. However, the increase of t_{ap} will also increase the on-resistance due to the longer carrier travel distance. Figure 7.18b shows the influence of the length of the current aperture on the on-resistance and the threshold voltage by simulation. The increase of the current aperture length can effectively reduce the on-resistance due to the larger cross section of the current flow but will weaken the gate electrostatic control on the carriers in the aperture and negatively shift the threshold voltage. When L_{ap} is smaller than 1.0 μm, the threshold voltage becomes negative and the device loses the ability of normally off operation. The simulation shows that when electron concentration in N^+-GaN is 10^{18} cm^{-3}, electron concentration in N—GaN is 10^{16} cm^{-3}, t_{dr} is 11 μm, hole concentration in P—GaN is 3×10^{17} cm^{-3}, L_{ap} is 0.9 μm, and t_{ap} is 2 μm, the device can demonstrate high performance with V_{th} = 0.8 V, V_{BR} = 1260 V, and R_{on} = 5.2 mΩcm^2. Till now, the reported best device performance for the fabricated GaN vertical JFET was obtained by Diduck from Avogy in 2013 [44]. A high V_{BR} = 1000 V, V_{on} = ~1.0 V, and $R_{on,sp}$ = 4.8 mΩcm^2 were achieved.

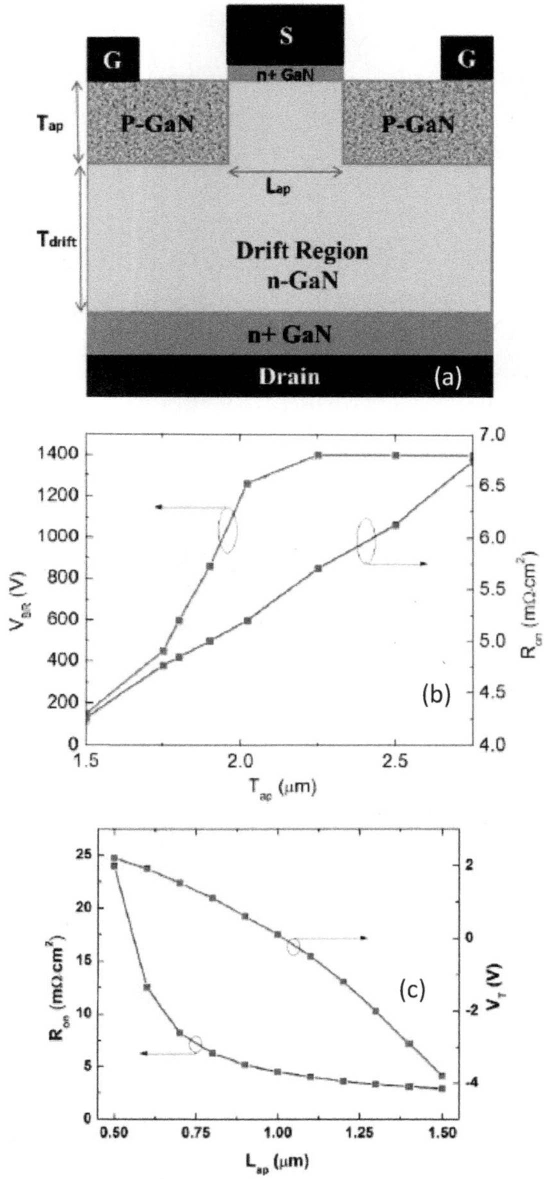

Figure 7.18 (a) Schematic of the structure of a GaN vertical JFET. The dimensions used in the simulation are also shown here. (b) Simulated breakdown voltage (V_{BR}) and on-resistance (R_{on}) as a function of the thickness of the current aperture (T_{ap}). (c) Simulated on-resistance (R_{on}) and threshold voltage (V_T) as a function of the aperture length L_{ap}. From Ref. [43]. Copyright @ 2015 IEEE.

7.3.3 Trench Field-Effect Transistors

Besides the vertical JFET, a structure named "trench field-effect transistor" is also popular in three-terminal Si and SiC power devices. The schematic of a GaN trench metal-oxide-semiconductor FET (MOSFET) is shown in Fig. 7.19. As in the former two types of vertical transistors, VFET and JFET, the source and gate electrodes of the trench MOSFETs are fabricated on the front side and the drain electrode is fabricated on the back side. From bottom to top, a trench MOSFET includes an N^+-GaN substrate, an N^--GaN drift layer, a P^--GaN layer, and an N^+-GaN on the top. The N^+-GaN layers on the top and bottom help to form source and drain ohmic contacts, respectively. A trench is etched down below the P^--GaN layer, and the gate dielectric and the gate metal are fabricated on the vertical sidewall of the etched trench. When the gate is biased positively enough, an inversion layer is formed at the interface between the gate dielectric/P^--GaN sidewall. The electrons flow from the source electrode through the top N^+-GaN layer, the inversion layer at the sidewall of the P^--GaN, and N^--GaN drift layer and are finally collected by the bottom N^+-GaN layer and the drain electrode. Similar to the vertical diode, CAVET and JFET, the off-state breakdown voltage is determined by the doping concentration and the thickness of the N^--GaN drift layer. The threshold voltage can be higher than 3 V by adjusting the hole doping level of the P^--GaN layer.

The N^--GaN drift layer, the P^--GaN layer, and N^+-GaN layers can be grown on a freestanding GaN bulk substrate or a foreign substrate by a MOCVD. The doping level and the thickness of the N^--GaN drift layer is designed on the basis of the expected device breakdown level, typically with $<1 \times 10^{16}$ cm^{-3} and <20 μm for a smaller than 1200 V breakdown voltage. The hole doping concentration of the P^--GaN is typically 1×10^{17} cm^{-3}, and it will influence the V_{BR} and the turn-on voltage V_{on}. If the doping level is too low in this P^--GaN region, the V_{BR} of the device will also be low. And if the doping level is too high, then the channel will be difficult to inverse and the turn-on voltage will be too high.

The gate trench can be formed by dry etch using Cl_2/BCl_3 plasmas. The etch recipes need to be optimized to minimize the etch damage on the trench sidewalls, especially on the sidewall of the P^--

GaN region. Typically the sidewall becomes rougher after plasma etch and lots of surface states and/or fixed charge are induced. There are a few approaches to remove the etch damage and recover the surface quality, such as high-temperature annealing, tetramethyl ammonium hydroxide (TMAH) treatment, and plasma treatment. The gate dielectric can be SiO_2 or SiN or high-k oxides such as Al_2O_3, HfO_2, and HfO_2. It is desired for the gate dielectric to have high breakdown field strength, low interface-state density, low fixed interface charge and bulk charge densities, etc. Ti/Al is typically used for source and drain contact metals. Body contact is made by recess etching down into the P^--GaN layer and filling with p-contact metals, such as Pd. The source electrode is stacked above the p-body electrode to save the area and miniaturize the cell pitch. Post-metal-annealing of 500°C–600°C is carried out to form a good ohmic contact for source, drain, and P^--GaN body. The device isolation can be realized by mesa etch. It is important to include edge termination at the mesa edges to avoid the electrical field crowding at the mesa corners. For an example, as illustrated by Fig. 7.20, a field plate was adopted to cover the P^--GaN layers at the mesa edges, as reported by Oka et al. [45], which has improved the breakdown voltage from 775 V to 1605 V.

Oka et al. have demonstrated GaN trench MOSFETs with excellent performance in Refs. [45, 46]. Their devices have an N^+-GaN substrate with TDD of 3–4 × 10^6 cm^{-2}, on which a layer of 15 μm thick N^--GaN, 1 μm thick p-GaN, and a 0.2 μm thick N^+-GaN were grown. The Si doping concentrations are 3 × 10^{18} cm^{-3} and 8 × 10^{15} cm^{-3} for N^--GaN and N^+-GaN, respectively. The Mg doping concentration in P^--GaN is 4 × 10^{18} cm^{-3}. A layer of SiO_2 with a thickness of 100 nm deposited by an atomic layer deposition (ALD) system was used as the gate dielectric. The width of the source and p-body electrodes is 6 μm and 2 μm, respectively. The device has a regular rectangular shape with a size of 100 × 2 μm and a cell pitch of 15 μm. A field plate was used for edge termination, and 100 nm ALD Al_2O_3 and 600 nm PECVD SiO_2 were deposited as the dielectric below the field plate. The field plate is located 10 μm far from the isolation mesa. The device demonstrated a high breakdown voltage of 1.6 kV, V_{on} of 7 V, and R_{on} of 12.1 mΩcm^2. Then Oka et al. further optimized the design of dimensions to reduce the on-state resistance for a 1.2 kV level power transistor. The N⁻-GaN layer thickness is reduced to 13 μm, and the doping concentration was increased a bit to 9×10^{15} cm^{-3}. The Mg doping concentration in the P⁻-GaN layer was reduced to 2×10^{18} to decrease the turn-on voltage. Another import improvement was the device layout. A hexagonal shape was used for the device (as shown in Fig. 7.21) and such a layout increased the gate width per unit up to two times compared to that of a rectangular shape. Finally the fabricated device demonstrated a much lower $R_{on,\ ap}$ of 1.8 mWcm^2, V_{on} of 3.5 V, and V_{BR} of 1200 V.

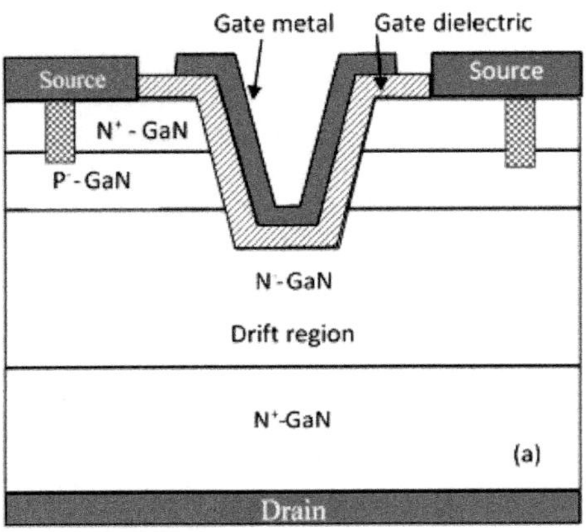

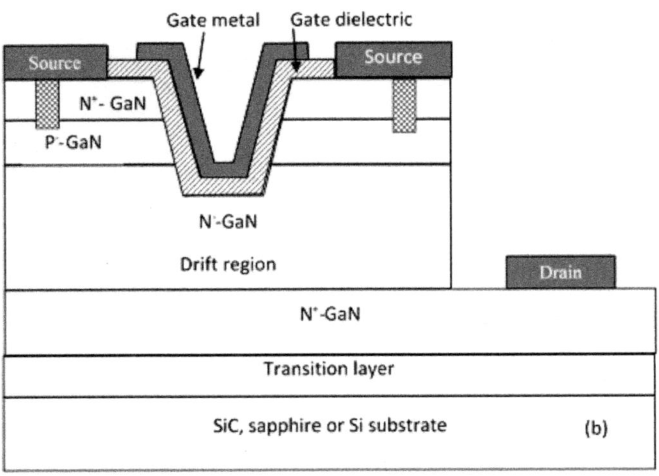

Figure 7.19 Schematic of the cross section of a GaN vertical trench MOSFET (a) on a bulk GaN substrate and (b) on a foreign substrate.

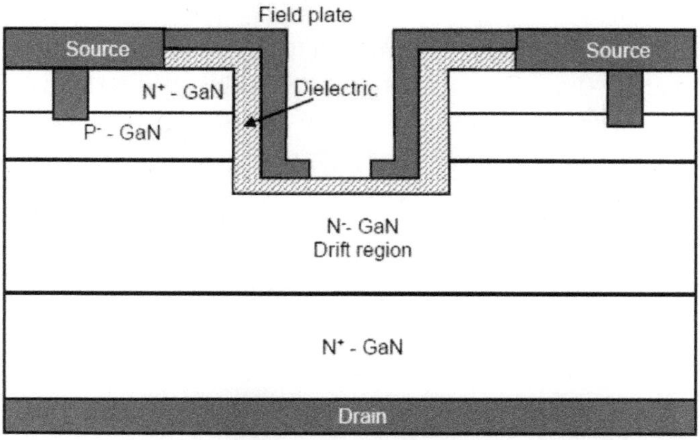

Figure 7.20 Schematic of the field plate edge termination for mesa isolation.

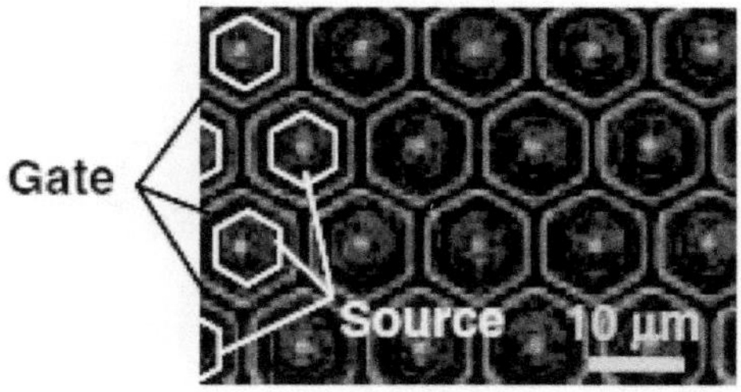

Figure 7.21 Micrograph of a hexagonal shape GaN vertical trench MOSFET [46].

Okada et al. from Sumitomo Electric Industries have reported a GaN vertical trench HEMT using a laterally regrown AlGaN/GaN heterostructure as the current channel [47]. The schematic of their device is shown in Fig. 7.22. Its vertical epilayers are the same as a traditional trench GaN MOSFET. Different from the fabrication process described above, after plasma etching to form a trench, an AlGaN/GaN heterostructure was regrown on the sidewalls followed by gate metal deposition on the AlGaN, instead of depositing a dielectric layer and metal to form the gate electrode directly. When the gate is biased above the threshold voltage, the device is turned on and the electrons flow from source into the N^- drift layer through the 2DEG formed between the regrown AlGaN and GaN, while for a traditional GaN trench MOSFET (as shown in Fig. 7.19), the electrons flow through the inversion layer formed on the P^--GaN sidewall. As the 2DEG formed between AlGaN and GaN has a much higher electron mobility and higher electron density than that in an inversion layer, such a trench HEMT device is expected to have a lower on-resistance. In [47] the device demonstrated a V_{BR} of 672 V, V_{on} of −1.1 V, and R_{on} of 7.5 mΩcm^2. The N^- drift layer thickness was thin and only 5 μm. With the increase of the drift region thickness, the trench GaN HEMT can be expected to hold a higher breakdown voltage.

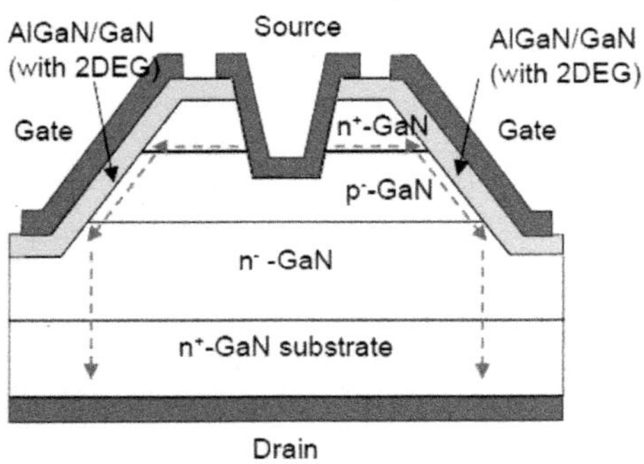

Figure 7.22 Schematic of a GaN vertical trench HEMT with regrown AlGaN/GaN at the sidewall of the trench [47].

Table 7.3 lists the reported GaN vertical three-terminal power devices fabricated on various substrates, that is, bulk GaN and sapphire. The V_{BR} and specific on-resistance ($R_{on,sp}$) of these fabricated GaN vertical transistors are summarized in Fig. 7.23. The theoretical limits of Si, SiC, and GaN are also plotted in Fig. 7.23 for comparison. Till now both CAVETs and trench MOSFETs grown on freestanding GaN substrates have shown excellent performance at a 1500 V breakdown level; however, unlike the GaN vertical diodes, the performance of the GaN vertical transistors is far away from the theoretical material limit. There are still lots of space for the material quality and fabrication process to further improve to enhance the device performance, including the p-GaN CBL quality after high-temperature regrowth in CAVETs, and the p-GaN/gate interface to reduce the sidewall leakage in the trench MOSFETs, etc.

Table 7.3 Summary of the reported GaN vertical three-terminal power devices in the literature

Device type	Sub.	V_{BR} (V)	$R_{on,sp}$ (mΩm^2)	V_{on} (V)	$N_{D,drf}$ (cm^{-3})	t_{drf} (μm)	Research group/ company	Year	Ref.
CAVET	Sapphire	65	NA	−6	NA	2	UCSB	2004	[48]
CAVET	GaN	NA	2.6	−16	1e16	3	Toyota	2007	[49]
CAVET	GaN	200	2.5	−7	2e16	3	UCSB	2010	[50]
CAVET	GaN	250	2.2	−5.5	2e16	3	UCSB	2012	[40]
CAVET	GaN	1000	4.8	1.0	NA	NA	Avogy	2013	[51]
CAVET	GaN	1500	2.2	0.5	1e16	15	Avogy	2014	[42]
CAVET	GaN	20	0.4	−7	NA	3	UCSB	2015	[52]
Trench FET	Sapphire	NA	30.6	5.1	1e17	1	Rohm	2007	[53]
Trench FET	GaN	240	9.3	3.7	1e17	1	Rohm	2008	[54]
Trench FET	GaN	180	1.6	10	2e16	6	Toyota	2008	[55]
Trench FET	GaN	672	7.6	−1.1	7e15	5	Sumitomo	2010	[47]
Trench FET	GaN	1600	12.1	7	8e15	15	Toyoda	2014	[45]
Trench FET	Sapphire	80	14	3	5e16	1	SK Hynx	2014	[56]
Trench FET	GaN	1200	1.8	3.5	9e15	13	Toyoda	2015	[46]

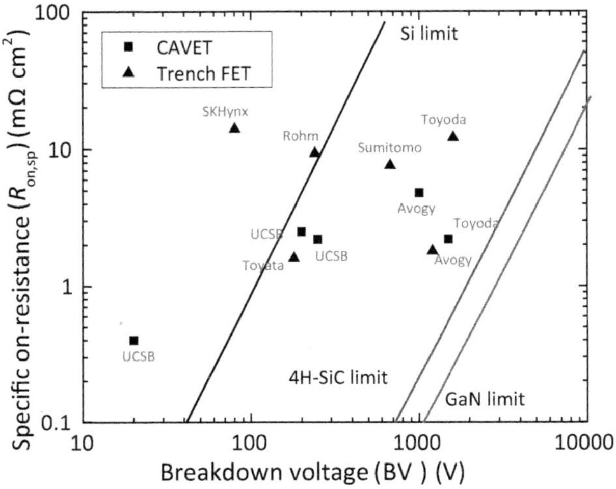

Figure 7.23 Characteristics of the state-of-the-art reported GaN vertical transistors. The theoretical limits of Si, 4H-SiC, and GaN are also plotted for comparison.

References

1. Ueda, T., Uemoto, Y., Tanaka, T., and Ueda, D. (2009). GaN transistors for power switching and millimeter-wave applications, *Int. J. High Speed Electron. Syst.*, **19**, 145–152.
2. Williams, A. D., and Moustakas, T. D. (2013). Materials issues for vertical gallium nitride power devices, *ECS Trans.*, **58**, 427–438.
3. Kizilyalli, I. C., Prunty, T., and Aktas, O. (2015). 4-kv and 2.8- mΩ -cm^2 vertical GaN p-n diodes with low leakage currents, *IEEE Electron Device Lett.*, **36**, 1073–1075.
4. Zhang, Y., Sun, M., Wong, H.-Y., Lin, Y., Srivastava, P., Hatem, C., Azize, M., Piedra, D., Yu, L., Sumitomo, T., de Braga, N. A., Mickevicius, R. V., and Palacios, T. (2015). Origin and control of off-state leakage current in GaN-on-Si vertical diodes, *IEEE Trans. Electron Devices*, **62**, 2155–2161.
5. Avrutin, V., Silversmith, D. J., Mori, Y., Kawamura, F., Kitaoka, Y., and Morkoc, H. (2010). Growth of bulk GaN and AlN: progress and challenges, *Proc. IEEE*, **98**, 1302–1315.
6. Kucheyeva, S. O., Williamsa, J. S., and Pearton, S. J. (2001). Ion implantation into GaN, *Mater. Sci. Eng.*, **33**, 51–107.

7. Chynoweth, A. G. (1960). Uniform silicon p-n junctions: II-Ionization rates for electrons, *J. Appl. Phys.*, **31**, 1161–1165.
8. Baliga, B. J. (2013). Gallium nitride devices for power electronic applications, *Semicond. Sci. Technol.*, **28**, 074011.
9. Khemka, V., Patel, R., Chow, T. P., and Gutmann, R. J. (1999). Design considerations and experimental analysis for silicon carbide power rectifiers, *Solid-State Electron.*, **43**, 1945–1962.
10. Pearton, S. J., Abernathy, C. R., and Ren, F. (2006). *Gallium Nitride Processing for Electronics, Sensors and Spintronics* (Springer Science & Business Media), p. 182.
11. Baliga, B. J. (2008). *Fundamentals of Power Semiconductor Devices* (Springer).
12. Kizilyalli, I. C., Edwards, A., Bour, D., Shah, H., Disney, D., and Nie, H. (2013). Very high performance GaN-on-GaN diodes, in *The 1st IEEE Workshop on Wide Bandgap Power Devices and Applications*, Columbus, OH, pp. 1–5.
13. Laroche, J. R., Ren, F., Baik, K. W., Pearton, S. J., Shelton, B. S., and Peres, B. (2005). Design of edge termination for GaN power Schottky diodes, *J. Electron. Mater.*, **34**, 370–374.
14. Liu, Z., Huang, C.-C., Xing, W., Fahle, D., Heuken, M., Ng, G. I., and Palacios, T. (2015). Si ion implantation and activation in NH_3 for CMOS-compatible fabrication of GaN HEMTs on Si substrates, in *International Conference on Nitride Semiconductors*, Beijing, China.
15. Ozbek, A. M., and Baliga, B. J. (2011). Finite-zone argon implant edge termination for high-voltage GaN Schottky rectifiers, *IEEE Electron Device Lett.*, **32**, 1361–1363.
16. Sundaramoorthy, V. K., and Nistor, I. (2011). Study of edge termination structures for high power GaN Schottky diodes, *Phys. Status Solidi C*, **8**, 2270–2272.
17. Kobayashi, K., Hatakeyama, S., Yoshida, T., Yabe, Y., Piedra, D., Palacios, T., Otsuji, T., and Suemitsu, T. (2014). Improved breakdown voltage and RF characteristics in AlGaN/GaN high-electron-mobility transistors achieved by slant field plates, *Appl. Phy. Express*, **7**, 096501-1–096501-4.
18. Dora, Y., Chakraborty, A., McCarthy, L., Keller, S., DenBaars, S. P., and Mishra, U. K. (2006). Effect of ohmic contacts on buffer leakage of GaN transistors, *IEEE Electron Device Lett.*, **27**, 529–531.
19. Chu, R., Li, Z., Boutros, K. S., and Burnham, S. (2013). *High Current High Voltage GaN Field Effect Transistors and Method of Fabricating Same*, US Patent No. 8530978B1.

20. Zhang, A. P., Dang, G., Ren, F., Han, J., Cho, H., Pearton, S. J., Chyi, J.-I., Nee, T.-E., Lee, C. M., Chuo, C. C., and Chu, S. N. G. (2000). Forward turn-on and reverse blocking characteristics of GaN Schottky and p-i-n rectifiers, *Solid-State Electron.*, **44**, 1157–1161.

21. Cao, X. A., Lu, H., LeBoeuf, S. F., Cowen, C., Arthur, S. D., and Wang, W. (2005). Growth and characterization of GaN PiN rectifiers on free-standing GaN, *Appl. Phys. Lett.*, **87**, 053503.

22. Nishikawa, A., Kumakura, K., Akasaka, T., and Makimoto, T. (2006). p-InGaN/n-GaN vertical conducting diodes on n+-SiC substrate for high power electronic device applications, *Jpn. J. Appl. Phys.*, **45**, 3387–3390.

23. Limb, J. B., Yoo, D., Ryou, J.-H., Shen, S.-C., and Dupuis, R. D. (2007). Low on-resistance GaN pin rectifiers grown on 6H-SiC substrates, *Electron. Lett.*, **43**, 366–367.

24. Yoshizumi, Y., Hashimoto, S., Tanabe, T., and Kiyama, M. (2007). High-breakdown-voltage pn-junction diodes on GaN substrates, *J. Cryst. Growth*, **298**, 875–878.

25. Hatakeyama, Y., Nomoto, K., Kaneda, N., Kawano, T., Mishima, T., and Nakamura, T. (2011). Over 3.0 GW/cm^2 figure-of-merit GaN p-n junction diodes on free-standing GaN substrates, *IEEE Electron Device Lett.*, **32**, 1674–1676.

26. Disney, D., Nie, H., Edwards, A., Bour, D., Shah, H., and Kizilyalli, I. C. (2013). Vertical power diodes in bulk GaN, in *25th International Symposium on Power Semiconductor Devices & IC's (ISPSD)*, Kanazawa, pp. 59–62.

27. Kizilyalli, I. C., Edwards, A. P., Nie, H., Disney, D., and Bour, D. (2013). High voltage vertical GaN p-n diodes with avalanche capability, *IEEE Trans. Electron Devices*, **60**, 3067–3070.

28. Zhang, Y., Sun, M., Piedra, D., Azize, M., Zhang, X., Fujishima, T., and Palacios, T. (2014), GaN-on-Si vertical Schottky and p-n diodes, *IEEE Electron Device Lett.*, **35**, 618–620.

29. Kizilyalli, I. C., Edwards, A. P., Nie, H., Quang, P. B., Disney, D., and Bour, D. (2014). 400-A (pulsed) vertical GaN p-n diode with breakdown voltage of 700 V, *IEEE Electron Device Lett.*, **35**, 654–656.

30. Kizilyalli, I. C., Edwards, A. P., Nie, H., Bour, D., Prunty, T., and Disney, D. (2014). 3.7 kV vertical GaN PN diodes, *IEEE Electron Device Lett.*, **35**, 247–249.

31. Dang, G. T., Zhang, A. P., Mshewa, M. M., Ren, F., Chyi, J.-I., Lee, C.-M., Chuo, C. C., Chi, G. C., Han, J., Chu, S. N. G., Wilson, R. G., Cao, X. A., and

Pearton, S. J. (2000). High breakdown voltage Au/Pt/GaN Schottky diodes, *J. Vac. Sci. Technol. A*, **18**, 1135–1138.

32. Zhu, T. G., Lambert, D. J. H., Shelton, B. S., Wong, M. M., Chowdhury, U., and Dupuis, R. D. (2000). High-voltage mesa-structure GaN Schottky rectifiers processed by dry and wet etching, *Appl. Phys. Lett.*, **77**, 2918.
33. Zhang, A. P., Johnson, J. W., Luo, B., Ren, F., Pearton, S. J., Park, S. S., Park, Y. J., and Chyi, J.-I. (2001). Vertical and lateral GaN rectifiers on free-standing GaN substrates, *Appl. Phys. Lett.*, **79**, 1555–1557.
34. Johnson, J. W., Zhang, A. P., Wen-Ben Luo, Fan Ren, Pearton, S. J., Park, S. S., Park, Y. J., and Jenn-Inn Chyi (2002). Breakdown voltage and reverse recovery characteristics of free-standing GaN Schottky rectifiers, *IEEE Trans. Electron Devices*, **49**, 32–36.
35. Zhou, Y., Wang, D., Ahyi, C., Tin, C.-C., Williams, J., Park, M., Williams, N. M., and Hanser, A. (2006). High breakdown voltage Schottky rectifier fabricated, *Solid-State Electron.*, **50**, 1744–1747.
36. Hashimoto, S., Yoshizumi, Y., Tanabe, T., and Kiyama, M. (2007). High-purity GaN epitaxial layers for power devices on low-dislocation-density GaN substrates, *J. Cryst. Growth*, **298**, 871–874.
37. Saitoh, Y., Sumiyoshi, K., Okada, M., Horii, T., Miyazaki, T., Shiomi, H., Ueno, M., Katayama, K., Kiyama, M., and Nakamura, T. (2010). Extremely low on-resistance and high breakdown voltage observed in vertical GaN Schottky barrier diodes with high-mobility drift layers on low-dislocation-density GaN substrates, *Appl. Phys. Express*, **3**, 081001.
38. Ozbek, A. M., and Baliga, B. J. (2011). Planar nearly ideal edge-termination technique for GaN devices, *IEEE Electron Device Lett.*, **32**, 300–302.
39. Ozbek, A. M., and Baliga, B. J. (2011). Finite-zone argon implant edge termination for high-voltage GaN Schottky rectifiers, *IEEE Electron Device Lett.*, **32**, 1361–1363.
40. Chowdhury, S., Wong, M. H., Swenson, B. L., and Mishra, U. K. (2012). CAVET on bulk GaN substrates achieved with MBE-regrown AlGaN/GaN layers to suppress dispersion, *IEEE Electron Device Lett.*, **33**, 41–43.
41. Chowdhury, S. (2015). Gallium Nitride based power switches for next generation of power conversion, *Phys. Status Solidi A*, **212**, 1066–1074.
42. Nie, H., Diduck, Q., Alvarez, B., Edwards, A. P., Kayes, B. M., Zhang, M., Ye, G., Prunty, T., Bour, D., and Kizilyalli, I. C. (2014). Remarkable reduction of on-resistance by ion implantation in GaN/AlGaN/GaN HEMTs with low gate leakage current, *IEEE Electron Device Lett.*, **35**, 939–941.

43. Ji, D., and Chowdhury, S. (2015). Design of 1.2 kV power switches with low Ron using GaN-based vertical JFET, *IEEE Trans. Electron Devices*, **62**, 2571–2578.

44. Diduck, Q., Nie, H., Alvarez, B., Edwards, A., Bour, D., Aktas, O., Disney, D., and Kizilyalli, I. C. (2013). 1000 V vertical JFET using bulk GaN, *ECS Trans.*, **58**, 295–298.

45. Oka, T., Ueno, Y., Ina, T., and Hasegawa, K. (2014). Vertical GaN-based trench metal oxide semiconductor field-effect transistors on a freestanding GaN substrate with blocking voltage of 1.6 kV, *Appl. Phys. Express*, **7**, 021002-1–021002-4.

46. Oka, T., Ina, T., Ueno, Y., and Nishii, J. (2015), 1.8 mΩ·cm2 vertical GaN-based trench metal–oxide–semiconductor field-effect transistors on a free-standing GaN substrate for 1.2-kV-class operation, *Appl. Phys. Express*, **8**, 054101-1–054101-4.

47. Okada, M., Saitoh, Y., Yokoyama, M., Nakata, K., Yaegassi, S., Katayama, K., Ueno, M., Kiyama, M., Katsuyama, T., and Nakamura, T. (2010). Novel vertical heterojunction field-effect transistors with re-grown AlGaN/GaN two-dimensional electron gas channels on GaN substrates, *Appl. Phys. Express*, **3**, 054201-1–054201-3.

48. Yaacov, I. B., Seck, Y.-K., Mishra, U. K., and DenBaars, S. P. (2004). AlGaN/GaN current aperture vertical electron transistors with regrown channels, *J. Appl. Phys.*, **95**, 2073–2077.

49. Kanechika, M., Sugimoto, M., Soejima, N., Ueda, H., Ishiguro, O., Kodama, M., Hayashi, E., Itoh, K., Uesugi, T., and Kachi, T. (2007). A vertical insulated gate AlGaN/GaN heterojunction field-effect transistor, *Jpn. J. Appl. Phys.*, **46**, L503–L505.

50. Chowdhury, S., Wong, M. H., Swenson, B. L., and Mishra, U. K. (2010). Dispersion-free AlGaN/GaN CAVET with low Ron achieved with plasma MBE regrown channel with Mg-ion-implanted current blocking layer, in *68th Device Research Conference*, South Bend, IN, pp. 201–202.

51. Diduck, Q., Nie, H., Alvarez, B., Edwards, A., Bour, D., Aktas, O., Disney, D., and Kizilyalli, I. C. (2013). 1000 V vertical JFET using bulk GaN, *ECS Trans.*, **58**, 295–298.

52. Yeluri, R., Lu, J., Hurni, C. A., Browne, D. A., Chowdhury, S., Keller, S., Speck, J. S., and Mishra, U. K. (2015). Design, fabrication, and performance analysis of GaN vertical electron transistors with a buried p/n junction, *Appl. Phys. Lett.*, **106**, 183502.

53. Otake, H., Egami, S., Ohta, H., Nanishi, Y., and Takasu, H. (2001). GaN-based trench gate metal oxide semiconductor field effect transistors

with over 100 cm2/(V s) channel mobility, *Jpn. J. Appl. Phys.*, **46**, L599–L601.

54. Otake, H., Chikamatsu, K., Yamaguchi, A., Fujishima, T., and Ohta, H. (2008), Vertical GaN-based trench gate metal oxide semiconductor field-effect transistors on GaN bulk substrates, *Appl. Phys. Express*, **1**, 011105-1–011105-3.

55. Kodama, M., Sugimoto, M., Hayashi, E., Soejima, N., Ishiguro, O., Kanechika, M., Itoh, K., Ueda, H., Uesugi, T., and Kachi, T. (2008). GaN-based trench gate metal oxide semiconductor field-effect transistor fabricated with novel wet etching, *Appl. Phys. Express*, **1**, 021104-1–021104-3.

56. Won, C.-H., Kim, K.-W., Kim, D.-S., Kang, H.-S., Im, K.-S., Jo, Y.-W., Kim, D.-K., Kim, R.-H., and Lee, J.-H. (2014). Normally-off vertical-type mesa-gate GaN MOSFET, *Electron. Lett.*, **50**, 1749–1751.

Chapter 8

Reliability of GaN HEMT Devices

Pengkun Xia
Department of Electrical and Electronic Engineering, Southern University of Science and Technology, 1088 Xueyuan Road, Nanshan District, Shenzhen, Guangdong 518055, China
xiapk@mail.sustc.edu.cn

Nowadays, GaN HEMT devices still cannot reach the theoretical brilliant performance due to reliability issues. Firstly, though GaN is a high-temperature-tolerant material, for a Schottky contact device, the semiconductor–metal contact suffers degradation under high temperature. The performance of GaN devices also varies with different temperatures. Secondly, the current collapse effect significantly limits the current performance, which is mainly reflected in the fact that current performance of a GaN HEMT deteriorates after stress or at an AC condition compared to a DC condition. Finally, the prebreakdown phenomenon of a GaN HEMT limits further improvement of the breakdown voltage in a high-power field, which can be attributed to an electric field peak, bulk leakage, and the reverse piezoelectric effect.

Gallium Nitride Power Devices
Edited by Hongyu Yu and Tianli Duan

ISBN 978-981-4774-09-3 (Hardcover), 978-1-315-19662-6 (eBook)
www.panstanford.com

8.1 Temperature-Related Degradation

In practical applications, power devices usually operate at high temperatures. One of the advantages of a GaN material for high-power applications is its excellent high-temperature performance. However, the thermal performance of GaN devices can be limited by poor quality of material, change of surface charge distribution, Au interdiffusion, etc. A thermal storage test that can simulate the harsh thermal environment is conducive to investigating these issues. Some previous results are summarized in this section. Performance variation with different temperatures is also discussed.

8.1.1 Degradation under Thermal Storage

After thermal storage, degradation often occurs at the contact between metal and semiconductor in a GaN high-electron-mobility transistor (HEMT). Under temperatures below 300°C, ohmic and Schottky contacts both have excellent stability [1, 2]. For the prevailing ohmic contact with metal Ti/Al/Ni/Au, no significant degradation is observed at 400°C thermal storage for 48 h in N_2 [3]. However, the Schottky gate consisting of Ni/Au reveals instability at this condition. The I_G–V_G curve before and after the thermal storage is shown in Fig. 8.1. The reduction of the gate leakage indicates the increase of the Schottky barrier. Other performances such as maximum drain current, transconductance, and threshold voltage are also affected after thermal storage, which are shown in Fig. 8.2 and Fig. 8.3. The degradation can be attributed to the change of surface charge distribution between the nitride passivation and the AlGaN barrier layer or imperfect material quality on AlGaN/GaN HEMTs subjected to thermal storage [3].

For the degradation of the ohmic contact, Piazza et al. reported the Au interdiffusion and Ga out-diffusion phenomenon in a Ti/Al/Ni/Au ohmic contact under high temperature [4]. Au interdiffusion results in the formation of Au-rich intermetallic compounds and strong out-diffusion of Ga. The former can introduce larger grains and, thereby, crack the passivation. Figure 8.4 shows Ga-rich (gray areas) and Au-rich (white areas) compounds and voids after 2000 h at 340°C. The increase of ohmic resistance and passivation film cracking are detected at 340°C for 100-hour thermal storage. At annealing temperature 200°C, nickel nitrides on GaN are formed, which leads to the reduction of the Schottky barrier height [4].

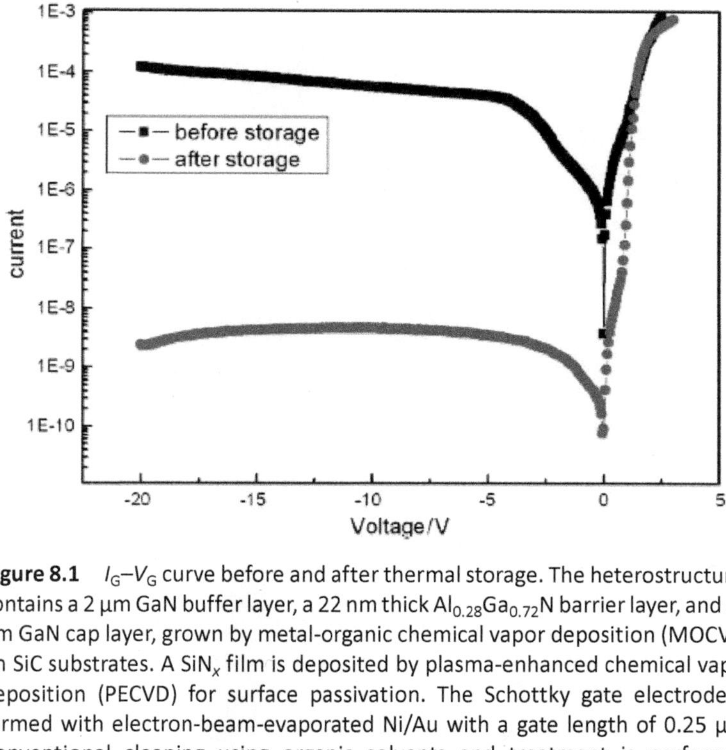

Figure 8.1 I_G–V_G curve before and after thermal storage. The heterostructures contains a 2 μm GaN buffer layer, a 22 nm thick $Al_{0.28}Ga_{0.72}N$ barrier layer, and a 2 nm GaN cap layer, grown by metal-organic chemical vapor deposition (MOCVD) on SiC substrates. A SiN_x film is deposited by plasma-enhanced chemical vapor deposition (PECVD) for surface passivation. The Schottky gate electrode is formed with electron-beam-evaporated Ni/Au with a gate length of 0.25 μm. Conventional cleaning using organic solvents and treatment is performed prior to the deposition. Gate recessing is performed using inductively coupled plasma–reactive ion etching (ICP-RIE). From Ref. [3]. Copyright @ 2010 IEEE.

The thermal stability of the Schottky contact can be optimized by inserting a layer of Pt between Ni and Au and thermal treatment subsequently. The gate leakage can be significantly reduced without sacrificing the transconductance. Experiments show that the barrier height depends on the work function of the inserted metal at the Schottky diodes on n-GaN after rapid thermal annealing (RTA) at 500°C for 5 min. [5].

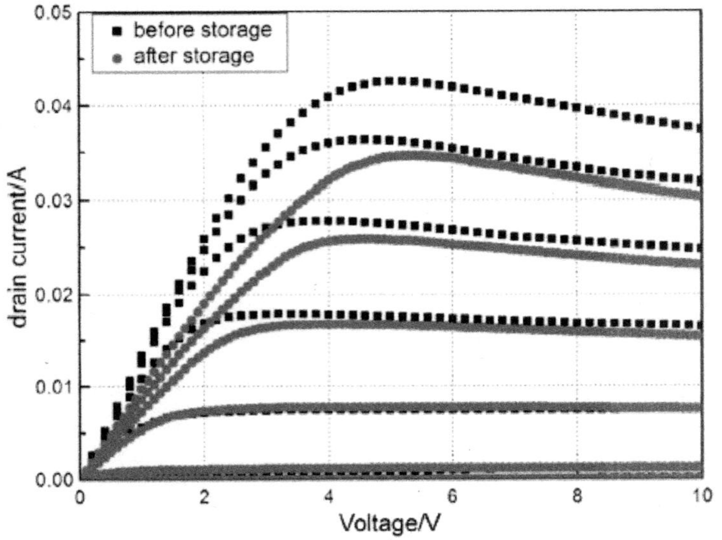

Figure 8.2 Output characteristics before and after thermal storage. The maximum drain current decreases after thermal storage. From Ref. [3]. Copyright @ 2010 IEEE.

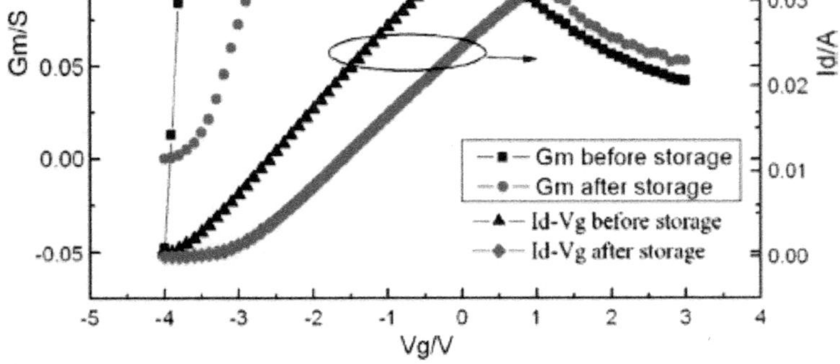

Figure 8.3 Transfer characteristics before and after thermal storage. After thermal storage, the transconductance decreases and the threshold voltage increases. From Ref. [3]. Copyright @ 2010 IEEE.

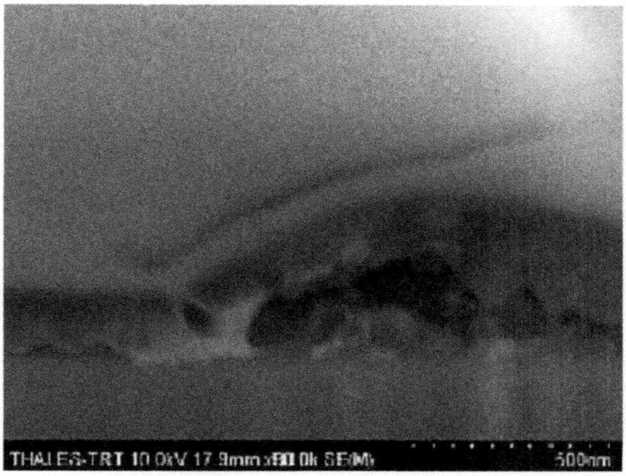

Figure 8.4 Focused-ion-beam (FIB) cross section showing Ga-rich (gray areas) and Au-rich (white areas) compounds and voids after 2000 h at 340°C. Reprinted from Ref. [4], Copyright (2009), with permission from Elsevier.

8.1.2 Variation of Device Performance under Different Temperatures

The temperature in a GaN HEMT is not always constant while it is working. It varies with the environment temperature and the operating current. Also, for GaN HEMT devices working in different applications, the temperatures in the device are also different due to different device sizes and output power. Therefore, the investigation of device performance under different temperatures is very important for confirming and improving standard working temperatures.

Experiments have shown that the time to breakdown of the gate of a GaN HEMT has a weak dependence on temperature when the gate is stressed [6]. Figure 8.5 shows the relation between the time to breakdown and temperature under a –55 V gate stress. The η values (63.2% failure) of the time-dependent breakdown (t_{BD}) distributions (Fig. 8.6) follow the Arrhenius relation (Fig. 8.7) with a 0.12 eV activation energy [6].

However, the generation of leakage hot spots at the edge of the gate near the drain is found to be strongly accelerated by higher temperature [7, 8], while for the average leakage current per leakage hot spot, no significant increase is observed when temperature rises [7].

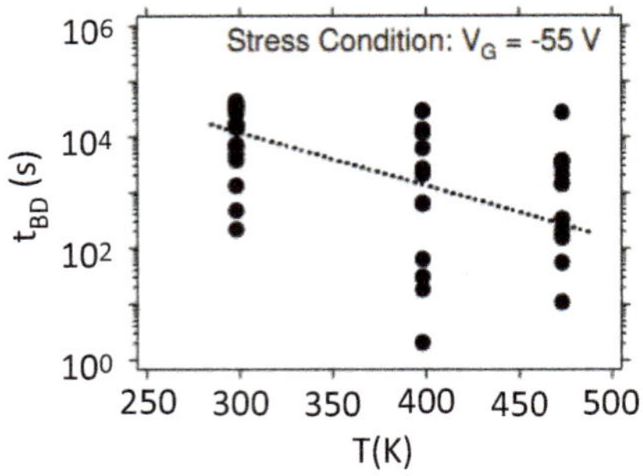

Figure 8.5 Time-dependent breakdown (t_{BD}) in relation to the temperature at $V_G = -55$ V. From Ref. [6]. Copyright @ 2010 IEEE.

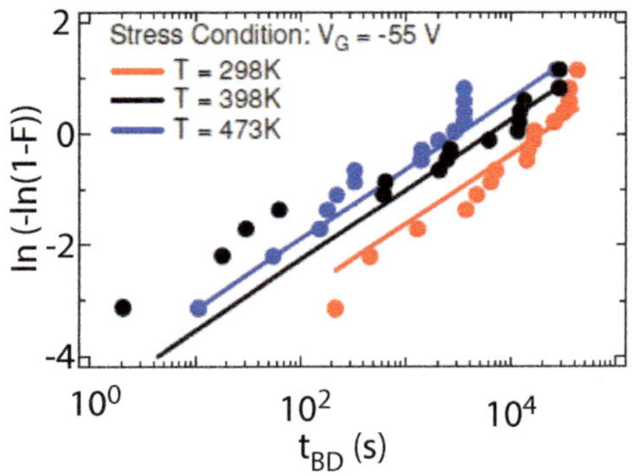

Figure 8.6 Weibull plot of the time-dependent breakdown (t_{BD}) distributions for three temperatures used during t_{DB} experiments at $V_G = -55$V. Vertical columns of data are time-outs (and not failures), which are consistently included in the Weibull fitting algorithm. From Ref. [6]. Copyright @ 2010 IEEE.

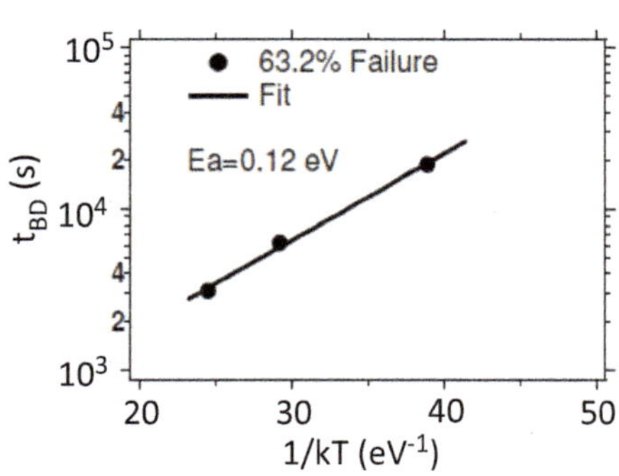

Figure 8.7 Arrhenius relation of the t_{BD} on the 63.2% of the population (*h* of the Weibull distributions). The activation energy is 0.12 eV. From Ref. [6]. Copyright @ 2010 IEEE.

At room temperature, the gate leakage current may be dominated by the conductive dislocations in AlGaN and at the interface of AlGaN/GaN, whereas at higher temperature, Frenkel–Poole emission in Schottky diodes will dominate the leakage current [9].

8.2 Current Collapse

One of the most severe reliability issues of a GaN HEMT is current collapse. It has been found that under large DC stress measurement, the knee voltage (V_{knee}) and the on-resistance will increase and the maximum current density (I_{Dmax}) as well as the maximum transconductance will decrease [10]. A schematic of output characteristics of a GaN HEMT before and after current collapse is shown in Fig. 8.8. Besides, after AC stress, current frequency dispersion is observed, that is, the output current is dramatically reduced compared to that under DC conditions [11]. Figure 8.9 demonstrates the radio frequency (RF) dispersion in GaN HEMTs.

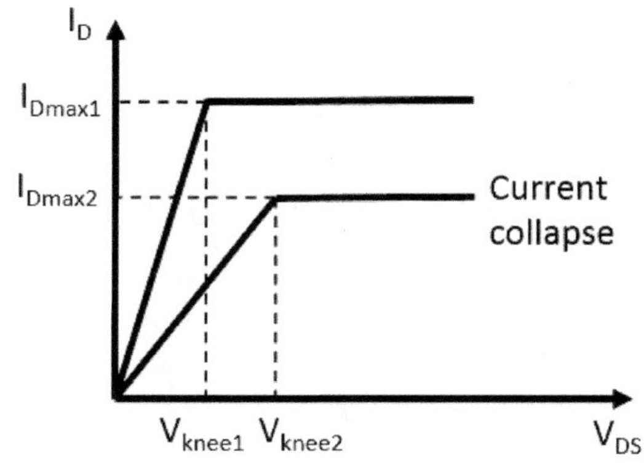

Figure 8.8 A schematic of output characteristics of a GaN HEMT before and after current collapse. The knee voltage (V_{knee}) and the on-resistance increase and the maximum current density (I_{Dmax}) decreases after current collapse.

Current collapse is mainly caused by traps in the device. The traps located under the gate or in the gate–drain access region can be separately identified by pulsed I_D–V_D and I_D–V_G characterization [12]. If the trapping exists under the gate, there will be a dynamical shift in the threshold voltage since the net polarization charge and two-dimensional electron gas (2DEG) sheet charge density are correlated to the negative charge trapped under the gate [13]. Experiments indicate that the shift is strongly correlated to the current collapse in the saturation region, suggesting charge trapping under the gate region, into the semiconductor epilayers [14]. If the trapping takes place in the channel between gate and drain, on-resistance will increase. Figure 8.10 shows the possible trapping mechanisms in semi-on-state and off-state bias operating points. In the semi-on-state, the trapping of negative charge may dominate and hot electrons can promote the trapping. In the off-state, the charge trapping may be promoted by the high reverse gate voltage and high gate leakage current levels [14]. Under the condition of a high electric field, electrons are accelerated to be hot electrons with high energy. These electrons may spill over the channel and be trapped by the traps outside the channel [15]. To avoid a high electric field, and thereby hot electrons, a field plate technology can be adopted to inhibit an electric field peak [16].

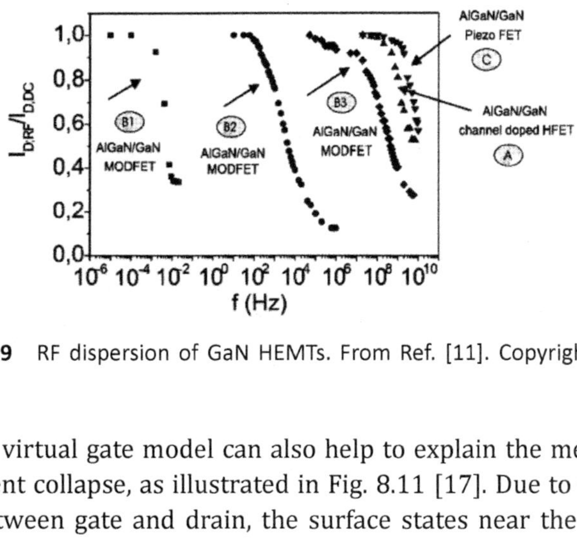

Figure 8.9 RF dispersion of GaN HEMTs. From Ref. [11]. Copyright @ 2001 IEEE.

The virtual gate model can also help to explain the mechanism of current collapse, as illustrated in Fig. 8.11 [17]. Due to the large bias between gate and drain, the surface states near the gate are negatively charged as a virtual gate. Therefore, the electrons are depleted under the virtual gate since the polarization is reduced in this region. Consequently, the maximum drain current decreases and current collapse occurs. Trew et al. also proposed a similar degradation mechanism [18]. They found that electrons could quantum-mechanically tunnel from the gate electrode to the semiconductor surface near the gate on the drain side, forming a negative charge layer. The negative charge layer on the AlGaN surface could deplete the conducting channel, resulting in current degradation. Surface passivation can reduce the current collapse prominently, which further proves the applicability of the virtual gate model [17, 19].

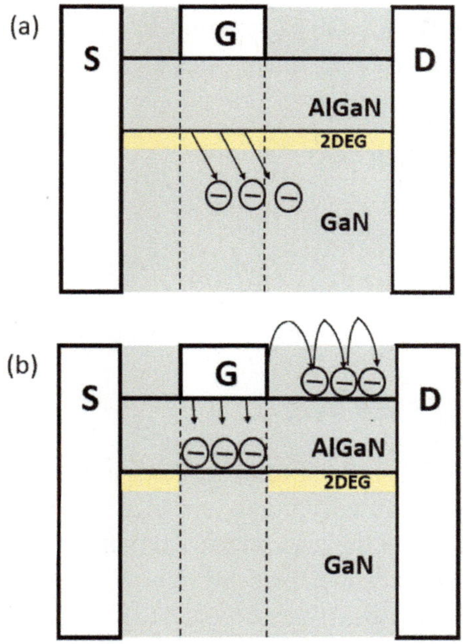

Figure 8.10 Possible trapping mechanisms in (a) semi-on-state and (b) off-state bias operating points.

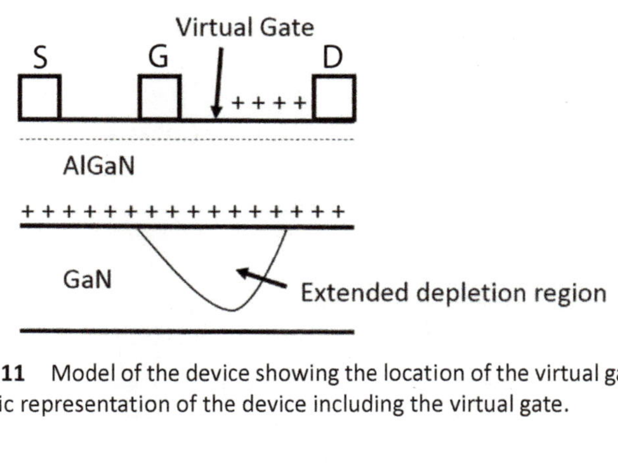

Figure 8.11 Model of the device showing the location of the virtual gate and a schematic representation of the device including the virtual gate.

Besides field plate technology and surface passivation, to inhibit current collapse, the material growth technique needs to be improved. Experiments show that a good growth technique of III nitride materials is strongly related to good reliability of the devices [20]. A high-quality material offers less traps, so current reduction can be inhibited under high current. And the optimization of the Schottky contact is also very important to improve the reliability of the device to avoid current collapse [21].

8.3 Prebreakdown

A GaN material has a high theoretical breakdown electric field, which can reach 3.3×10^6 V/cm, ten times higher than that of Si [22]. However, there are still some factors limiting the breakdown voltage, so GaN HEMT devices often have a prebreakdown.

8.3.1 Electric Field Peak

Under the reverse-bias condition, the electric field of a GaN HEMT is not uniformly distributed along the channel between gate and drain. It is extremely high at the drain side of the gate, so tunneling leakage will increase significantly once the electric field at this location reaches the breakdown electric field [23]. Therefore, making the electric field between gate and drain uniform is meaningful for enhancing the breakdown voltage.

A field-modulating plate can effectively mitigate the electric field peak [24]. The structure is illustrated in Fig. 8.12. The field plate is electrically connected with the gate so that the electric field under the gate can be modulated. Furthermore, multiple-grating field plates (MGFPs) have been shown to be more effective in inhibiting the electric field peak effect [25]. A structure of an MGFP is shown in Fig. 8.13. However, a field plate brings parasitic capacitance between gate and drain, which can degrade the device performance.

Another technology to reduce the peak electric field is doping a magnesium layer under the 2DEG channel [26]. Figure 8.14 gives the structure of a GaN HEMT with a magnesium layer. The Mg layer here functions as a charge-balanced floating field plate without parasitic capacitance. The modulation of the electric field using different lengths of magnesium layer is demonstrated in Fig. 8.15.

L is the length of the magnesium layer. The electric field peak can be effectively inhibited when the length of the magnesium layer exceeds 1.5 μm in this case.

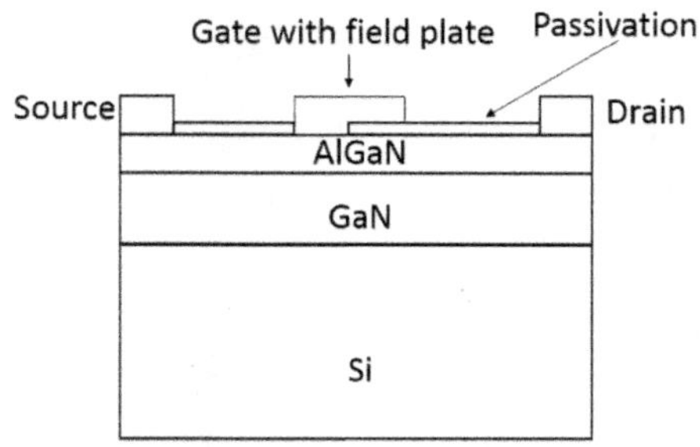

Figure 8.12 Schematic of a GaN HEMT with a field plate.

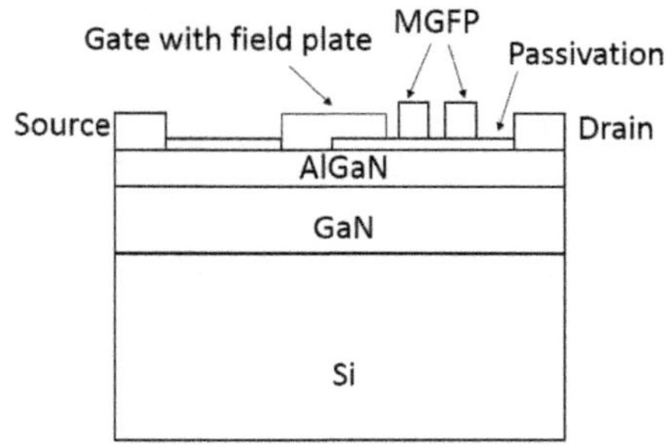

Figure 8.13 Schematic of a GaN HEMT with an MGFP.

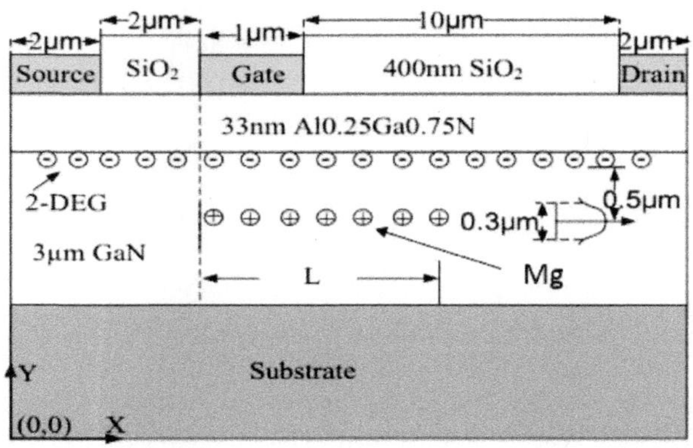

Figure 8.14 GaN HEMT with a magnesium layer. Reprinted with permission from Ref. [26].

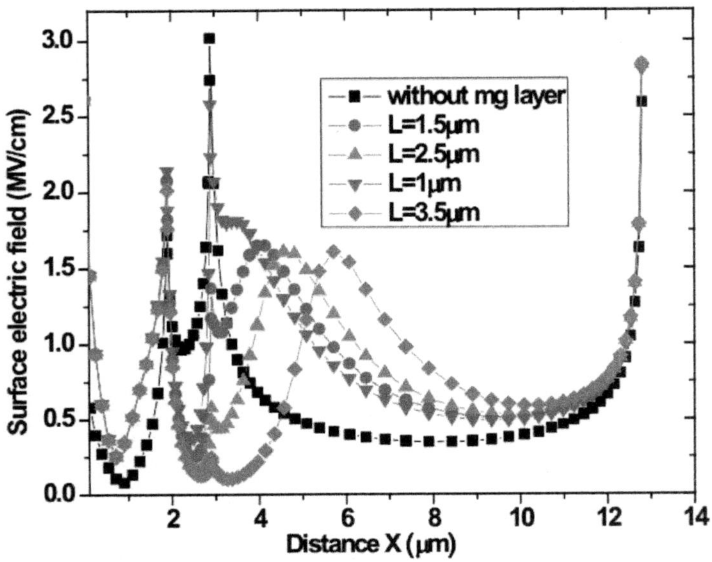

Figure 8.15 Simulated surface electric field distribution along the AlGaN/GaN interface for different *L*s, $V_{GS} = -5$ V, $T = 300$ K (*Y* axis: 8 μm, *X* axis: 2–15 μm). Reprinted with permission from Ref. [26].

8.3.2 Bulk Leakage

For a Si substrate GaN HEMT, the large lattice mismatch between Si and GaN causes obvious bowing and severe cracks. Defects originated from the bowing and cracking can reduce the insulation performance of the GaN buffer layer and the Si substrate, leading to severe bulk leakage and a low vertical breakdown voltage. In addition, the good conductivity of Si compared to III nitrides also contributes to the bulk leakage a lot [27]. One solution to this problem is increasing the thickness of the GaN buffer layer, on which the i-GaN is grown [28]. Figure 8.16 shows the effect of increasing the GaN buffer thickness. It demonstrates strong correlation of the thickness of the GaN buffer layer with leakage. The basic device structure is demonstrated in the inset of Fig. 8.16. Besides, using AlGaN instead of GaN as the buffer layer due to a higher bandgap, and thereby higher resistivity of AlGaN is also helpful for improving the vertical breakdown voltage [27].

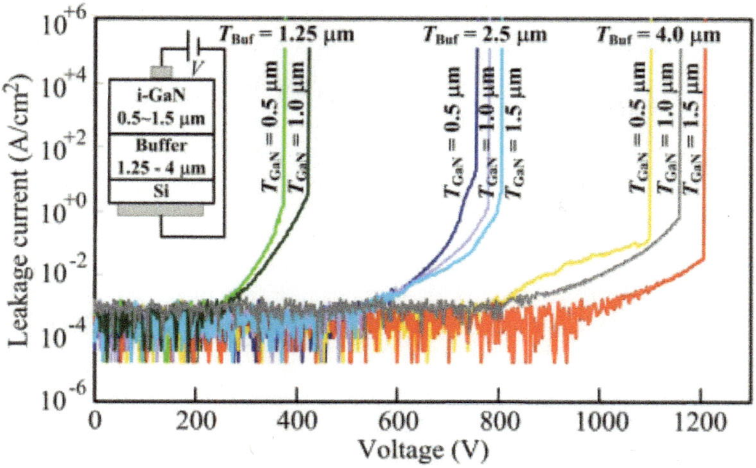

Figure 8.16 Vertical *I–V* measured across bulk GaN/Si with bias applied at the top electrode. The breakdown enhancement is small (11%) for increasing the GaN thickness and large (50%) for increasing the buffer thickness. From Ref. [28]. Copyright @ 2011 IEEE.

Additionally, though the AlN nucleation layer between a GaN buffer and a Si substrate can reduce the GaN-Si lattice mismatch, under a high drain–gate voltage, simulation shows there are AlN/Si interface inversion layer channel and local trap tunneling effect in the buffer layer, which is demonstrated in Fig. 8.17 [29]. This can also result in a severe current leakage path from drain to gate or drain to source. The device dominated by this effect would break down at a very low voltage [29]. Figure 8.18 provides the schematic band diagram and the buffer leakage model across the AlGaN/GaN/AlN/Si multiple layers. With the increase of the drain voltage, electron tunneling occurs due to the increase of the vertical electric field. The tunneling effect booms as the electric field continuously rises up. Also, the tunneling efficiency is strongly correlated to the band difference as well as the barrier thickness [29]. The problem can be solved by the addition of a high-barrier P-type buried layer between the AlN Layer and the Si substrate [29], which is shown in Fig. 8.19. After the treatment, the concentration of electrons in the AlN/Si interface inversion layer is reduced. The breakdown voltage of the devices can be enhanced significantly.

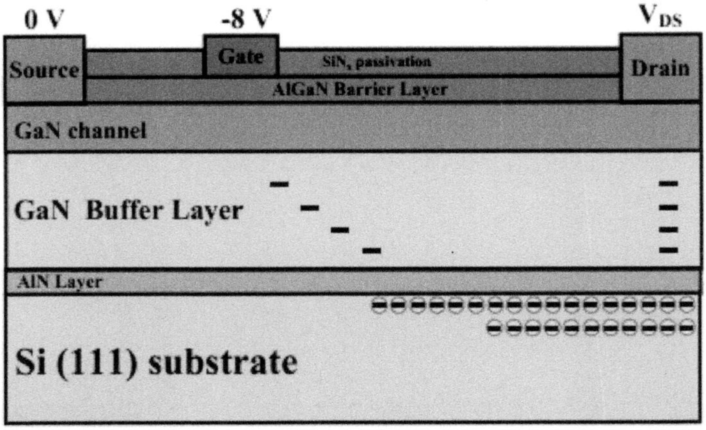

Figure 8.17 Schematic of a GaN HEMT with the inversion electron channel and the trap distributions in the buffer layer. From Ref. [29]. Copyright @ 2013 IEEE.

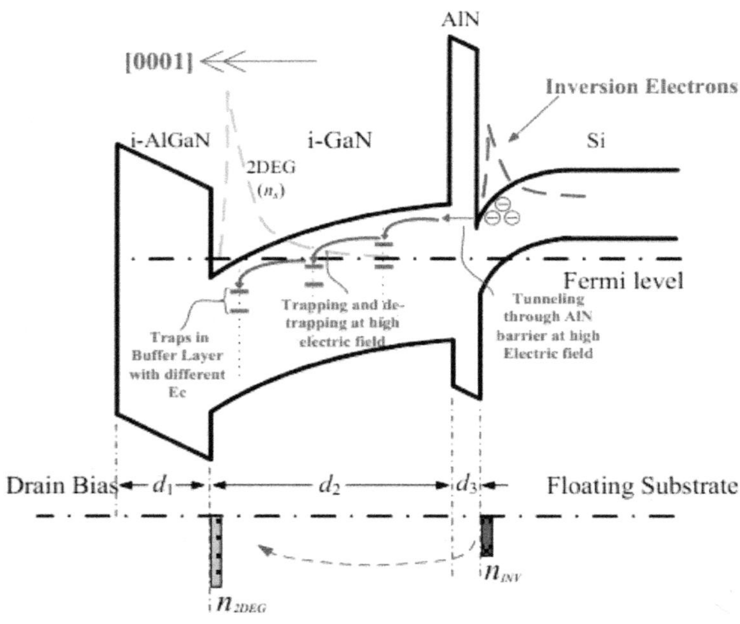

Figure 8.18 Schematic band diagram and buffer leakage model across AlGaN/GaN/AlN/Si multiple layers. The dimensional parameters include i-AlGaN thickness d_1 = 33 nm, i-GaN thickness d_2 = 3 µm, and AlN thickness d_3 = 10 nm. The green column and the red column are the 2DEG and the inversion electron channel concentrations, respectively. From Ref. [29]. Copyright @ 2013 IEEE.

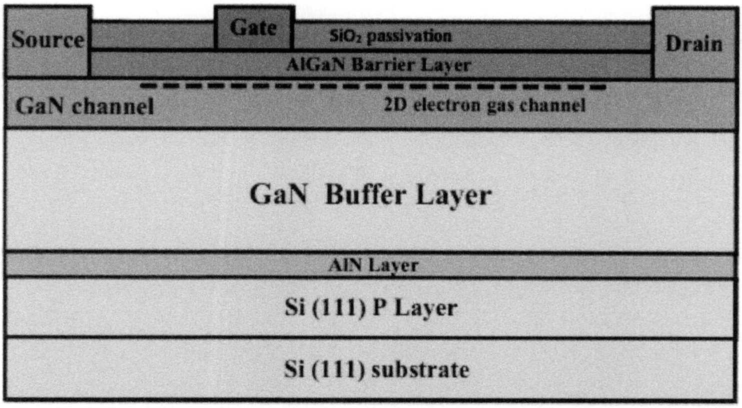

Figure 8.19 Cross-sectional view of a device with a high-barrier P-type buried layer. From Ref. [29]. Copyright @ 2013 IEEE.

8.3.3 Inverse Piezoelectric Effect

An inverse piezoelectric effect was also proposed for the breakdown mechanism at the strong reverse bias by Jungwoo and del Alamo [30]. They pointed out that high voltage, instead of high current density, is the main reason for degradation. When the device is under high voltage stress, an inverse piezoelectric effect may lead to the generation of defects and breakdown. Strain would be generated in the lattice of III nitrides under the electric field due to an inverse piezoelectric effect. Then the lattice may relax at some critical point when the electric field is strong enough. Cracks will be generated following the relaxation. The critical voltage and the abruptly increased leakage current (I_{Dmax}) are shown in Fig. 8.20. Figure 8.21 shows the transmission electron microscopy (TEM) images of the pits, cracks, and degradation at the edge of the gate, including the pits on both the source- and drain-side edges of the gate (Fig. 8.21a), a crack at the drain-side edge of the gate (Fig. 8.21b), and a severe case of degradation where the gate metal (Pt) has diffused into the crack formed (Fig. 8.21c) [31].

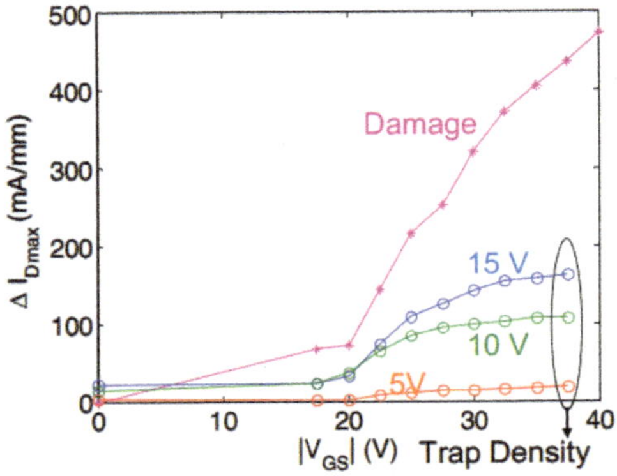

Figure 8.20 Cumulative damage and change in leakage current (I_{Dmax}) in response to diagnostic pulses (V_{GS} = –5, –10, and –15 V) in the step-stress-recovery experiment start beyond 20 V. From Ref. [30]. Copyright @ 2006 IEEE.

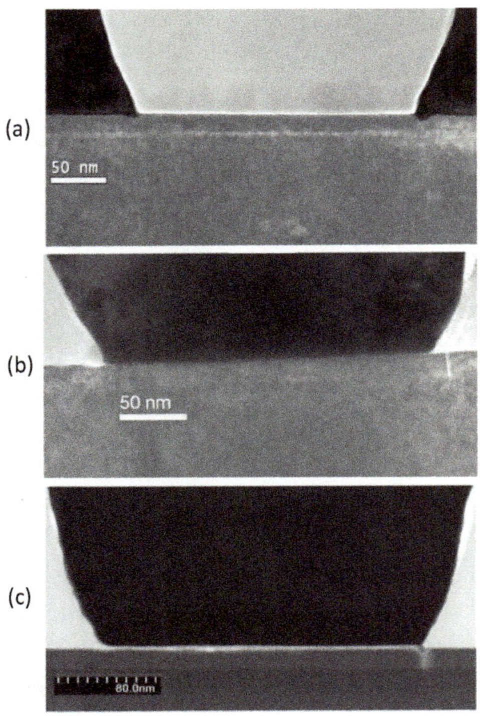

Figure 8.21 Cumulative damage and change in leakage current (I_{Dmax}) in response to diagnostic pulses (V_{GS} = –5, –10, and –15 V) in the step-stress-recovery experiment start beyond 20 V. From Ref. [31]. Copyright @ 2008 IEEE.

8.4 Conclusion

The reliability of GaN devices is one of the biggest issues that are an obstacle in its commercialization. Several degradation phenomena and mechanisms are summarized in this chapter. For future widespread application of GaN HEMT devices, further optimizations are still strongly required, such as optimizing growth technology of materials to avoid defects and traps, novel device structure design like optimized Schottky contacts, addition of a new interlayer or a new material, and surface passivation to enhance the performance of the devices.

References

1. Cheney, D. J., Douglas, E. A., Liu, L., Lo, C. F., Xi, Y. Y., Gila, B. P., et al. (2013). Reliability studies of AlGaN/GaN high electron mobility transistors, *Semicond. Sci. Technol.*, **28**.
2. Danesin, F., Tazzoli, A., Zanon, F., Meneghesso, G., Zanoni, E., Cetronio, A., et al. (2008). Thermal storage effects on AlGaN/GaN HEMT, *Microelectron. Reliab.*, **48**, 1361–1365.
3. Zhao, M., Wang, X., Liu, X., Huang, J., Zheng, Y., and Wei, K. (2010). Thermal storage of AlGaN/GaN high-electron-mobility transistors, *IEEE Trans. Device Mater. Reliab.*, **10**, 360–365.
4. Piazza, M., Dua, C., Oualli, M., Morvan, E., Carisetti, D., and Wyczisk, F. (2009). Degradation of TiAlNiAu as ohmic contact metal for GaN HEMTs, *Microelectron. Reliab.*, **49**, 1222–1225.
5. Miura, N., Nanjo, T., Suita, M., Oishi, T., Abe, Y., Ozeki, T., et al. (2004). Thermal annealing effects on Ni/Au based Schottky contacts on n-GaN and AlGaN/GaN with insertion of high work function metal, *Solid-State Electron.*, **48**, 689–695.
6. Marcon, D., Kauerauf, T., Medjdoub, F., Das, J., Van Hove, M., Srivastava, P., et al. (2010). A comprehensive reliability investigation of the voltage-, temperature- and device geometry-dependence of the gate degradation on state-of-the-art GaN-on-Si HEMTs, in *International Electron Devices Meeting*, San Francisco, CA, pp. 20.3.1–20.3.4.
7. Sun, H. R., Bajo, M. M., Uren, M. J., and Kuball, M. (2015). Progressive failure site generation in AlGaN/GaN high electron mobility transistors under OFF-state stress: Weibull statistics and temperature dependence, *Appl. Phys. Lett.*, **106**, 20–23.

8. Heller, E., Choi, S., Dorsey, D., Vetury, R., and Graham, S. (2013). Electrical and structural dependence of operating temperature of AlGaN/GaN HEMTs, *Microelectron. Reliab.*, **53**, 872–877.

9. Zhang, H., Miller, E. J., and Yu, E. T. (2006). Analysis of leakage current mechanisms in Schottky contacts to GaN and $Al_{0.25}Ga_{0.75}N$/GaN grown by molecular-beam epitaxy, *J. Appl. Phys.*, **99**, 023703.

10. Khan, M. A., Shur, M. S., Chen, Q. C., and Kuznia, J. N. (1994). Current-voltage characteristic collapse in AlGaN/GaN heterostructure insulated gate field-effect transistors at high drain bias, *Electron. Lett.*, **30**, 2175–2176.

11. Daumiller, I., Theron, D., Gaquiere, C., Vescan, A., Dietrich, R., Wieszt, A., et al. (2001). Current instabilities in GaN-based devices, *IEEE Electron Device Lett.*, **22**, 62–64.

12. Meneghini, M., Ronchi, N., Stocco, A., Meneghesso, G., Mishra, U. K., Pei, Y., et al. (2011). Investigation of trapping and hot-electron effects in GaN HEMTs by means of a combined electrooptical method, *IEEE Trans. Electron Devices*, **58**, 2996–3003.

13. Kumar, S. P., Agrawal, A., Kabra, S., Gupta, M., and Gupta, R. (2006). An analysis for AlGaN/GaN modulation doped field effect transistor using accurate velocity-field dependence for high power microwave frequency applications, *Microelectron. J.*, **37**, 1339–1346.

14. Bisi, D., Meneghini, M., de Santi, C., Chini, A., Dammann, M., Bruckner, P., et al. (2013). Deep-level characterization in GaN HEMTs-part I: advantages and limitations of drain current transient measurements, *IEEE Trans. Electron Devices*, **60**, 3166–3175.

15. Binari, S. C., Klein, P. B., and Kazior, T. E. (2002). Trapping effects in GaN and SiC microwave FETs, *Proc. IEEE*, **90**, 1048–1058.

16. Hwang, I., Kim, J., Chong, S., Choi, H. S., Hwang, S. K., Oh, J., et al. (2013). Impact of channel hot electrons on current collapse in AlGaN/GaN HEMTs, *IEEE Electron Device Lett.*, **34**, 1494–1496.

17. Vetury, R., Zhang, N. Q. Q., Keller, S., and Mishra, U. K. (2001). The impact of surface states on the DC and RF characteristics of AlGaN/GaN HFETs, *IEEE Trans. Electron Devices*, **48**, 560–566.

18. Trew, R. J., Liu, Y., Kuang, W. W., and Bilbro, G. L. (2006). The physics of reliability for high voltage AlGaN/GaN HFET's, in *IEEE Compound Semiconductor Integrated Circuit Symposium*, San Antonio, TX, pp. 103–106.

19. Kunii, T., Totsuka, M., Kamo, Y., Yamamoto, Y., Takeuchi, H., Shimada, Y., et al. (2004). A high reliability GaN HEMT with SiN passivation by Cat-CVD, in *IEEE Compound Semiconductor Integrated Circuit Symposium*, pp. 197–200.

20. Alexeev, A. N., Chaly, V. P., Krasovitsky, D. M., Mamaev, V. V., Petrov, S. I., and Sidorov, V. G. (2014). Features and benefits of III-N growth by ammonia-MBE and plasma assisted MBE, *J. Phys. Conf. Ser.*, **541**, 012030.

21. Jung, H., Behtash, R., Thorpe, J. R., Riepe, K., Bourgeois, F., Blanck, H., et al. (2009). Reliability behavior of GaN HEMTs related to Au diffusion at the Schottky interface, *Phys. Status Solidi C*, **6**, S976–S979.

22. Gaudenzio, M., Matteo, M., and Enrico, Z. (2014). Breakdown mechanisms in AlGaN/GaN HEMTs: an overview, *Jpn. J. Appl. Phys.*, **53**, 100211.

23. Sathaiya, D. M. and Karmalkar, S. (2007). Edge effects on gate tunneling current in HEMTs, *IEEE Trans. Electron Devices*, **54**, 2614–2622.

24. Ando, Y., Okamoto, Y., Miyamoto, H., Nakayama, T., Inoue, T., and Kuzuhara, M. (2003). 10-W/mm AlGaN-GaN HFET with a field modulating plate, *IEEE Electron Device Lett.*, **24**, 289–291.

25. Bahat-Treidel, E., Hilt, O., Brunner, F., Sidorov, V., Wurfl, J., and Trankle, G. (2010). AlGaN/GaN/AlGaN DH-HEMTs breakdown voltage enhancement using multiple grating field plates (MGFPs), *IEEE Trans. Electron Devices*, **57**, 1208–1216.

26. Xie, G., Zhang, B., Fu, F. Y., and Ng, W. T. (2010). Breakdown voltage enhancement for GaN high electron mobility transistors, in *22nd International Symposium on Power Semiconductor Devices & IC's (ISPSD)*, Hiroshima, pp. 237–240.

27. Domenica, V., Marleen V., H., Joff, D., Stefan, D., Maarten, L., Kai, C., et al. (2009). AlGaN/GaN/AlGaN double heterostructures on silicon substrates for high breakdown voltage field-effect transistors with low on-resistance, *Jpn. J. Appl. Phys.*, **48**, 04C101.

28. Rowena, I. B., Selvaraj, S. L., and Egawa, T. (2011). Buffer thickness contribution to suppress vertical leakage current with high breakdown field (2.3 MV/cm) for GaN on Si, *IEEE Electron Device Lett.*, **32**, 1534–1536.

29. Tang, C., Sheng, K., and Xie, G. (2013). Buffer leakage induced pre-breakdown mechanism for AlGaN/GaN HEMTs on Si, in *International Conference on Communications, Circuits and Systems (ICCCAS)*, Chengdu, pp. 353–357.

30. Jungwoo, J., and del Alamo, J. A. (2006). Mechanisms for electrical degradation of GaN high-electron mobility transistors, in *International Electron Devices Meeting*, San Francisco, CA, pp. 1–4.

31. Chowdhury, U., Jimenez, J. L., Lee, C., Beam, E., Saunier, P., Balistreri, T., et al. (2008). TEM observation of crack- and pit-shaped defects in electrically degraded GaN HEMTs, *IEEE Electron Device Lett.*, **29**, 1098–1100.

Chapter 9

The Packaging Technologies for GaN HEMTs

Dong Bin
Department of Electrical and Electronic Engineering, South University of Science and Technology of China, No 1088, Xueyuan Road, Shenzhen, Guangdong, 518055, China
dongb@mail.sustc.edu.cn

In this chapter, the packaging technologies for GaN HEMTs are introduced, including the advances of packaging technologies and the commonly used package types in GaN device packaging.

9.1 Introduction

With an increasing demand for various electronic applications under high-frequency, high-power-density, and high-temperature working conditions, conventional Si-based power devices face their inherent limitation. GaN-based high-electron-mobility transistors (HEMTs) are considered a promising candidate to work in a harsher environment by virtue of wider bandgaps, higher electron mobility, and higher breakdown voltages. For power devices, packages

Gallium Nitride Power Devices
Edited by Hongyu Yu and Tianli Duan

ISBN 978-981-4774-09-3 (Hardcover), 978-1-315-19662-6 (eBook)
www.panstanford.com

exert a significant impact on their performance, such as electrical characteristics, robustness, and thermal dissipation. As for GaN HEMTs, greater demands are placed on the packaging technologies because the harsher working conditions bring a series of difficulties in the packaging design.

The assembling and packaging process often introduces external resistance and capacitance [1] to the devices, which usually have a negligible impact on low-power systems. However, it would be of difference in high-power applications. For example, in high-power modules, like inverters for the photovoltaic systems, a tiny inductor could cause high heat generation. Thus, the packages for GaN HEMTs, which are mainly used in high-frequency and high-voltage environments, are of great importance. Additionally, thermal managing challenges for GaN HEMTs cannot be adequately addressed using a conventional packaging design [2]. GaN HEMTs often work in much higher temperatures than Si metal-oxide-semiconductor field-effect transistors (MOSFETs), and the junction temperatures of HEMTs can reach up to 300°C [3]. In addition, the heat flux of GaN devices can be as large as several hundreds of W/cm^2. However, a GaN material has a relatively low thermal conductivity. Therefore, the thermal dissipation for GaN devices is a big challenge [4]. On the other hand, the thermal degradation of GaN HEMTs could be serious. As the operating temperature increases, the current density and efficiency of GaN HEMTs are degraded. High-temperature environments also damage the lifetime of devices [5]. Additionally, because there is a severe mismatch of thermal expansion in different epitaxial layers, high operating temperature may cause severe mechanical deformation of GaN HEMTs [6].

For the above reasons, the packages for GaN HEMTs have attracted much attention. Related research interests mainly focus on solving the most vexing issues, such as parasitic inductors during assembling and thermal dissipation of packaged products. The detailed challenges and advances are reviewed in the next section.

Though facing such packaging challenges, some commercial GaN power products have been launched. Most of the products adopt packaging technologies modified from some conventional packages for power devices. Moreover, some leading corporations in GaN power electronics also lay emphasis on new packaging technologies and launched some commercial products using new packages. These state-of-the-art packaging technologies combined with commercial products are reviewed in this chapter.

9.2 Challenges and Advances in Packages for GaN HEMTs

Apart from providing a mechanical support, packages should also provide excellent electrical interconnection between devices and external circuits. Wire- and ribbon-bonding methods are commonly used for interconnection in power electronics because of their low-cost advantage [7]. However, the large parasitic inductors introduced by wire bonding degrade the electrical performance of devices. Especially, it cannot take the advantage of GaN HEMTs in applications requiring high power density and high frequency [8]. On the other hand, a long-time operation in high-temperature environments may cause the performance degradation of HEMTs and reduce their lifetime. In particular, the high thermal mismatch between substrate and epilayers may cause the formation of defects and even fractures in devices [9], which becomes a big challenge in mass production for GaN HEMTs.

The challenges in terms of parasitic inductors and thermal dissipation for power electronics also drive the development of packaging technologies. Some new packaging technologies, such as new bonding materials and novel interconnecting methods, have been proposed to eliminate parasitic inductors. Two-dimensional and three-dimensional packaging technologies have been demonstrated to induce less parasitic inductors during assembling process [10]. In addition, in contrast to typical assembling technologies, 2D or 3D packages could significantly reduce the size and weight of products (even 40 to 50 times smaller than conventional products [11]). Regarding thermal-managing issues, the most effective way to manage heat dissipation is immobilizing a heat sink with the power device. Besides, packaging materials also have a significant influence on the heat-transferring capacity. In particular, because thermal interface materials (TIMs) offer heat-spreading routes by attaching devices to cold plates or substrates, great research interest is invested in the choosing of TIMs [12].

9.2.1 Advanced Technologies for Lower Parasitic Inductors

To decrease package-induced parasitic resistance, some advanced assembling solutions have been proposed. Conventional wire bonding was invented in the 1990s and was widely used in chip-on-board (COB) packages. COB packages usually include three steps: die attachment, wire bonding, and encapsulation. Due to the strengths of simple making and low cost, COB packages are widely used for power devices [13]. However, COB packages with wire-bonding interconnections occupy a large space. Meanwhile, bonding wires in packages introduce plenty of parasitic inductors. Nowadays, the miniaturization of products and energy saving are a popular trend in the power industry. Therefore, conventional COB packages are adverse to the developing trend in the industry. Recently, some new bonding technologies and assembling methods have been introduced to satisfy the demands of industry.

Adopting better bonding materials can help to decrease parasitic inductors [14]. Gold wire bonding could effectively reduce the package on-resistance. However, in high-temperature environments, the mechanical stability of gold wire may be affected and the package robustness would decline [15]. Palladium is an alternative material for wire bonding. This bonding material has been successfully used in the package for SiC-based sensors due to its superior mechanical stability in high temperature [4]. Recently, Zeiser et al. made comparative research on the gold and palladium wire bonding for GaN HEMTs and demonstrated that current palladium-bonding technologies were still lagging behind that of gold bonding (five times larger resistance and nearly the same mechanical performance) [16].

Flip-chip assembling technologies are introduced instead of COB for GaN HEMTs to achieve a smaller size, better electrical performance, and better thermal dissipation. In flip-chip packages, the bonding contacts are replaced with metalized bumps on a substrate, without bonding wire. [17]. Therefore, a direct electrical connection between chips and the carrier is created. Figure 9.1 shows a process of the flip-chip assembling technology [18]. Recently, GaN Systems announced that it utilized a 2.5D die stack for GaN HEMT packaging. The HEMTs are directly mated to complementary metal-oxide-semiconductor (CMOS) driver chips. These packaged HEMTs could work with a much higher switching speed as well as nearly no current crowding [19].

HEMT packages for high-efficiency and high-frequency applications, like power amplifiers, now get individual attention. Package-induced parasitic inductors are considered as a significant factor limiting the switching characteristics in high-efficiency and high-frequency applications [20]. In particular, GaN HEMTs are commonly packaged with cascode circuits to exhibit a normally off property [21]. However, cascode circuits introduce plenty of inductors, which significantly degrade the electrical performance of HEMTs. Recently, Zhang et al. packaged GaN HEMTs in features of a stacked-die structure, an embedded external capacitor, and a flip-chip configuration. This packaging design was demonstrated to reduce parasitic ringing and switching losses [8]. Pavlidis et al. established a review of GaN HEMTs packaged by different methods and demonstrated that flip-chipped GaN HEMTs on AlN exhibited an improved bandwidth of a hybrid power amplifier [22].

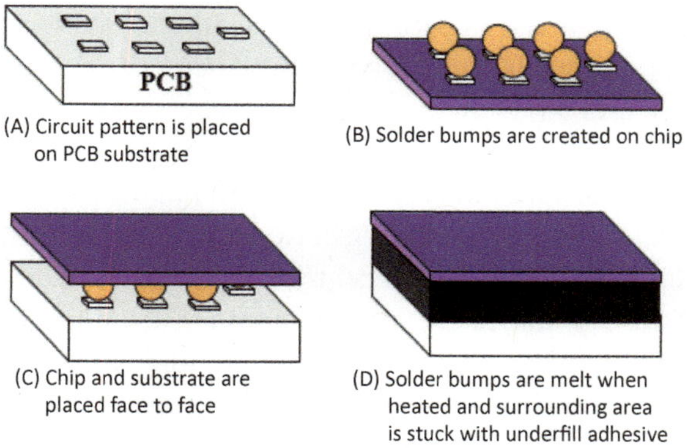

Figure 9.1 Flip-chip assembling process.

9.2.2 Thermal Management

A GaN material with a larger bandgap could work in higher operating temperatures because of its lower intrinsic carrier concentration in contrast to Si. However, the increased temperatures bring a significant degradation of performance in GaN HEMTs.

Because the two-dimensional electron gas (2DEG) density in GaN HEMTs significantly decreases in a high junction temperature, the carrier mobility of HEMTs would be reduced, which would cause a dramatic fall of electrical performance [23]. In addition, mechanical deformation in high temperatures also damages the performance and even the service life of GaN devices. The thermal effects on performances of GaN HEMTs were reviewed in Ref. [24]. In this work, a 12-finger HEMT with L_G = 0.25 μm and W_G = 150 μm was fabricated to analyze the performances of GaN HEMTs from 295 K to 540 K. With the temperature changing from 295 K to 540 K, a fall of output power (−0.138 dB/10°C), an increment of minimum noise (0.068 dB/10°C), and an increment of noise resistance (2.57 Ω/10°C) were observed when temperature started increasing.

One key role of packages is to provide heat dissipation paths for the devices. The operating temperatures of new-type power devices like GaN HEMTs are nearly reaching out of safety regulation in traditional packages. Therefore, some advanced packages for GaN HEMTs are adopted to manage thermal dissipation.

To take advantage of GaN devices, packaging materials with great influence on thermal dissipation capacity attracted much interest in the past decade. In particular, a proper selection of the TIM has been demonstrated to allow much better thermal dissipation capacity of GaN HEMTs, because it influences the thermal resistance of both devices and external cooling components [25]. Polymeric TIMs such as greases and gels are widely used because of their low cost and good thermal dissipation capacity. Recently, nanoparticles and nanotubes are introduced in the TIMs. Carbon nanotubes (CNTs) are of great concern because they have high thermal conductivity. However, the large interface thermal resistance limits their application. Recently, Prasher et al. demonstrated that a multiwalled CNT (MWCNT) with vertical contact has a much smaller contact resistance than that with horizontal contact. Therefore, a vertically grown CNT looks like a more promising TIM candidate [26, 27]. In addition to CNTs, nanoparticle-based TIMs were also reported [28]. These nanoparticles could significantly improve thermal conductance of filled materials.

The packaging configuration is another significant factor that determines the thermal dissipation capacity of devices. Recently, some specific configuration designs were proposed at the laboratory level to improve thermal dissipation. Cheng et al. adopted a V-groove copper base in transistor outline (TO)-3P packages for GaN HEMTs. The thermal resistance could be significantly reduced (19.6% lower than dual in-line [DIP] packages) [5]. Moreover, in 2010, Narayan et al. utilized nanoporous alumina structures achieve a thin-film evaporative cooling device with very high heat removal efficiency [29]. With the aid of nanoengineering, the thermal dissipation of packages could be improved significantly. However, these technologies are still facing great challenges. High cost, poor reproducibility, and complex design make it hard for mass production.

9.3 Package Types for GaN Products

On the basis of applications, GaN commercialized products with different standard packages have been available. Discrete GaN field-effect transistors (FETs) (HEMTs) are usually launched in transistor outline (TO) packages or surface-mount packages, which are commonly used for conventional Si power devices. Since GaN power devices usually work in a harsh environment, the conventional packages should be modified to improve the mechanical stability and thermal transfer capacity. These commercialized packaging technologies adopted by related companies are introduced in this section.

In conventional power electronics, the package types of semiconductors are dependent on the power density of the system. For electrical applications in the range of small power, discrete device packages with a single power electronic switch are prevailing, since discrete devices with a compact size are convenient for application and installation in the system design. To design a sophisticated circuit system, electrical engineers usually immobilize the various discrete devices in a printed circuit board (PCB) by soldering and build the entire circuit on the basis of the PCB. Therefore, the installation and substitution of discrete devices could be very easy. The discrete packages are commonly used for low-power semiconductors because of the poor thermal dissipation capacity. However, for high-power applications, power modules are packaged to improve the heat dissipation. In contrast to discrete devices, power modules provide much higher density by connecting multiple semiconductor dies. The circuit design and connection in a power module are called topology. The power modules can connect to other modules or discrete devices by means of soldering to a PCB. In addition, a heat sink is often packaged with the PCB to improve the heat dissipation capacity of the system. Similarly, in GaN electronics, the discrete packages and module packages are both available. Discrete GaN HEMTs have been launched with some conventional packages like TO-220 and power-quad-flat "no-lead" (PQFN), and they have been utilized instead of Si MOSFETs in some switching power supply circuits. In addition, a parallel operation of GaN HEMTs in a module could be achieved for some applications with a high power requirement.

In conventional power electronics, three-lead packages, like TO-220, are commonly used for packaging discrete power transistors. As for power integrated circuit (IC) applications, PQFN packages are mostly available. However, these conventional packages cannot take advantages of GaN HEMTs. Therefore, some enhanced packages like DirectFET and Land Grid Array (LGA) are adopted. DirectFET is a packaging technology created by International Rectifier. Nowadays, this technology is used to package the GaN radio frequency (RF) power transistors, and the packaged products have been introduced to some base station manufactures. These devices with larger bandwidths, higher power density, and higher efficiency could pave the path to 5G technology. Another major corporation in the GaN power field, Efficient Power Conversion (EPC), is pushing its LGA packages for GaN HEMTs. By adopting this method, EPC has launched its fourth-generation commercial products, Gen4 eGaN® FETs, ranging from 30 V to 200 V [30].

Driven by the advances in GaN devices and a great demand of high-efficiency and high-frequency applications, circuit design based on GaN is developing fast. GaN power switches with lower switching loss and higher efficiency are drawing much attention. AC–AC, AC–DC, and DC–DC converters based on GaN HEMTs are all available, which offers a number of potential benefits, including extremely high efficiency and much smaller volume of complete commercial products. In addition, all GaN power modules are of most concern. In the RF field, GaN amplifiers with high output power, large bandwidth, and high efficiency are available. The commercialization of these modules has already started. However, the large thermal dissipation, the stray inductances and capacitances, and the reliability of circuits limit the commercialization of GaN power modules. All of these issues are influenced by the packaging technologies, so system packaging is becoming one of the top priorities in the system design of GaN modules.

9.3.1 An Introduction to Commercialized GaN Products

In contrast to Si MOSFETs, the implementation of GaN FETs in the circuit faces some challenges because of the special characteristics of GaN FETs. The current biggest problem in the implementation is the normally on property of GaN FETs, which results in the unserviceability of conventional driver circuit design. A commercialized solution is to develop a voltage regulator circuit and connect it with the GaN FET. However, the additional circuit components increase both the cost and the size of the product. In addition, the external inductors introduced by the circuit degrade the electrical performance and increase the energy wastage. Therefore, the power supply firms do not appreciate this design. Considering the market promotion, the normally off products are preferred. In the laboratory, several methods have been used to achieve normally off GaN products. However, most of these methods are not likely to be applied in mass production because the uniformity is hard to control. Some commercialized normally off GaN products are summarized in the following content. Particularly, the packaging types and technologies for these commercialized products are discussed.

Cascode GaN HEMTs are demonstrated to exhibit a normally off property. This kind of device is composed of a cascode structure in terms of a depletion-mode GaN HEMT and a low-voltage Si MOSFET. The package for cascode devices is more challenging than that for a single depletion-mode GaN HEMT because this circuit design increases the difficulty of the thermal management. On the one hand, GaN power switches are proposed to operate in converters for high-power-density and high-operating-temperature applications. The harsh working conditions put obstacles in the way of packages for GaN HEMTs. However, the addition of a Si MOSFET reduces the conversion efficiency and causes more heat production. Thus, the thermal burden of the GaN HEMT is exacerbated. On the other hand, the package size should be minimized due to the downsizing requirement of the electronics industry. Thus, the thermal design should not cover too much room in the package. The trade-off between thermal consideration and size consideration sharpens the difficulty of designing the package for cascode HEMTs. Therefore, the package and thermal management become a key factor in the production of cascode GaN HEMTs. Recently, by using a modified packaging technology (this modified package is named Quiet TabTM by Transphorm), Transphorm has launched 600 V/650 V normally off HEMTs based on the cascode structure. The minimum gate threshold voltage is larger than 1.6 V.

In fact, the industry and researchers are trying to fabricate enhancement-mode (E-mode) GaN HEMTs in order to eliminate external inductors brought by the Si MOSFET. Nowadays, E-mode HEMTs are also available. Fujitsu and Panasonic have launched sampling products. EPC and GaN Systems also launched E-mode products, which are the main competitors of cascode HEMTs from Transphorm in GaN power markets. In addition to the optimization of device design, the packaging technologies are also a focus of competition.

9.3.2 Transistor Outline Packages

The TO family dominates the package market discrete transistors today. Particularly, the TO-220 package is a typical type in this package family. A TO-packaged discrete power device is usually with three leads: drain, source, and gate. However, with an increasing demand for a large die, more leads are introduced for the TO package. Nowadays, TO-220 packages with five or seven leads and even more are available. In a TO package, the die is attached to a pad by solder paste and the chip is placed on the leadframe. By wire bonding, the electrodes of the die are connected to the leads. In particular, in three-lead TO packages, the middle lead is usually connected to the exposed dap of the leadframe. To maintain mechanical stability, epoxy mold compound (EMC) is used as an encapsulant in the package. However, the poor thermal performance of the EMC limits its application [31]. Therefore, a heat sink is often required to improve the heat dissipation capacity, since the heat sink could ensure an unblocked heat transfer route. As Fig. 9.2 shows, there is a tab with a hole on the top of the TO package, and this tab is used to mount the packaged device to the heat sink.

Figure 9.2 (a) TO-220 packaging structure. (b) The specially designed heat sink for heat transfer.

Nowadays, TO-packaged GaN HEMTs are available. The 600 V/650 V GaN HEMT products from Transphorm are based on a modified TO package (Quiet Tab™ package). These products have passed the certification by the Joint Electron Device Engineering Council (JEDEC). To match with the general requirement in the circuit in terms of a normally off property of a power MOSFET, Transphorm utilizes a cascode design to fabricate its GaN products. Figure 9.3 shows a schematic of a circuit configuration in packaged cascode GaN HEMTs. Though the normally off characteristic is achieved, the additional Si MOSFET introduces a series of parasitic inductors, in particular the large common source inductor, which cause a huge energy loss [32]. Though still facing such a trouble, Quiet Tab™ adopted by Transphorm could overcome a big challenge in Si devices for high-power applications. In conventional TO packages for Si MOSFETs, only the drain tab could be available, which is limited by the structure of the Si material. However, the large transient current may bring a big disturbance to V_{gs}, especially in extremely high-power environments. The drain tab cannot address this problem, so the device faces a certain amount of risk in some high-power circuits. However, there is no such restriction in a GaN material. In fact, Quiet Tab™–packaged GaN HEMTs could use a source tab. The metal source tab helps to decrease the ringing of V_{gs} by carting current, and a Kelvin source pin is connected to the metal tab for the drive circuit. Besides, adopting a package with source tab is more suitable for low-side transistors in circuit because source could connect to ground in low-side transistors in such a package. Meanwhile, a TO package with a drain tab is suitable for high-side transistors since the drain side could connect to the DC terminal in high-side transistors (see Fig. 9.4). Particularly, to achieve a shorter wire in the packages, the pin-out arrangement of Quiet Tab^{TM} is G, S, and D (see Fig. 9.4) because driver signals could enter from the first pin and output from the second pin using such a sequence, which decreases the inner wire in packages. By means of these improvements, Quiet Tab^{TM}–packaged HEMTs could provide a higher switch speed than conventional packaged HEMTs. Thanks to the advanced packaging technologies, Transphorm's 650 V GaN FET product (TPH3207) in a TO-247 package now enjoys the lowest on-resistance in contrast to the counterparts in the same package. As the firm indicates, the on-resistance of this GaN FET is as low as 41 mΩ, while the reverse recovery charge (Q_{rr}) is only 175 nC. According to the estimation of Transphorm, the whole volume of the system adopting this GaN FET is able to be reduced as much as 50% [33].

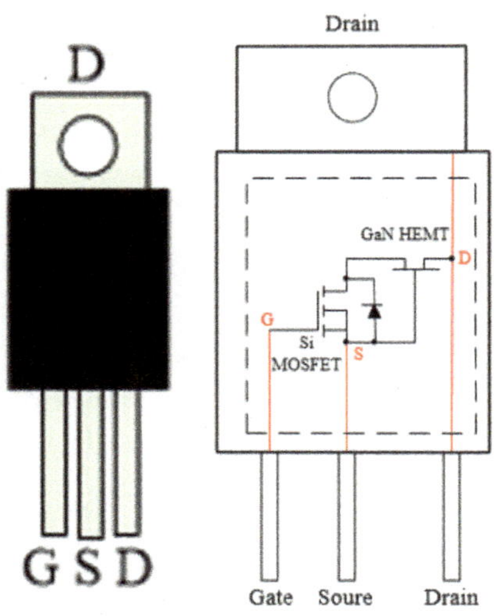

Figure 9.3 A schematic of the circuit configuration in packaged cascode GaN HEMTs.

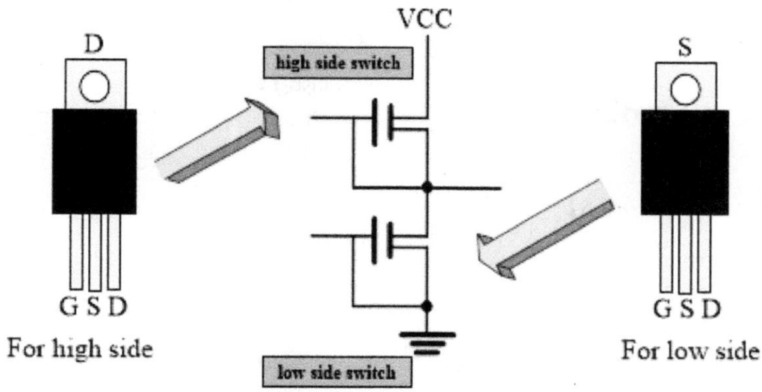

Figure 9.4 A schematic of the circuit configuration in packaged cascode GaN HEMTs.

Fujitsu and Panasonic also launched some TO-packaged GaN HEMTs for sale, but these products are still under trial manufacture. In 2013, Fujitsu Semiconductor Limited announced sampling of its 600 V GaN power device (using the TO-247 package) and low-voltage (150 V and 30 V) GaN devices (using wafer-level chip-scale packaging). The 600 V TO-247 GaN transistor exhibits a normally off property with a threshold voltage of 1.8 V. Fujitsu may have adopted recessed-gate technology for achieving the normally off property, because some researchers from Fujitsu once reported that a recessed-gate structure could push the threshold voltage of GaN HEMTs to 3 V [34]. However, due to the fabrication limitation of recessed-gate structures, the mass production may be a big challenge, which might account for the slow implementation of these products. In addition, Fujitsu announced that its research team and intellectual property rights were transferred to Transphorm in 2013. In exchange, Fujitsu invested in Transphorm and obtained an amount of stock. Nowadays, Transphorm authorizes the foundry owned by Fujitsu (the Aizu-Wakamatsu facility) for the fabrication of its GaN products, and Fujitsu acts as an agent for the GaN products of Transphorm. Another Japanese company, Panasonic Semiconductor Solutions, also announced the tentative sampling of an E-mode GaN device with a p-type gate in 2013. The firm claimed that this 600 V GaN device was the first commercial product free from current collapse. In addition, the TO-220 package was adopted for this transistor. In 2015, Panasonic announced that it would launch the industry's smallest E-mode GaN transistor using a dual-flat no-lead (DFN) package and the mass production of GaN power devices would be finished in 2016. These above-mentioned products all appeared in the exhibition of the Asia-Pacific Economic Cooperation (APEC) 2016. However, the mass production has not been achieved up to now.

Although TO packages are popular in the power industry due to its advantage of easy implementation, these types of packages cannot adequately take advantage of GaN power devices. The bonding wire in the packages induces copious parasitic inductors, which decrease the device efficiency and causes a huge power waste in high-power applications. In addition, the big size of TO packages is adverse to a thin type and miniaturizing trend of electronic products. Therefore, the GaN power companies prefer using surface-mount packaging solutions for GaN transistors.

9.3.3 Surface-Mount Technology

Unlike TO-220, surface-mount technology (SMT) is a leadless packaging technology intended for board-mounted power electronics. In contrast to the through-hole technology, SMT-packaged devices can be mounted to the PCB directly with smaller or even no leads. The leadless design of SMT packages further reduces the parasitic inductors introduced by packages, hence achieving a better electrical performance. In addition, the leadless design helps with increasing component density and reducing the whole size of electrical products. Besides, automated assembly solutions in industry could be simpler in SMT packages than in TO packages. Therefore, the assembly efficiency is improved with a reduction of cost at the same time. Because of these benefits, SMT has brought about a revolution to the electronic packaging market.

Particularly, the quad-flat no-lead (QFN) package is a commonly used SMT package for power electronics. QFN is a quad-in-line SMT, which is usually based on a planar copper leadframe substrate. In a QFN package, an exposed thermal pad could facilitate the heat transfer, and the perimeter lands under the pad provide a good electrical connection to the PCB. Modified from a QFN package, International Rectifier launched an advanced power-packaging technology with better electrical characteristics and thermal performance and named it PQFN (see the front view of a typical PQFN package from International Rectifier and Infineon [35] in Fig. 9.5). In contrast to the common QFN package, the firm optimized the design of the leadframe and thermal management. In a PQFN package, a bare-die pad is introduced to reduce the thermal resistance between the die and the PCB. The PQFN package with a single pad is most popular, and it has been registered in the JEDEC now. For some special applications, multiple pads are also available. In addition, IR pushed another package type in power electronics, the DirectFET package, which is a dual-side-cooled packaging technology. As a variant type of PQFN, DirectFET can further eliminate unnecessary elements for lower inductance and resistance. Figure 9.6 shows a cross-sectional view of both PQFN and DirectFET. As the figure shows, the DirectFET package does not adopt any wire bonding and leadframe, in contrast to the PQFN package. Therefore, parasitic inductors are further reduced in the DirectFET package [36].

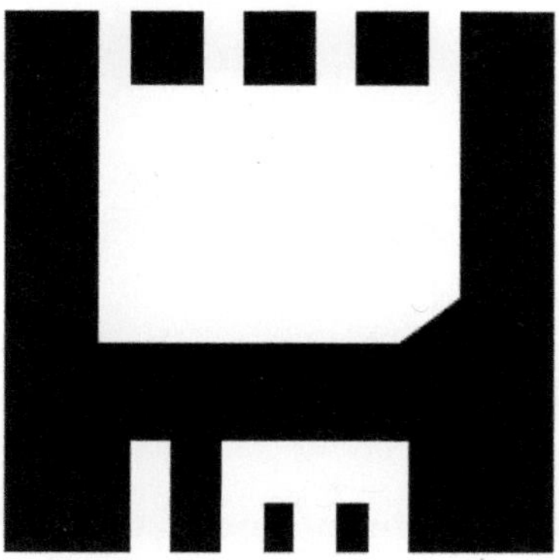

Figure 9.5 The front view of a typical PQFN package from International Rectifier & Infineon.

LGA has been hailed as a technological revolution in the package for microprocessors. LGA-packaged components can use a grid of contacts rather than pins for the connection to the PCB, which helps with avoiding the signal noise brought by pins. Therefore, Ball Grid Array (BGA) gradually replaced the conventional Socket 478 package for CPU chips in INTEL. Recently, EPC launched both single GaN transistors and modules based on the LGA package. Figure 9.7 shows a typical schematic of the LGA package. The source and drain terminals are placed alternately. Compared to PQFN and DirectFET, LGA can significantly reduce the overall size of the package. Though LGA may usually have a higher package resistance than DirectFET, it is better than the PQFN package. Utilizing its GaN FET products, EPC demonstrated some application demos, such as DC-DC converters and wireless power circuits. The firm claimed that eGaN® FETs could provide significant performance enhancements in contrast to Si devices, which is mainly because of the brilliant properties of GaN devices and also partly due to the fine characteristics of LGA packages.

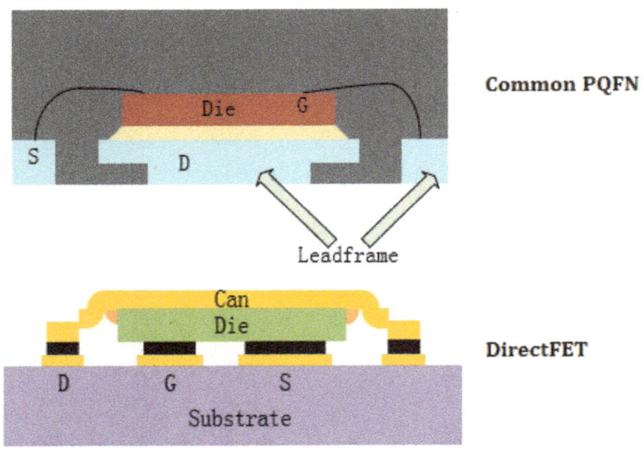

Figure 9.6 The cross-sectional view of common PQFN and DirectFET packages.

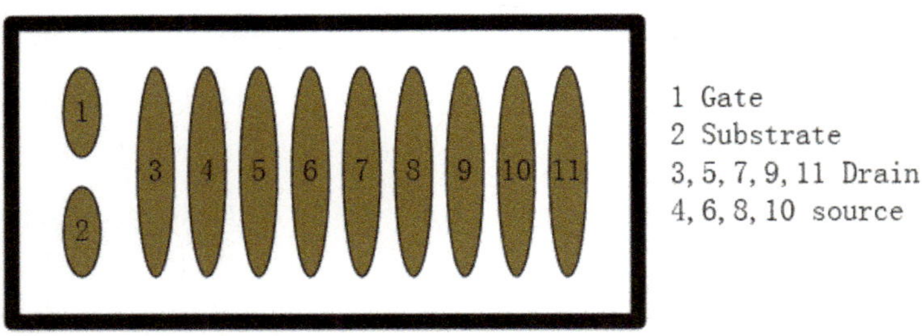

Figure 9.7 A typical schematic of the LGA package.

GaN Systems is a fabless producer of GaN power devices in Canada. This firm has launched 100 V and 600 V E-mode GaN HEMT transistors. In particular, GaN Systems adopts GaN-on-Si technology and the near-chip-scale package. The firm claims that it adopts special island technology in the fabrication of its GaN transistors. It uses the island structure rather than traditional fingers in the devices for a reduction in the size as well as better thermal performance. In addition, in its exclusive GaN*PX* package, the GaN die is embedded in a laminate without any wire bonds and molding compounds. All of these techniques are replaced by galvanic processes [37]. The firm claims that this package results in optimal thermal performance and extremely low inductance.

Due to the high-frequency capacity, GaN HEMTs also draw special attention in an RF field. In particular, many research interests are invested in the packages for GaN HEMTs and modules. According to the requirement in different applications, Cree provides different packaged products for commercial applications, including bare-die, packaged transistors, and packaged RF circuits. For example, Cree has launched some GaN HEMT–based products with functional circuits as an entity like monolithic microwave integrated circuits (MMICs). These MMIC products utilize surface-mount packages or Flange and Pill packages according to their demands. Another enterprise, TriQuint, has also launched GaN HEMT–based switches, power amplifiers, and low-noise amplifiers. The packaging methods for its products include QFN, PQFN, and Flange packages.

References

1. Lee, C.-H., Billings, D., and Wen, Y. (2015). New MOSFET packaging techniques for high power density, in *PCIM Asia 2015; International Exhibition and Conference for Power Electronics, Intelligent Motion, Renewable Energy and Energy Management*, Shanghai, China, pp. 1–8.
2. Wyk, J. D., and Lee, F. C. (2013). On a future for power electronics, *J. Power Electron.*, **1**, 59–72.
3. Bajwa, A., Qin, Y., Wilde, J., Reiner, R., Waltereit, P., and Quay, R. (2014). Assembly and packaging technologies for high-temperature and high-power GaN HEMTs, in *IEEE 64th Electronic Components and Technology Conference (ECTC)*, Orlando, FL, pp. 2181–2188.
4. Zhao, Y., Semenic, T., and Bhunia, A. (2013). Advanced packaging and thermal management for high power density GaN devices, in *IEEE Compound Semiconductor Integrated Circuit Symposium (CSICS)*, Monterey, CA, pp. 1–4.

5. Cheng, S., Chou, P.-C., Chieng, W.-H., and Chang, E. (2013). Enhanced lateral heat dissipation packaging structure for GaN HEMTs on Si substrate, *Appl. Therm. Eng.*, **51**, 20–24.
6. Choi, S., Heller, E., Dorsey, D., Vetury, R., and Graham, S. (2013). The impact of mechanical stress on the degradation of AlGaN/GaN high electron mobility transistors, *J. Appl. Phys.*, **114**, 164501.
7. Xu, T., Walker, T., Chen, R., Fu, J., and Luechinger, C. (2014). Advanced interconnect equipment and process development, in *IEEE 16th Electronics Packaging Technology Conference (EPTC)*, Singapore, pp. 564–569.
8. Lee, F. C., Zhang, W., Huang, X., Liu, Z., Du, W., and Li, Q. (2015). A new package of high-voltage cascode gallium nitride device for high-frequency applications, in *IEEE International Workshop on Integrated Power Packaging (IWIPP)*, Chicago, IL, pp. 9–15.
9. Liu, H., Dolmanan, S., Zhang, L., Chua, S., Chi, D., Heuken, M., et al. (2013). Influence of stress on structural properties of AlGaN/GaN high electron mobility transistor layers grown on 150 mm diameter Si (111) substrate, *J. Appl. Phys.*, **113**, 023510.
10. Zheng, L., Zhang, Y., Huang, G., and Bakir, M. S. (2014). Novel electrical and fluidic microbumps for silicon interposer and 3-D ICs, in *IEEE Trans. Compon. Packag. Manuf. Technol.*, **4**, 777–785.
11. Al-Sarawi, S. F., Abbott, D., and Franzon, P. D. (1998). A review of 3-D packaging technology, *IEEE Trans. Compon. Packag. Manuf. Technol. Part B*, **21**, 2–14.
12. Moore, A. L., and Shi, L. (2014). Emerging challenges and materials for thermal management of electronics, *Mater. Today*, **17**, 163–174.
13. Lau, J. H. (1994). *Chip on Board: Technology for Multichip Modules* (Springer).
14. Plößl, A., and Kräuter, G. (1999). Wafer direct bonding: tailoring adhesion between brittle materials, *Mater. Sci. Eng., R*, **25**, 1–88.
15. Burla, R. K., Chen, L., Zorman, C. A., and Mehregany, M. (2009). Development of nickel wire bonding for high-temperature packaging of SiC devices, *IEEE Trans. Adv. Packag.*, **32**, 564–574.
16. Zeiser, R., Wagner, P., and Wilde, J. (2012). Investigation of ultrasonic platinum and palladium wire bonding as interconnection technology for high-temperature SiC-MEMS, in *4th Electronic System-Integration Technology Conference*, Amsterdam, the Netherlands, pp. 1–6.
17. Amalu, E. H., and Ekere, N. N. (2012). High temperature reliability of lead-free solder joints in a flip chip assembly, *J. Mater. Process. Technol.*, **212**, 471–483.

18. Ding, F. (2006). Flip chip and lid attachment assembly process development (Auburn University Press, USA).

19. Longford, A., and Klowak, G. (2013). Utilising advanced packaging technologies to enable smaller, more efficient GaN power devices, in *European Microelectronics Packaging Conference (EMPC)*, Grenoble, pp. 1–5.

20. Zhang, W., Huang, X., Liu, Z., Lee, F. C., She, S., Du, W., et al. (2016). A new package of high-voltage cascode gallium nitride device for megahertz operation, *IEEE Trans. Power Electron.*, **31**, 1344–1353.

21. Huang, X., Liu, Z., Li, Q., and Lee, F. C. (2014) Evaluation and application of 600 V GaN HEMT in cascode structure, *IEEE Trans. Power Electron.*, **29**, 2453–2461.

22. Pavlidis, S., Ulusoy, A. C., Khan, W. T., Chlieh, O. L., Gebara, E., and Papapolymerou, J. (2014). A feasibility study of flip-chip packaged gallium nitride HEMTs on organic substrates for wideband RF amplifier applications, in *IEEE 64th Electronic Components and Technology Conference (ECTC)*, Orlando, FL, pp. 2293–2298.

23. Liu, H.-Y., Hsu, W.-C., Lee, C.-S., Chou, B.-Y., Liao, Y.-B., and Chiang, M.-H. (2014). Investigation of temperature-dependent characteristics of AlGaN/GaN MOS-HEMT by using hydrogen peroxide oxidation technique, *IEEE Trans. Electron Devices*, **61**, 2760–2766.

24. Nuttinck, S., Wagner, B. K., Banerjee, B., Venkataraman, S., Gebara, E., Laskar, J., et al. (2013). Thermal analysis of AlGaN-GaN power HFETs, *IEEE Trans. Microwave Theory Tech.*, **51**, 2445–2452.

25. Lu, D., and Wong, C. (2009). *Materials for Advanced Packaging* (Springer).

26. Prasher, R. S., Hu, X. J., Chalopin, Y., Mingo, N., Lofgreen, K., Volz, S., Cleri, F., and Keblinski, P. (2009). Turning carbon nanotubes from exceptional heat conductors into insulators, *Phys. Rev. Lett.*, **102**, 105901.

27. Nan, C.-W., Liu, G., Lin, Y., and Li, M. (2014). Interface effect on thermal conductivity of carbon nanotube composites, *App. Phys. Lett.*, **85**, 3549–3551.

28. Irwin, P., Cao, Y., Bansal, A., and Schadler, L. (2003). Thermal and mechanical properties of polyimide nanocomposites, in *Annual Report Conference on Electrical Insulation and Dielectric Phenomena*, pp. 120–123.

29. Narayanan, S., Fedorov, A. G., and Joshi, Y. K. (2010). Experimental characterization of a micro-scale thin film evaporative cooling device, in *12th IEEE Intersociety Conference on Thermal and Thermomechanical Phenomena in Electronic Systems*, Las Vegas, NV, pp. 1–10.

30. Efficient Power Conversion (EPC). Moore's Law Revival: Gen4 eGaN® FETs [online]. Available at http://epc-co.com/epc/Products/eGaNFETs/Gen4eGaNFETs.aspx (accessed May 5, 2016).

31. Liu, Y. (2012). *Power Electronic Packaging: Design, Assembly Process, Reliability and Modeling* (Springer).

32. Liu, Z., Huang, X., Zhang, W., Lee, F. C., and Li, Q. (2014). Evaluation of high-voltage cascode GaN HEMT in different packages, in *IEEE Applied Power Electronics Conference and Exposition, APEC 2014*, Fort Worth, TX, pp. 168–173.

33. Transphorm, TPH3207: World's only fully-qualified 650V GaN FET with the lowest $R_{(on)}$ in a TO-247 [online]. Available at http://www.transphormusa.com/products/ (accessed May 5, 2016).

34. Kanamura, M., Ohki, T., Kikkawa, T., Imanishi, K., Imada, T., Yamada, A., et al. (2010). Enhancement-mode GaN MIS-HEMTs with n-GaN/i-AlN/n-GaN triple cap layer and high-gate dielectrics, *IEEE Electron Device Lett.*, **31**, 189–191.

35. International Rectifier, *Discrete Power Quad Flat No-Lead (PQFN) Board Mounting Application Note, Appl. Note AN-1136*, p. 2.

36. International Rectifier, *DirectFET® Technology Board Mounting Application Note, Appl. Note AN-1035*, p. 2.

37. Semiconductor Today, GaN Systems' power semiconductors gain top-side cooling for simpler PCB design [online]. Available at http://www.semiconductortoday.com/nwas_items/2015/mar/gansystems_160315.shtml (accessed March 16, 2016).

Index